Ian McBride.

Spon's Railways Construction Price Book

2nd Edition

Spon's Railways Construction Price Book

This unique reference, based on years of specialist experience, will provide an understanding of the key drivers and components which affect the cost of railway projects. Any company, whether designers, contractors or consultants, looking to participate in the regeneration of the UK's railway network will find the guidance provided here an essential strategic asset.

Spon's Railways Construction Price Book

2nd Edition

Edited by Franklin + Andrews Ltd
Construction Economists

Spon Press
Taylor & Francis Group

LONDON AND NEW YORK

First published 1999 by Spon Press
11 New Fetter Lane, London EC4P 4EE

Simultaneously published in the USA and Canada by Routledge
29 West 35th Street, New York, NY 10001

Spon Press is an imprint of the Taylor & Francis Group

© 2003 Franklin + Andrews Ltd

The right of Franklin + Andrews Ltd to be identified as the Authors of this work has been asserted by them in accordance with the Copyright, Designs and Patents Act 1988

Printed and bound in Great Britain by TJ International Ltd, Padstow, Cornwall

All rights reserved. No part of this book may be reprinted or reproduced or utilised in any form or by electronic, mechanical or other means, now known or hereafter invented, including photocopying and recording, or in any information storage and retrieval system, without permission in writing from the publishers.

The publisher makes no representation, express or implied, with regard to the accuracy of the information contained in this book and cannot accept any legal responsibility or liability for any omissions that may be made.

Publisher's note
This book has been produced from camera-ready copy supplied by the authors.

British Library Cataloging in Publication Data
A catalogue record for this book is available from the British Library.

Library of Congress Cataloging in Publication Data
Spon's railways construction price book. – 2nd ed.
    p. cm.
  Includes bibliographical references and index.
  ISBN 0-415-32623-0 (alk. paper)
  1. Railroads–Great Britain–Design and construction–Costs. 2. Railroads–Equipment and supplies–Prices–Great Britain.

TF193.S66 2003
338.4'36251'0941–dc21   2003054261

ISBN 0-415-32623-0

Contents

Preface to Second Edition. vii

Acknowledgements ix

Introduction xi

General Items xvii

General and Sector Preambles xxi

Railway Construction Measured Rates

   Signalling 3

   AC Electrification. 47

   DC Electrification 79

   Permanent Way 97

   Telecommunications 139

   Property – New Build 149

   Property – Refurbishment 159

   Level Crossings 235

   Bridges 253

   Tunnels 287

   General Civils. 297

Appendices

   Useful Addresses 395

   Acronyms and Common Terms 400

   Location Factors. 407

   Indicative International Location Factors 409

Index. 413

Preface to the Second Edition

This, the second edition of Spon's Railways Construction Price Book, builds on the library from the first edition and addresses subsequent issues and feedback received from those working in this sector. It is the result of a substantial update to reflect those costs and technical changes necessitated by new and enhanced new and mandatory safety issues following recent concerns in the industry.

As before it follows the industry standard for construction works within the rail sector and every endeavour has been made to add more activities and to adjust cost information in order to reflect current rates. Included in this issue are a more extensive range of switches and crossings, a larger library of cables, terminations and running joints, electric points heating, additions to signalling and telecoms equipment, electrification structures, piling and tunnelling work in Civils and General Works.

Railways infrastructure works throughout the world have at their roots a common base and construction and technical issues and systems remain the same. Costs may differ but the actual work elements are more or less the same using similar materials and specialist labour. The Spon's Railways Construction Price Book allows this work to be identified within an international context.

André Holden
Franklin + Andrews Ltd

Acknowledgements

The publishers express their appreciation to those who have made a valuable contribution to the content of this, the second edition, of the Spon's Railways Construction Price Book and without whose input this publication could not have been written.

Particular acknowledgement is due to those working within the rail industry for their valued and experienced assistance with the content.

A number of manufacturers, merchants and specialist contractors have also provided invaluable input, as have many of the consultants involved with railways projects at all levels.

Finally special thanks are due to Franklin + Andrews for enhancing another ground-breaking publication.

Introduction

The rates which have been compiled for this publication are intended as a guide to the cost of carrying out major project works within the railways infrastructure and reflect costs current at a base date of 1Q2003. Each of the rates presented can only be considered as representative of an average value for work carried out under normal site conditions. Indicated costs should be viewed in consideration of possible cost enhancements to allow for the many other contributory influences which can add considerably to the base cost. All rates are intended to form a basis for works of a significant nature and should not be used for minor works without suitable modifications.

POSSESSIONS

Works within the railways infrastructure are, more often than not, carried out in close proximity to a fully operational and highly organised transportation business. Essentially, carrying out works of construction in order to accommodate the railways operations with the least disruption will be the order of the day.

In order to protect personnel and to allow for the railways to continue at maximised operational capacity during any construction or services activities, it will be necessary for disruption to be fully controlled to ensure minimal disorder to the rail schedule. All work has to be carried out within an Operational Safety Zone. All works within this OSZ must also be carried out during a 'possession'.

'Possessions' are a means of controlling construction activity safely. A possession arrangement actually means the possession by the Engineer of the line to the exclusion of all operational train movements. It is not permissible to run an operational train – passenger, freight, light engine or as engineer's plant – in traffic through a possession. This 'possession' arrangement results in on-costs for carrying out construction activities. Such a regime can add considerably to the cost of any activity.

Possessions, therefore, are windows when construction activity may take place safely. In situations where a possession is continuous the construction activity is relatively straight-forward. However if the possession is of an intermittent nature to enable main line services to continue then the construction process will suffer considerable disruption.

Hence, possessions can dramatically affect construction cost. The cost of disjointed working, additional management and safety staff required, electrical isolations and the like, will add greatly to the cost of the works. Labour productivity will vary considerably depending on the length of a possession. For example, an eight hour possession may be reduced by two hours in order to allow for the electrical supply services to be suspended and for re-energising the supply at the end of the possession. This would, therefore, result in an actual site labour production period of six hours. Similarly, a three hour possession could be reduced to an actual working time slot of just one hour.

Those preparing estimates of costs will need to carefully consider all aspects of working during a possession, not least the on-cost of the additional personnel involved.

POSSESSION STAFFING PERSONNEL

Person In Charge Of Possessions (PICOP):

Responsibilities include co-ordinating with the Signaller for taking on and for giving up a possession, arranging for all necessary protection to be implemented, authorising Engineer's trains into a possession and between work sites and sanctioning the Engineering Supervisor in charge of each work site to commence work.

Engineering Supervisor (ES):

Responsible for: the safety of all personnel within the work site, safe movement of trains entering the working area and erecting and maintaining site limit defining marker boards.

Person In Charge Of Work (PICOW):

Responsible for organising a safe working regime for the protection of working personnel from train movements.

Hand Signaller:

In charge of giving hand signals to train drivers on entering, leaving and within the site possession area where other signalling would be ineffective.

Site Warden:

Responsible for warning straying personnel of the site safe working limits whether entering or leaving the site and ensuring that they remain within safe limits.

Lookout:

Appointed by the PICOW to advise a work group of an approaching train or of a train movement.

Strap Man:

A person, certified in Electrified Lines instructions, who fits earthing straps to the electrified lines during an isolation to short circuit the system in the event of any inadvertent re-energising of the lines, to protect any site personnel.

POSSESSION COSTS

The following factors need to be taken into account in the calculation of costs for work to be carried out under a possession:

SAFETY

Training costs	
Signalling staff	Handsignalman/Lookout costs
Supervision	Site warden costs

Introduction

ISOLATIONS
Cut off / Re-energise   Electrification services

STAFFING and PRODUCTIVITY
Output restrictions   Unsequenced working
Unsociable working hours   Non-productive overtime
Possession staff costs   Safety requirements

TRAIN OPERATING COMPANY'S COSTS
Service interruption payments   Costs of possessions
Arranging possessions   Management costs

PROTECTION
Security and fencing   Cost for provision

POSSESSION COSTS

All works executed within the Operational Safety Zone (OSZ) must be carried out during a possession and this could involve disjointed working, additional safety and management requirements and general other conditions which will add greatly to the costs of the works.

Works carried out under a possessions regime will restrict site working and labour output can be seriously affected with non-productive time, disrupted working, overtime working and additional management costs all adding to the basic cost of the work.

Typical uplifts in respect of labour costs for carrying out work to accommodate a possession may result in the following adjustments:

Weekday Nights	+25%
Saturdays	+40%
Sundays	+40%
Bank Holidays	+55%

The Rules of the Route are normally laid down by the owner of the infrastructure and are pre-arranged with the Train Operating Companies in order to agree the periods of possessions. The contractor is obliged to concur with these arrangements and payments will have to be made to the TOCs if additional periods are required, these charges can be expensive, extremely so should they need to be arranged at short notice. Possessions need to be carefully assessed and the work element which is to be carried out during such will require careful consideration.

All personnel working within a possessions regime, on or near a live railway, have to have extensive training and must attend training courses before being allowed to carry on their respective trades within the site confines.

PRICING GUIDELINES

The reader will now be aware of the considerable on-costs associated with railways construction activities and the need to carefully assess each project in order to determine the effect on the base cost of construction and to arrive at unit values for pricing purposes.

These on-cost factors affect each and every project in varying degrees. All will be project specific and it is probably reasonable to assume from inception that each project will be unique.

Scheduled below are a number of the cost drivers that will add on-costs to the base rates and guidance on the possible cost enhancements that should be made to the indicated rates.

MEASURED RATES

To assist the reader in understanding the range and effect of the cost drivers for railways construction the measured rates pages display a base rate for each library description, illustrating labour resource costs, plant hire costs, material costs and an all-in unit rate for the unit of measure. A labour productivity output is also shown to enable the reader to adjust for own staff productivity information. In addition to these Base rates two further rates encompassing Green Zone and Red Zone working have been provided. These two rates (Green and Red Zone) provide National average rates inclusive of allowances for typical possession costs but exclude contractor's overhead and profit mark-up.

Green Zone working is defined as an environment where construction activities are carried out under a regime where trains and people are kept apart either by distance or a physical barrier and all work is carried out within a possessions regime

Red Zone working is defined as works carried out where trains and people are working in close proximity and additional safety requirements may be in force.

All rates exclude the contractor's overheads and profit mark-up and exclude VAT. They do, however, include some allowances for the on-costs associated with railways construction. An explanation for each of the three rates is given below.

The measured work section within this publication provides the reader with three rates, to represent the net cost of resources for each description plus grossed up rates inclusive of project specific costs for Green and Red Zone working.

BASE COST UNIT RATES

The Base Cost Unit Rate assumes and includes for labour working at optimum utilisation with materials being easily supplied to an accessible site and plant available as required. The rate does not include for the contractor's general overheads or for a profit element, neither does it take into account project specific on-costs or for any Value Added Tax element.

Introduction

GREEN ZONE UNIT RATES

The rates for Green Zone working include an assumption on possession costs as a percentage uplift on Base Cost Unit Rate and take into account a proportion of isolated working arrangements and the subsequent material and plant delivery problems that this type of location presents. The rates indicated at this level assume that all work may be carried out under a continuous and uninterrupted possession. No account has been taken in respect of Value Added Tax.

The rates within the green zone allow for the following uplifts to the net rates:

    General Items (Preliminaries)   +10%

the adjusted rate is then further uplifted by:

    Freight Costs (Engine Power)   +8%
    Design Costs (Contractor)   +10%
    Project Management   +10%

The resultant adjustment has the cumulative effect of an uplift of 40.8% on net rate indicated, this figure has been rounded to 40% for Green Zone working rates in this publication.

RED ZONE UNIT RATES

Red zone working rates take an average view of working under a restrictive regime. They include for work carried out during the continued operation of the rail industry and generally allow for an element of night, weekend and Bank Holiday working and for works of isolation of services, Value Added Tax has not been included. The range of uplifting percentages for carrying out a contract under these conditions is extremely wide and it can only be adequately assessed by considering each and every contributory factor applicable to the extent of the works. For the purposes of this publication Red Zone rates have assumed the following average calculations:

    General Items (Preliminaries)   +15%

the adjusted rate is then further uplifted by:

    Freight Costs (Engine Power)   +12%
    Design Costs (Contractors)   +12%
    Project Management   +15%
    Restricted Working Environment   +35%

The resultant adjustment has the cumulative effect of an uplift of 100.1%, on base rates indicated, this has been rounded to 100% for Red Zone working rates in this publication.

GENERAL ITEMS

Preliminary and general items relate directly to the whole of the works. This would include the many and varied costs which the contractor would need to allow for in order to set-up, manage and conclude the works in respect of the whole contract.

Unlike the measured rates which contain resources affected by quantity changes, certain costs are not affected by quantity change and require a different approach to recovery. Examples might be site accommodation or the provision of a telephone, both of which are required for the works but are not affected by small changes in work quantity. Similarly, plant may be utilised in a general capacity on several site operations making it difficult to cost against a specific measured item. Such costs, therefore, may be more appropriately recovered by inserting a sum in the General Items section of the Tender Documents.

The following are examples of the range of items that may be included in the General Items section. The list is not exhaustive but will identify those common items which, if not included in the tender elsewhere, should be included, if required, under the section headed General Items.

Employer's Requirements

The Employer may impose certain obligations and restrictions including:
1. Access to and possession or use of the site.
2. Working space limitations.
3. Working hours limitations.
4. Protection screens and hoardings.
5. Display signs and name boards.
6. Maintenance and protection of existing services.
7. Communications installations.
8. Temporary office accommodation.
9. Work scheduling.

Contractor's Administrative Arrangements

An assessment of the contractor's site management requirements will normally include allowances for the following:

1. Site administration.
2. Site supervision.
3. Security and site protection.
4. Safety, health and welfare of site staff.
5. Transport of site staff.
6. Construction Design and Management (CDM)
7. Isolations management
8. Possessions management and supervision

Contractor's Facilities

The contractor should establish the extent of facilities required for it's own operations and the cost of providing them against the following headings:

1. Site accommodation including offices, compounds, stores, mess-rooms, toilet facilities and living accommodation. Include for any fees and charges on building structures as appropriate.
2. Temporary fencing, hoardings, screens, guardrails and general protection.
3. Water for the works including any temporary plumbing and services distribution and subsequent removal on completion.
4. Temporary lighting and power for the works including temporary installations and removals.
5. Temporary telephones and other means of communications.

Constructional Plant

A programme will often be prepared in order to establish the quantity and types of plant required on site, as well as the hire periods involved, items of plant that are directly operationally related will generally be included within the measured rates. Items of plant required for general usage and not already included in the measured rates can be included under the following headings, together with an allowance for transport and maintenance costs:

1. Small plant and tools.
2. Scaffolding and access equipment.
3. Cranes and lifting gear.
4. Site transport.
5. Plant required for specific trades.

Temporary Works

It may also be necessary to identify any temporary works costs which can add considerably to the cost of the project in certain locations. Temporary works can best be considered under the following headings:

1. Traffic diversions and road closures.
2. Access roads including reinstatement on completion of the contract.
3. Bridges.
4. Coffer dams and temporary restrictions and diversions.
5. Pumping and de-watering .
6. Compressed air provision and maintenance.

Sundry Items

There will be certain items which do not fall easily under any of the foregoing headings and are best acknowledged as sundry work, for example:

1. Material testing.
2. Testing of the works.
3. Protection of the works against inclement weather.
4. Protective casings and coverings, debris removal and general cleaning of the works.
5. Traffic regulations.
6. Maintenance of roads and service installations.
7. Drying the works.
8. Control of noise and pollution.
9. All statutory obligations.

Additional Commentary

Construction activities relating to the railways have additional requirements which need to be identified and allowed for as necessary in order to comply with special requirements of the Employer. The following list is not specific to railways project working neither is it intended to be exhaustive but is given here as a guide to possible further considerations:

1. Medical fitness of persons competent in track safety.
2. Alcohol and drugs policy.
3. Method statement for working in the vicinity of the running line.
4. Photograph identity card.
5. Project safety plan.
6. Project quality plan.
7. On Demand retention bond.
8. Specialist training requirements.

General Items

Pricing Example

Each project will have its own unique requirements in respect of general items and should be carefully assessed.

General items and preliminary costs for, say, a bridge rebuilding project of three months duration and with a contract value of £1,000,000 might be priced as follows:

Supervision and administration	£40,000.00
Site accommodation	£10,500.00
Lighting and Power	£2,500.00
Bonds	£4,000.00
Insurances	£5,000.00
Water for the Works	£1,500.00
Labour accommodation facilities	£15,000.00
Temporary works	£25,000.00
Small plant	£2,500.00
Crane	£11,500.00
Transport and travelling costs	£9,500.00
Total Preliminaries	**£127,000.00**

This represents 12.7% of the tender sum.

Project Overheads and Profit

Every contractor will require to recover from his activities sufficient income to fund general office overheads and administrative costs. In addition the contractor will look to receive a fair return (profit) upon capital invested in the project. This is in addition to the measured rates cost and the general.

This mark-up on costs is a commercial decision making progress and is dictated by economic pressures. No allowance has been made for the recovery of overheads and mark-up in the rates

General and Sector Preambles

General Preambles

1.0 Generally

In order to standardise measurement of work carried out within the rail industry it is necessary to draw on more than one standard method for measurement of the construction activities. These include: Civil Engineering Standard Method of Measurement, Third Edition, referred to as CESMM3; Standard Method of Measurement of Building Works, Seventh Edition, referred to as SMM7, and Standard Method of Measurement for Industrial Engineering Construction, referred to as SMMIEC. Items not falling within one of these three specific measurement rule categories are fully described and are generally in accordance with similar classifications.

1.1   Work sections in accordance with Civil Engineering Standard Method of Measurement Third Edition (CESMM3)

The following sections have been compiled wholly or in part in accordance with CESMM3 standard measurement rules:

Signalling	civils work element
AC Electrification	civils work element
DC Electrification	civils work element
Permanent Way	civils work element
Telecommunications	civils work element
Level Crossings	civils work element
Bridges	
Tunnels	
General Civils	

1.2   Work sections in accordance with Standard Method of Measurement of Building Works, Seventh Edition (SMM7)

The following sections have been compiled wholly or in part in accordance with SMM7 standard measurement rules:
Property – New Build
Property – Refurbishment

1.3   Work sections in accordance with Standard Method of Measurement of Industrial Engineering Construction (SMMIEC)

The following sections have been compiled wholly or in part in accordance with SMMIEC standard measurement rules:

Signalling	cabling and the like
AC Electrification	cabling and the like
DC Electrification	cabling and the like
Telecommunications	cabling and the like

1.4 All rates and sums entered in the publication are exclusive of Value Added Tax (VAT).

1.5 Due allowance is made within the sums and unit rates for material procurement. Material costs include procurement, transport to work area, packaging, handling, storage, protection and inclusion in the measured work item.

1.6 All sums and unit rates entered against the respective items are deemed to be fully inclusive for finished work described under such items.

1.7 The following abbreviations have been used within the publication:-

Millimetre	mm
Metre	m
Square millimetre	mm^2
Square metre	m^2
Hectare	ha
Cubic metre	m^3
Kilogramme	kg
Tonne	tonne
Sum	sum
Number	Nr
Item	Item
Pair	pr
Hour	hr
Day	day
Week	wk

1.8 Lengths, heights, widths, depths and girths expressed, for example, as 1.50–3.00m shall mean exceeding the former figure but not exceeding the latter.

1.9 The term 'up to' shall mean 'up to and including'.

1.10 Where the text refers to singular this denotes both in the singular and the plural and vice versa.

1.11 Figures within descriptive text are deemed to be in millimetres unless qualified by a unit of measurement indicating otherwise or by a word or words signifying a reference.

1.12 The term 'structure' when used in the description of fixing or installing shall mean any background including concrete, masonry, steelwork and the like in respect of the relevant sections of work.

1.13 Work described to be fixed with screws, bolts and similar fixings assumes the provision of holes for fixing.

1.14 The term 'bolts' is deemed to include ragbolts, wedgebolts, rawlbolts, holding down bolts and the like as appropriate to the work.

1.15 The term 'fix/install only' includes for ordering, taking delivery of the item or items either on-site or off-site as appropriate, handling, loading and unloading, protecting, storing, double-handling, hoisting, lowering, checking, sorting, assembling, use of temporary supports including bracing, positioning, trial and permanent erection, fixing complete. It also includes for returning all packaging and surplus materials.

1.16 Any method of fixing so described includes the provision of all necessary fixing devices, whether or not specifically identified, in order to complete the installation. It includes all necessary drilling, abortive holes and the like and general making good to disturbed works as and where required.

1.17 Surface finishings are deemed to include the preparation of surfaces to which those finishes are to be applied.

1.18 The terms 'included elsewhere' and 'measured separately' where used to qualify any description means that provision has been made elsewhere for the work element referred to.

2.0 Pricing

Rates and prices generally include for all temporary and permanent materials unless expressly described as being supplied by others.

Rates include for all costs in connection with the sequence of working necessary to complete the works.

Rates include for accommodating and working around and in conjunction with other direct and indirect contractors and subcontractors.

Rates include for protection of the works during installation and up to handover.

Rates include for all marking out including marking position of holes, chases, mortices and the like. Position of penetrations, cast in supports and the like are to be established.

Rates include for all necessary cleaning, maintaining and protection of the works including any special protection as required throughout the period of construction up to handover.

Rates include pre-installation tests, manufacturer's tests and for test certificates as and where required by the job specification.

Rates and prices include for all temporary cables, lighting, weather and other protection and facilities required for the proper, safe and timely execution of the works and for clearing away all temporary works on completion.

Unless otherwise specifically stated the following are deemed to be included with all items:

- (a) Labour and all costs in connection therewith
- (b) Materials, goods and all costs in connection therewith
- (c) Assembling, fitting and fixing materials and goods in position
- (d) Plant and all costs in connection therewith
- (e) Square cutting

Part 2 Sector Preambles

The preamble notes for each sector refer to the specific pricing requirements and measurement rules for the particular discipline. As and where appropriate the notes shall apply to all other sectors to the extent that they are relevant.

Definitions

The following definitions shall be used for the purposes of interpretation and all such definitions will apply to the extent that they are relevant:-

'Supply' shall include, but not be limited to, procurement, expediting, inspection, packing, protecting, preservations, haulage, storage and delivery to site of materials and equipment, taking delivery, unloading, examining, handling, accepting into Contractor's control and all waste and overage and removal of surplus and scrap material on completion.

'Fabricate' shall include, but not be limited to, taking delivery, unloading, examining, removal of protective materials and wrappings, handling, accepting material into Contractor's control, maintaining records, preparation of necessary documentation for fabrication drawings, sketches and calculations, identifying and removing material from store, setting out, preparing, cutting, prefabricating into sub-units, supply of all consumables, inspection, testing, other than where specifically measured, and transporting to specific location.

'Erect' or 'Install' or 'Fix' shall include, but not be limited to, taking delivery, unloading, examining, removal of protective materials and wrappings, handling, accepting material into Contractor's control, maintaining records, preparation of necessary documentation for fabrication drawings, sketches and calculations, identifying and removing material from store, setting out, preparing, cutting, prefabricating into sub-units, supply of all consumables, inspection, testing, other than where specifically measured, and dismantling and clearing away temporary support structures.

Where the definitions are used together, e.g. 'Supply and Erect', the item will be deemed to include the interpretation and meaning as if each definition was used independently and the definition will include everything necessary to carry out all the work described.

Schedule 10 Signalling

General and Specific Preambles

Generally: Rates shall include for carrying out the following items of work during all types of possessions and as appropriate:

- a) transport of all materials to site for bases to Location Casings and Relocatable Equipment Buildings (REBs) and for foundations to signal posts
- b) construction of bases for Location Casings and Relocatable Equipment Buildings and foundations for signal posts and the like
- c) transporting to site and erecting signal posts and all signal equipment
- d) transporting to site and erecting Location Casings and Relocatable Equipment Buildings
- e) laying cables, ducts and the like
- f) commissioning signalling and equipment
- g) taking out, removing and disposing of all redundant signals and signalling equipment

All signalling and ancillary equipment shall comply with the relevant and specific industry standards for supply and installation including integration with existing layouts as necessary.

Casings, Cubicles and Buildings

Cubicles generally:

Rates for interlocking cubicles and technicians' terminals shall be deemed to include for all equipment located therein and all cabling and wiring and connections between the equipment located therein and for subsequent testing.

Location Casings:

Rates for Location Casings shall include location casing modules, components and associated wiring required for trackside equipment therein and, in addition, allowance for Automatic Train Protection (ATP).

Rates for Location Casings shall be deemed to include 650 volt spur unit, 650–110 volt transformer, 50 volt pulse generator, 110/50 volt transformer/rectifier, busbar, power distribution rack/block, telecoms termination rack/block, relay racks, trackside equipment modules, track relays, fuses, data link modules, field engineers telephone socket, maintenance circuit plug points and all wiring and connections between the equipment within the Location Casing and earthing to Location Casing.

Relocatable Equipment Building:

Rates for Relocatable Equipment Buildings shall include location casing modules, components and associated wiring required for trackside equipment therein and, in addition, allowance for Automatic Train Protection (ATP).

Rates for Relocatable Equipment Buildings shall be deemed to include 650 volt spur unit, 650–110 volt transformer, busbar, power distribution rack/block, relay racks, telecoms termination rack/block, trackside equipment modules, relays, fuses, data link modules, 24 volt battery charger, field engineers telephone socket, maintenance circuit plug points and all wiring and connections between the equipment within the Location Casing and earthing to Location Casing.

Foundation bases:

Platform bases to Location Casings and Relocatable Equipment Buildings are deemed to include provision and erection of precast concrete base and corner units, block fillers and associated fittings and fixings on levelled and prepared site, measured separately, and including holding down bolts and ancillary fittings.

Signal Posts and Signals

Rates for signal posts shall be deemed to include provision of access ladders.

Rates for signals shall be deemed to include for signal identification plates and black PVC covers placed prior to commissioning and for subsequent removal on commissioning.

Signal Post Foundation Bases

Rates for foundation bases to signal posts are deemed to include excavation, backfilling with arisings and imported material, disposal of surplus excavated material, earthwork support, trimming, compacting, levelling, provision and laying of concrete and reinforcement, tamping and levelling, formwork, cable ducts, provision and setting of plates, bolts and the like.

Telephones

Rates for Point Zone and ground signal telephones shall be deemed to include for fixing to and including 50 mm diameter steel posts, 1000 mm high above ground, cast in concrete foundation together with associated excavations, disposals and the like.

Rates for Level Crossing telephones and housings shall be deemed to include for fixing to and including 50 mm diameter steel posts, 2000 mm high above ground, cast in concrete foundation together with associated excavations, disposals and the like.

Automatic Warning Systems (AWS)

Rates for AWS components shall be deemed to include permanent magnet, electromagnets, suppressed magnets where necessary, and for drilling holes in sleepers and bolting to same, for protective housing and ramp installation and for placing pre-commissioning shield over magnets within protective housing.

Train Operated Warning Systems (TOWS)

Rates for TOWS switches shall be deemed to include for plastic conduit connecting switch to Yodalarm position and for provision of appropriate Allen key.

Rates for TOWS Yodalarms and switches shall be deemed to include for fixing to and including 50 mm diameter steel posts, 2000 mm high above ground, cast in concrete together with all associated excavations, disposals and other ancillary works, with and including necessary mounting signs and fixing brackets

Redundant Signalling and Equipment

Removal of redundant signalling and equipment is deemed to include, as described and where appropriate, removal of the following:

- a) redundant signal posts, attached associated signals and telephones
- b) redundant signals fitted to gantries together with associated telephones
- c) redundant AWS, treadles, track circuits and other trackside equipment and associated tail cables
- d) redundant signalling and power cables within cable troughs
- e) redundant location casings
- f) defective and/or redundant cable troughing including concrete asbestos on and including posts
- g) ground frames
- h) level crossing gates, barriers and associated equipment
- i) resultant debris from the works

Cables and Wiring

Cables:

Rates for tail cables shall be deemed to include for forming 'knock out' openings in cable troughing and shall also allow for laying adjacent to cable troughing route initially, where appropriate, and for relaying within cable troughing during a subsequent visit during removal of redundant signal cables after commissioning.

Descriptions in accordance with relevant section of SMMIEC

Conduit, Trunking and the like

Conduit:

Conduit shall be measured over all fittings and boxes and shall be deemed to include bending, cutting, threading, jointing and all conduit fittings and fixings, boxes, tees, bends, bushes, locknuts and stopping plugs.

Forming openings for conduit entries in trunking, accessories, fittings, switchgear, distribution gear and equipment, providing associated components including bushes, locknuts, nipples and the like shall be deemed to be included with the items in which they occur.

Trunking:

Trunking shall be measured on the centre line over all fittings and shall be deemed to include all cutting and jointing.

Stop ends, bends, tees, crosses, offsets, reducers, internal fire barriers and the like shall be enumerated and measured as Extra Over the trunking.

Trays:

Trays shall be measured on the centre line over all fittings and shall be deemed to include all cutting and jointing.

Stop ends, bends, tees, crosses, offsets, reducers and the like shall be enumerated and measured as Extra Over the trays.

Cutting openings in trays shall be enumerated and grouped together irrespective of size.

Ladder racks:

Ladder racks and like supports shall be measured on the centre line over all fittings and shall be deemed to include all cutting and jointing.

Elbows, risers, tees, crosses, reducers and the like shall be enumerated and measured as Extra Over the trays.

Sundries:

Treatments to the cut edge of finished components shall be deemed to be included.

Earth continuity components and supporting steelwork shall be deemed to be included.

Cables and Conductors

Cables:

Cables shall be measured from gland to gland, allowance for tails shall be 0.30m for switches, socket outlets, light fittings, equipment outlets and the like and 1.00m for all other items.

Determining circuits, capping ends of cables (including end on drum or coil after cutting), providing draw wires and draw cables, cleaning out conduits, pipe ducts, trunking and trays, threading cables through sleeves and the like shall be deemed to be included.

Cables drawn into existing pipe ducts, conduits or trunking shall be so described and measured separately and rates shall be deemed to include removing and replacing covers, inspection lids and the like as applicable.

Dressing cables after installation shall be deemed to be included.

Cable joints:

Cable termination joints, branch joints, line taps, through joints and joints to existing cables where described shall be deemed to include all jointing items including end boxes, glands, seals, joint boxes, earth tags, identification ferrules and the like other than where described separately for switchgear and similar items.

The removal of cable insulation, sheaths and armouring, terminal covers, connecting conductors and the provision of heat insulating and joint material, locknuts, bushes, cable connecting lugs, cable and core identification markers and the like shall be deemed to be included.

Forming openings for cable entries in trunking, accessories, fittings and the like together with any bushing material shall be deemed to be included.

Openings for cable entries in switchgear, distribution gear and equipment following delivery shall be measured separately and adequately described.

Cable fixings:

Cable fixings including saddles, cleats, clips, hangers and the like shall be deemed to be included in cable items unless enumerated separately.

Accessories and Fittings

Accessories:

Switches, socket outlets, thermostats, indicating units, push buttons and the like are deemed to include ancillary boxes and earthing tails and so described.

Fittings:

Fittings generally are deemed to include ancillary boxes, cover plates and the like and so described.

Switchgear, Distribution Gear and Equipment

Generally:

Items of switchgear, distribution gear and equipment are deemed to include ancillary works including forming openings in trunking together with any bushing material required

Assembling and jointing composite items including the provision of all necessary items as described for completing the work item are deemed to be included.

Earthing Tapes and Fittings

Bends, twists, termination joints, branch joints and through joints for tape items together with all jointing materials, drilling, sweating and tinning shall be deemed to be included.

Driving earth rods is deemed to be included.

Supporting Steelwork

Generally:

All cutting, drilling, welding, bolting and the like for supporting steelwork items is deemed to be included. Welding of joints and connections in all positions together with all cutting, cleansing, bevelling, grinding, machining, pre-heating, tack welding, sealing runs at weld roots and temporary backstraps shall be deemed to be included.

Assembly of components shall be deemed to be included.

Sundries

Generally:

Boring or excavating of holes for poles and stays shall be deemed to be included.

Cable trenches:

Disposal of any displaced excavated material from cable trenches and the like shall be deemed to be included.

Excavating for concrete surrounds and the like shall be deemed to be included.

Testing and commissioning:

Provision of test equipment for inspecting and testing and the provision of associated reports and certificates shall be deemed to be included.

Schedule 20: AC Electrification

Description generally in accordance with relevant section of SMMIEC and as specifically described in Signalling section of these Preambles. Civil works elements are generally in accordance with CESMM3.

Schedule 30: DC Electrification

Description generally in accordance with relevant sections of SMMIEC and as specifically described in Signalling section of these Preambles. Civil works elements are generally in accordance with CESMM3.

Schedule 40: Permanent Way

Work elements are generally described in accordance with CESMM3.

Schedule 50: Telecommunications

Description generally in accordance with relevant section of SMMIEC and as specifically described in Signalling section of these Preambles. Civil works elements are generally in accordance with CESMM3.

General and Sector Preambles

Schedule 60: Property – New Build

Note: Work in the Property – New Build schedule is generally described in accordance with SMM7 except where specifically mentioned.

60.30 Platforms

60.30.10 Solid fill construction composite

Note: These items are composite work build-ups.

Includes: Excavation, disposal of redundant material, earthwork support, filling with excavated and imported material as appropriate, concrete foundations, brickwork and blockwork constructions, copings, surface paving material and including all labour, plant and materials required.

60.30.20 Cross wall construction composite

Note: These items are composite work build-ups.

Includes: Excavation, disposal of redundant material, earthwork support, backfilling with excavated material, concrete foundations, brickwork and blockwork constructions, beam and pot structural flooring, concrete topping, copings, surface paving material and including all labour, plant and materials required.

60.50 Facilities

60.50.10 Storage and Repair Facilities

Note: These facilities costings are cost composites of typical constructions presented in elemental format at an average cost for the particular construction type specified.

Schedule 65: Property – Refurbishment

Note: Work in the Property – Refurbishment Schedule is generally compiled in accordance with SMM7 except where specifically mentioned and as noted below.

Note: **The following items have not been described strictly in accordance with the measurement rules for SMM7**

65.01 Demolitions and Removals

65.01.10 General demolitions and removals

65.01.10.10　Concrete structural elements

Includes: Demolition of concrete elements by volume, in situ and reinforced, including walls, columns, slabs, piers and the like. Work deemed to include removal and off-site disposal of all redundant materials. Allow for temporary support incidental to demolitions and at the discretion of the Contractor and for temporarily diverting, maintaining or sealing off existing services as necessary.

65.01.10.20　Masonry structural elements

Includes: Demolition of masonry elements by area of specific dimensioned thickness or construction in brickwork and blockwork, including walls and partitions, attached piers, buttresses and the like. Work deemed to include removal and disposal off-site of all redundant materials. Allow for temporary support incidental to demolitions and at the discretion of the Contractor and for temporarily diverting, maintaining or sealing off existing services as necessary.

Generally: General removals of roofs, ceilings and timber floors, internal partitions; taking down, cleaning and setting aside materials for re-use; removal of building fabric fixtures and fittings; removal of windows, doors and frames; removal of sundry joinery items; removal of plumbing and heating installations; removal of electrical installations; general demolitions and removals are all described according to lineal or square measure or by number as appropriate to the work element. All work is deemed to include off-site disposal of all redundant materials and for cleaning and storing salvaged items described as set aside for re-use.

65.05　　　　Brickwork and Blockwork

65.05.20　　　New Work

65.05.20.40　Damp proof course

Note: Description of damp proof courses and membranes is at variance with SMM7, with alternative widths to standard rules.

65.06　　　　Asphalt Work

65.06.20　　　New Work

65.06.20.10　Mastic asphalt tanking BS 6925
65.06.20.20　Mastic asphalt tanking BS 6577
65.06.20.30　Mastic asphalt flooring BS 6925 and
65.06.20.40　Coloured mastic asphalt flooring BS 6577

Note: Descriptions for tanking, flooring, roofing and damp proofing are at variance with SMM7 with widths of '<150mm', '150–300mm' and '>300mm' as opposed to four separate width considerations in respect of SMM7.

Schedule 70: Level Crossings

Section 1: Definitions

Schedule 80: Bridges

Note: Bridge works are described generally in accordance with CESMM3 except where stated and as noted below.

Note: **The following items have not been described strictly in accordance with the measurement rules for CESMM3, SMM7 or SMMIEC**

80.20 Demolitions and Removals

80.20.10 Structures

80.20.10.10 Masonry structural elements

Includes: Demolition of masonry elements by area of specific dimensioned thickness in brickwork and blockwork, including walls and partitions, attached piers, buttresses and the like. Work deemed to include removal and disposal off-site of all redundant materials. Allows for temporary support incidental to demolitions and at the discretion of the Contractor and for temporarily diverting, maintaining or sealing off existing services as necessary.

80.20.10.20 Concrete structural elements

Includes: Demolition of concrete elements by volume, in situ and reinforced, including walls, columns, slabs, piers and the like. Work deemed to include removal and off-site disposal of all redundant materials. Allows for temporary support incidental to demolitions and at the discretion of the Contractor and for temporarily diverting, maintaining or sealing off existing services as necessary.

80.20.10.30 Steel columns and stanchions

Includes: Removal of steel elements by weight, including columns, stanchions, rails, purlins and the like. Work deemed to include removal and off-site disposal of all redundant materials. Allows for temporary support incidental to removals and at the discretion of the Contractor and for temporarily diverting, maintaining or sealing off existing services as necessary.

80.20.20   Removal of old work

80.20.20.10   General metalwork

Includes: Removal of general metalwork elements by item, including stairs, handrails and the like. Work deemed to include removal and off-site disposal of all redundant materials. Allows for temporary support incidental to removals and at the discretion of the Contractor.

                Removal of general signage elements by number, including road signs and the like. Work deemed to include removal and off-site disposal of all redundant materials.

                Removal of general signage elements by number, including road signs and the like. Work deemed to include removal and setting aside for re-use.

80.20.20.20   General elements

Includes: Removal of general elements by area, including timber decking and the like. Work deemed to include removal and off-site disposal of all redundant materials or for setting aside for re-use, as appropriate to the item description.

80.20.20.30   Fencing

Includes: Removal of fencing elements by length, including open post and rail, post and wire and the like. Work deemed to include removal and off-site disposal of all redundant materials.

80.20.20.40   Removal of hard access constructions

Includes: Breaking out hard surfacing elements by area, including asphalt and macadam wearing and base courses, concrete pavings and the like. Work deemed to include removal and off-site disposal of all redundant materials.

                Taking up kerb and edging elements by length, including foundations. Work deemed to include removal and off-site disposal of all redundant materials.

80.20.20.50   Pipelines and ducts

Includes: Removal of general drainage and duct pipework by length, including pipes described by internal diameter and demolition and removal of supports. Work deemed to include removal and off-site disposal of all redundant materials. Excavations deemed to include return, fill and ram with excavated material and making up deficit with similar approved materials procured at the Contractor's discretion and backfilled into excavation as before.

General and Sector Preambles

80.30 Earthworks

80.30.10 General Excavations

80.30.10.10 Excavation by machine in material other than rock

Includes: Oversite excavation to remove topsoil; by volume, in average depths of 150 mm and 300 mm.

80.30.10.20 Excavation by hand in material other than rock

Includes: Oversite excavation to remove topsoil; by area, in average depths of 150 mm and 300 mm.

80.49 Precast Concrete

80.49.70 Copings and sills

80.49.50.10 Precast units

Note: Notwithstanding standard measurement rules coping, sill and parapet units are described by specific cross section dimensions per metre run.

Note: **Work described in accordance with SMM7**

80.25 Site Clearance

80.25.10 General site clearance

80.25.10.10 Trees, stumps and general vegetation

Includes: Trees; including disposal, described by alternative girth. Deemed to include off-site disposal and backfilling excavation with selected excavated material including making up deficit with approved materials, compacted in layers.

Stumps; including disposal and backfilling with imported materials, described by alternative girth. Deemed to include off-site disposal of redundant materials.

Vegetation; undergrowth, bushes and the like, described by square metre.

Cut down hedges; grub up roots and remove from site, described by length in specific height range. Deemed to include backfilling resultant void with selected excavated material including making up deficit with approved materials, compacted in layers.

80.30.50	Filling to excavations
Includes:	Grading filled oversite to contours, embankments and the like; by machine; described to falls, crossfalls and slopes, by area
	Grading filled oversite to contours, embankments and the like; by hand; described to falls, crossfalls and slopes, by area
80.40	Concrete Work
Includes:	Plain and reinforced in situ concrete work items described generally in accordance with SMM7 rules in respect of foundations, ground beams and the like, excepting where amended to include a varied classification; i.e. '450–600mm' replacing 'over 450mm' and an additional classification; i.e. 'over 600mm'.
	Reinforced in situ concrete work items for columns, piers, beams and casings as described in CESMM3 with the exception the unit rates include all labour, plant and material notwithstanding the measurement rules for CESMM3 requiring provision and placing of concrete as separate rates.
80.45.10	Formwork
Note:	**Described in accordance with SMM7 measurement rules.**
80.45.30	Joints in concrete pavings
Note:	**Described in accordance with SMM7 measurement rules.**
80.65	Miscellaneous Metalwork
Note:	**Described in accordance with SMM7 measurement rules.**

Schedule 85: Tunnels

Note:	The work in this section is generally described in accordance with CESMM3 except where stated and as noted below.
Note:	**The following items are described in accordance with SMM7:**
85.80	Work to Existing
85.80.01	Brickwork

Note: Generally described in accordance with SMM7 measurement rules throughout the brickwork section. For remedial works to brickwork the item description describes the work to be carried out and is deemed to include all labour, small tools and material necessary to complete the works. Extensive access equipment or substantial temporary works, where required, may need to be included and are at the Contractor's discretion. The inclusion of such may require an addition to the Preliminaries section of the Bill.

Schedule 90: General Civils

Note: **The work in this section is generally described in compliance with the Civil Engineering Standard Method of Measurement, third edition, referred to as CESMM3.**

Railway Construction
Measured Rates

Signalling

Rail 2003 Ref		Unit	Labour Hours	Labour Cost £	Plant Cost £	Materials Cost £	Unit Rate Base Cost £	Unit Rate Green Zone £	Unit Rate Red Zone £
11	NEW INSTALLATIONS – RESIGNALLING WORKS								
1101	EQUIPMENT/SIGNALS								
110101	Posts; 168mm dia, 3000mm high, incl. holding down bolts (bases/foundations measured separately), BRS-SM drawing number:								
11010101	2052; comprising 612mm platform with central post; NRS catalogue number:								
11010101A	86/49269 (4 or 3 aspect)............	Nr	20.00	764.23	–	768.97	1533.20	2146.48	3066.40
11010101B	86/49271 (4 or 3 aspect)............	Nr	20.00	764.23	–	768.97	1533.20	2146.48	3066.40
11010106	2053; comprising 612mm platform with central post and foot step; NRS catalogue number:								
11010106A	86/49000 (4 aspect)................	Nr	20.00	764.23	–	1185.81	1950.04	2730.06	3900.09
11010106B	86/49001 (3 aspect)................	Nr	20.00	764.23	–	1185.81	1950.04	2730.06	3900.09
11010106C	86/49002 (4 aspect)................	Nr	20.00	764.23	–	1261.74	2025.97	2836.36	4051.94
11010106D	86/49003 (3 aspect)................	Nr	20.00	764.23	–	1185.81	1950.04	2730.06	3900.09
11010106E	86/49004 (4 aspect)................	Nr	20.00	764.23	–	1261.74	2025.97	2836.36	4051.94
11010106F	86/49005 (3 aspect)................	Nr	20.00	764.23	–	1166.47	1930.70	2702.98	3861.41
11010106G	86/49006 (4 aspect)................	Nr	20.00	764.23	–	1261.74	2025.97	2836.36	4051.94
11010106H	86/49007 (3 aspect)................	Nr	20.00	764.23	–	1185.81	1950.04	2730.06	3900.09
11010106I	86/49008 (4 aspect)................	Nr	20.00	764.23	–	1120.58	1884.81	2638.73	3769.62
11010106J	86/49009 (3 aspect)................	Nr	20.00	764.23	–	1116.44	1880.67	2632.94	3761.35
11010106K	86/49010 (4 aspect)................	Nr	20.00	764.23	–	1117.49	1881.72	2634.41	3763.44
11010106L	86/49011 (3 aspect)................	Nr	20.00	764.23	–	1110.90	1875.13	2625.17	3750.25
11010106M	86/49012 (4 aspect)................	Nr	20.00	764.23	–	1115.57	1879.80	2631.72	3759.61
11010106N	86/49013 (3 aspect)................	Nr	20.00	764.23	–	1116.56	1880.79	2633.10	3761.58
11010106O	86/49014 (4 aspect)................	Nr	20.00	764.23	–	1123.41	1887.64	2642.69	3775.28
11010106P	86/49015 (3 aspect)................	Nr	20.00	764.23	–	1119.26	1883.49	2636.88	3766.98
11010106Q	86/49016 (4 aspect)................	Nr	20.00	764.23	–	1090.30	1854.53	2596.33	3709.05
11010106R	86/49017 (3 aspect)................	Nr	20.00	764.23	–	1090.30	1854.53	2596.33	3709.05
11010111	2054; comprising 912mm platform with central post; NRS catalogue number:								
11010111A	86/49233 (3 or 4 aspect)............	Nr	20.00	764.23	–	837.32	1601.55	2242.17	3203.10
11010111B	86/49235 (3 or 4 aspect)............	Nr	20.00	764.23	–	934.77	1699.00	2378.59	3397.99

Signalling

Rail 2003 Ref		Unit	Labour Hours	Labour Cost	Plant Cost	Materials Cost	Unit Rate	Unit Rate	Unit Rate
				£	£	£	Base Cost £	Green Zone £	Red Zone £

11   NEW INSTALLATIONS – RESIGNALLING WORKS

110101   Posts; 168mm dia, 3000mm high, incl. holding down bolts (bases/foundations measured separately), BRS-SM drawing number:

11010116   2055; comprising 912mm platform with central post and footstep; NRS catalogue number:

Ref	Description	Unit	Hours	Labour	Plant	Materials	Base	Green	Red
11010116A	86/49018 (4 aspect)	Nr	20.00	764.23	–	1243.59	2007.82	2810.94	4015.64
11010116B	86/49019 (3 aspect)	Nr	20.00	764.23	–	1241.44	2005.67	2807.94	4011.35
11010116C	86/49020 (4 aspect)	Nr	20.00	764.23	–	1314.56	2078.79	2910.30	4157.58
11010116D	86/49021 (3 aspect)	Nr	20.00	764.23	–	1238.61	2002.84	2803.97	4005.67
11010116E	86/49022 (4 aspect)	Nr	20.00	764.23	–	1315.43	2079.66	2911.52	4159.32
11010116F	86/49023 (3 aspect)	Nr	20.00	764.23	–	1239.49	2003.72	2805.20	4007.44
11010116G	86/49024 (4 aspect)	Nr	20.00	764.23	–	1315.42	2079.65	2911.51	4159.30
11010116H	86/49025 (3 aspect)	Nr	20.00	764.23	–	1240.55	2004.78	2806.69	4009.56
11010116I	86/49026 (4 aspect)	Nr	20.00	764.23	–	1188.77	1953.00	2734.19	3905.99
11010116J	86/49027 (3 aspect)	Nr	20.00	764.23	–	1185.49	1949.72	2729.60	3899.43
11010116K	86/49028 (4 aspect)	Nr	20.00	764.23	–	1189.63	1953.86	2735.41	3907.73
11010116L	86/49029 (3 aspect)	Nr	20.00	764.23	–	1185.49	1949.72	2729.60	3899.43
11010116M	86/49030 (4 aspect)	Nr	20.00	764.23	–	1190.70	1954.93	2736.89	3909.85
11010116N	86/49031 (3 aspect)	Nr	20.00	764.23	–	1186.54	1950.77	2731.08	3901.55
11010116O	86/49032 (4 aspect)	Nr	20.00	764.23	–	1191.58	1955.81	2738.13	3911.62
11010116P	86/49033 (3 aspect)	Nr	20.00	764.23	–	1187.44	1951.67	2732.34	3903.34
11010116Q	86/49034 (4 aspect)	Nr	20.00	764.23	–	1158.45	1922.68	2691.75	3845.36
11010116R	86/49035 (3 aspect)	Nr	20.00	764.23	–	1158.45	1922.68	2691.75	3845.36

11010121   2056; comprising 912mm platform with off-set post; NRS catalogue number:

Ref	Description	Unit	Hours	Labour	Plant	Materials	Base	Green	Red
11010121A	86/49249 or 50 (3 or 4 aspect)	Nr	20.00	764.23	–	885.31	1649.54	2309.35	3299.07
11010121B	86/49253 or 54 (3 or 4 aspect)	Nr	20.00	764.23	–	982.75	1746.98	2445.78	3493.97

11010126   2057; comprising 912mm platform with off-set post and foot step; NRS catalogue number:

Ref	Description	Unit	Hours	Labour	Plant	Materials	Base	Green	Red
11010126A	86/49077 or 78 (4 aspect)	Nr	20.00	764.23	–	1336.56	2100.79	2941.10	4201.58
11010126B	86/49079 or 80 (3 aspect)	Nr	20.00	764.23	–	1334.42	2098.65	2938.10	4197.29
11010126C	86/49081 or 82 (4 aspect)	Nr	20.00	764.23	–	1372.01	2136.24	2990.73	4272.47
11010126D	86/49083 or 84 (3 aspect)	Nr	20.00	764.23	–	1327.30	2091.53	2928.14	4183.06
11010126E	86/49085 or 86 (4 aspect)	Nr	20.00	764.23	–	1383.40	2147.63	3006.68	4295.26
11010126F	86/49087 or 88 (3 aspect)	Nr	20.00	764.23	–	1334.87	2099.10	2938.74	4198.20
11010126G	86/49089 or 90 (4 aspect)	Nr	20.00	764.23	–	1383.41	2147.64	3006.69	4295.28
11010126H	86/49091 or 92 (3 aspect)	Nr	20.00	764.23	–	1335.12	2099.35	2939.09	4198.70

Signalling

Rail 2003 Ref		Unit	Labour Hours	Labour Cost	Plant Cost	Materials Cost	Unit Rate	Unit Rate	Unit Rate
				£	£	£	Base Cost £	Green Zone £	Red Zone £

11 NEW INSTALLATIONS – RESIGNALLING WORKS

110101 Posts; 168mm dia, 3000mm high, incl. holding down bolts (bases/foundations measured separately), BRS-SM drawing number:

11010126 2057; comprising 912mm platform with off-set post and foot step; NRS catalogue number:

11010126I	86/49093 or 94 (4 aspect)	Nr	20.00	764.23	–	1296.94	2061.17	2885.63	4122.33
11010126J	86/49095 or 96 (3 aspect)	Nr	20.00	764.23	–	1292.79	2057.02	2879.82	4114.03
11010126K	86/49097 or 98 (4 aspect)	Nr	20.00	764.23	–	1297.19	2061.42	2885.98	4122.84
11010126L	86/49099 or 100 (3 aspect)	Nr	20.00	764.23	–	1293.06	2057.29	2880.21	4114.59
11010126M	86/49101 or 102 (4 aspect)	Nr	20.00	764.23	–	1285.28	2049.51	2869.31	4099.02
11010126N	86/49103 or 104 (3 aspect)	Nr	20.00	764.23	–	1281.13	2045.36	2863.50	4090.72
11010126O	86/49105 or 106 (4 aspect)	Nr	20.00	764.23	–	1284.56	2048.79	2868.31	4097.58
11010126P	86/49107 or 108 (3 aspect)	Nr	20.00	764.23	–	1280.42	2044.65	2862.51	4089.31
11010126Q	86/49109 or 110 (4 aspect)	Nr	20.00	764.23	–	1251.44	2015.67	2821.93	4031.33
11010126R	86/49111 or 112 (3 aspect)	Nr	20.00	764.23	–	1251.44	2015.67	2821.93	4031.33

11010131 2058; comprising 1262mm platform with central post; NRS catalogue number:

11010131A	86/49237 or 39 (3 or 4 aspect)	Nr	20.00	764.23	–	1000.73	1764.96	2470.94	3529.92
11010131B	86/49241 or 43 (3 or 4 aspect)	Nr	20.00	764.23	–	975.60	1739.83	2435.76	3479.66
11010131C	86/49245 or 47 (3 or 4 aspect)	Nr	20.00	764.23	–	1073.05	1837.28	2572.19	3674.56

11010136 2059; comprising 1262mm platform with off-set post and foot step; NRS catalogue number:

11010136A	86/49045 or 46 (4 aspect)	Nr	20.00	764.23	–	1347.81	2112.04	2956.86	4224.08
11010136B	86/49047 or 48 (3 aspect)	Nr	20.00	764.23	–	1343.66	2107.89	2951.05	4215.78
11010136C	86/49049 or 50 (4 aspect)	Nr	20.00	764.23	–	1347.11	2111.34	2955.87	4222.67
11010136D	86/49051 or 52 (3 aspect)	Nr	20.00	764.23	–	1342.97	2107.20	2950.08	4214.40
11010136E	86/49061 or 62 (4 aspect)	Nr	20.00	764.23	–	1348.18	2112.41	2957.37	4224.82
11010136F	86/49063 or 64 (3 aspect)	Nr	20.00	764.23	–	1415.98	2180.21	3052.29	4360.42
11010136G	86/49065 or 66 (4 aspect)	Nr	20.00	764.23	–	1419.41	2183.64	3057.10	4367.29
11010136H	86/49067 or 68 (3 aspect)	Nr	20.00	764.23	–	1415.28	2179.51	3051.31	4359.01

Signalling

Rail 2003 Ref		Unit	Labour Hours	Labour Cost £	Plant Cost £	Materials Cost £	Unit Rate Base Cost £	Unit Rate Green Zone £	Unit Rate Red Zone £
11	**NEW INSTALLATIONS – RESIGNALLING WORKS**								
110101	Posts; 168mm dia, 3000mm high, incl. holding down bolts (bases/foundations measured separately), BRS-SM drawing number:								
11010141	2060; comprising 1262mm platform with off-set post; NRS catalogue number:								
11010141A	86/49257 or 58 (3 or 4 aspect)	Nr	20.00	764.23	–	956.15	1720.38	2408.53	3440.76
11010141B	86/49261 or 62 (3 or 4 aspect)	Nr	20.00	764.23	–	931.01	1695.24	2373.33	3390.47
11010141C	86/49265 or 66 (3 or 4 aspect)	Nr	20.00	764.23	–	1028.43	1792.66	2509.72	3585.32
11010146	2061; comprising 1262mm platform with central post and foot step; NRS catalogue number:								
11010146A	86/49113, 14, 18 or 21 (4 aspect)	Nr	20.00	764.23	–	1392.48	2156.71	3019.39	4313.42
11010146B	86/49115, 16, 20 or 23 (3 aspect)	Nr	20.00	764.23	–	1388.33	2152.56	3013.58	4305.12
11010146C	86/49129, 30, 34 or 37 (4 aspect)	Nr	20.00	764.23	–	1467.88	2232.11	3124.95	4464.22
11010146D	86/49131, 32, 36 or 39 (3 aspect)	Nr	20.00	764.23	–	1460.65	2224.88	3114.83	4449.76
11010151	2062; comprising 1500mm platform with central post; NRS catalogue number:								
11010151A	86/49209 or 10 (3 or 4 aspect)	Nr	20.00	764.23	–	1179.92	1944.15	2721.81	3888.30
11010151B	86/49213 or 14 (3 or 4 aspect)	Nr	20.00	764.23	–	1154.81	1919.04	2686.65	3838.07
11010151C	86/49217 or 18 (3 or 4 aspect)	Nr	20.00	764.23	–	1252.23	2016.46	2823.04	4032.92
11010156	2063; comprising 1500mm platform with central post and foot step; NRS catalogue number:								
11010156A	86/49145, 46, 50 and 53 (4 aspect)	Nr	20.00	764.23	–	1601.68	2365.91	3312.28	4731.83
11010156B	86/49147, 48, 52 and 55 (3 aspect)	Nr	20.00	764.23	–	1597.53	2361.76	3306.47	4723.53
11010156C	86/49161, 62, 66 and 69 (4 aspect)	Nr	20.00	764.23	–	1674.00	2438.23	3413.53	4876.47
11010156D	86/49163, 64, 68 and 71 (3 aspect)	Nr	20.00	764.23	–	1669.85	2434.08	3407.72	4868.17
11010161	2064; comprising 1500mm platform with off-set post; NRS catalogue number:								
11010161A	86/49221 (3 or 4 aspect)	Nr	20.00	764.23	–	1137.94	1902.17	2663.04	3804.34
11010161B	86/49225 (3 or 4 aspect)	Nr	20.00	764.23	–	1112.81	1877.04	2627.86	3754.08
11010161C	86/49229 (3 or 4 aspect)	Nr	20.00	764.23	–	1210.26	1974.49	2764.29	3948.98

Signalling

Rail 2003 Ref		Unit	Labour Hours	Labour Cost £	Plant Cost £	Materials Cost £	Unit Rate Base Cost £	Unit Rate Green Zone £	Unit Rate Red Zone £

11 NEW INSTALLATIONS – RESIGNALLING WORKS

110101 Posts; 168mm dia, 3000mm high, incl. holding down bolts (bases/foundations measured separately), BRS-SM drawing number:

11010166 2065; comprising 1500mm platform with off-set post and foot step; NRS catalogue number:

Ref	Description	Unit	Hours	Labour	Plant	Materials	Base	Green	Red
11010166A	86/49177, 178, 182 and 185 (4 aspect)	Nr	20.00	764.23	–	1599.40	2363.63	3309.08	4727.26
11010166B	86/49179, 180, 184 and 187 (3 aspect)	Nr	20.00	764.23	–	1595.25	2359.48	3303.27	4718.96
11010166C	86/49193, 194, 198 and 201 (4 aspect)	Nr	20.00	764.23	–	1671.71	2435.94	3410.31	4871.88
11010166D	86/49195, 196, 200 and 203 (3 aspect)	Nr	20.00	764.23	–	1667.57	2431.80	3404.52	4863.60

110121 Netting for structures; including tie rods, U-bolts and clamp brackets; to suit platform width:

11012101 612mm, 912mm with central post, 1262mm and 1500mm; NRS reference:

Ref		Unit	Hours	Labour	Plant	Materials	Base	Green	Red
11012101A	A	Nr	3.40	119.75	–	120.60	240.35	336.48	480.68
11012101B	B	Nr	6.60	232.45	–	241.31	473.76	663.25	947.51
11012101C	C	Nr	3.40	119.75	–	134.59	254.34	356.06	508.66
11012101D	D	Nr	6.60	232.45	–	269.17	501.62	702.26	1003.24
11012101E	E	Nr	3.60	126.79	–	309.48	436.27	610.77	872.54
11012101F	F	Nr	7.60	267.66	–	638.99	906.65	1269.31	1813.30
11012101G	G	Nr	7.00	246.53	–	444.08	690.61	966.86	1381.22
11012101H	H	Nr	3.40	119.75	–	150.25	270.00	378.00	540.00
11012101I	J	Nr	3.60	126.79	–	339.72	466.51	653.10	933.01
11012101J	K	Nr	7.60	267.66	–	699.45	967.11	1353.96	1934.23
11012101K	L	Nr	7.00	246.53	–	489.98	736.51	1031.12	1473.03

11012111 912mm with off-set post; NRS reference:

Ref		Unit	Hours	Labour	Plant	Materials	Base	Green	Red
11012111A	P	Nr	2.60	91.57	–	306.49	398.06	557.28	796.12
11012111B	Q	Nr	5.60	197.23	–	441.09	638.32	893.64	1276.62
11012111C	R	Nr	6.60	232.45	–	636.01	868.46	1215.84	1736.91
11012111D	S	Nr	2.60	91.57	–	328.74	420.31	588.44	840.62
11012111E	T	Nr	5.60	197.23	–	479.01	676.24	946.73	1352.46
11012111F	U	Nr	6.60	232.45	–	688.49	920.94	1289.30	1841.87

Signalling

Rail 2003 Ref		Unit	Labour Hours	Labour Cost £	Plant Cost £	Materials Cost £	Unit Rate Base Cost £	Unit Rate Green Zone £	Unit Rate Red Zone £
11	NEW INSTALLATIONS – RESIGNALLING WORKS								
110131	Signal heads; fixing to structure (structure measured separately)								
11013101	Long range high voltage; with filament switching relay and individual lamp transformer:								
11013101A	2 aspect	Nr	13.20	504.39	–	797.10	1301.49	1822.09	2602.98
11013101B	3 aspect	Nr	19.80	756.59	–	1053.72	1810.31	2534.43	3620.63
11013101C	4 aspect	Nr	26.40	1008.78	–	1436.50	2445.28	3423.41	4890.58
110151	Junction route indicators, fixing to structure (structure measured separately)								
11015101	Standard position combinations; BRS-SE 152 (issue C):								
11015101A	position 1	Nr	4.00	152.85	–	1613.05	1765.90	2472.25	3531.79
11015101B	position 1 and 2	Nr	4.00	152.85	–	3961.03	4113.88	5759.42	8227.75
11015101C	position 4	Nr	4.00	152.85	–	1613.05	1765.90	2472.25	3531.79
11015101D	position 4 and 5	Nr	4.00	152.85	–	3961.03	4113.88	5759.42	8227.75
11015101E	position 1 and 4	Nr	4.00	152.85	–	3961.03	4113.88	5759.42	8227.75
11015101F	position 1, 2 and 4	Nr	4.00	152.85	–	4635.92	4788.77	6704.27	9577.53
11015101G	position 1, 4 and 5	Nr	4.00	152.85	–	4635.92	4788.77	6704.27	9577.53
110161	Position light shunt signals; fix to structure (structure measured separately)								
11016101	Heads; fitted with B22D 3-lamp holder BRS-SE 160 (issue C):								
11016101A	one aspect; elevated mounting	Nr	4.00	152.85	–	478.92	631.77	884.47	1263.53
11016101B	two aspect; ground mounting	Nr	4.00	152.85	–	533.16	686.01	960.41	1372.02
11016101C	two aspect; limit of shunt	Nr	4.00	152.85	–	521.81	674.66	944.52	1349.31
11016151	Bases:								
11016151A	precast concrete, for Position Light Signal (BRS-SC36)	Nr	3.00	105.66	–	30.62	136.28	190.78	272.54

Signalling

Rail 2003 Ref		Unit	Labour Hours	Labour Cost	Plant Cost	Materials Cost	Unit Rate	Unit Rate	Unit Rate
				£	£	£	Base Cost £	Green Zone £	Red Zone £

11 NEW INSTALLATIONS – RESIGNALLING WORKS

110165 Axle counters

11016510 General equipment:

11016510A	SEL axle counter	Nr	–	–	–	24730.00	24730.00	34622.00	49460.00
11016510B	SEL balise counter	Nr	–	–	–	14842.00	14842.00	20778.80	29684.00
11016510C	Adtranz TEML32 controller	Nr	–	–	–	15460.00	15460.00	21644.00	30920.00
11016510F	Adtranz ELS-80 head	Nr	–	–	–	2319.00	2319.00	3246.60	4638.00
11016510H	TFM signal with AC ability	Nr	–	–	–	4638.00	4638.00	6493.20	9276.00
11016510J	TFM point	Nr	–	–	–	3092.25	3092.25	4329.15	6184.50
11016510K	cubicle installation complete	Nr	–	–	–	67750.00	67750.00	94850.00	135500.00
11016510L	TDM Telecode 80 Mk II	Nr	–	–	–	23190.00	23190.00	32466.00	46380.00

110171 Fibre optic route indicators; fix to structure; (structure measured separately)

11017101 Theatre type, to BR spec 1651 part 1:

11017101A	1 aspect	Nr	4.00	152.85	–	1608.38	1761.23	2465.71	3522.45
11017101B	2 aspect	Nr	4.26	162.78	–	1986.82	2149.60	3009.44	4299.21
11017101C	3 aspect	Nr	4.46	170.42	–	2226.50	2396.92	3355.70	4793.86
11017101D	4 aspect	Nr	4.80	183.42	–	2459.88	2643.30	3700.61	5286.58
11017101E	5 aspect	Nr	5.26	200.99	–	2617.56	2818.55	3945.98	5637.11
11017101F	6 aspect	Nr	6.00	229.27	–	2781.55	3010.82	4215.15	6021.65

11017111 Stencil type, to BR spec 1651 part 2:

11017111A	1 aspect	Nr	4.00	152.85	–	1645.49	1798.34	2517.67	3596.67
11017111B	2 aspect	Nr	4.26	162.78	–	1892.21	2054.99	2876.99	4109.98
11017111C	3 aspect	Nr	4.46	170.42	–	2207.58	2378.00	3329.20	4756.01
11017111D	4 aspect	Nr	4.80	183.42	–	2459.88	2643.30	3700.61	5286.58
11017111E	5 aspect	Nr	5.26	200.99	–	2838.32	3039.31	4255.04	6078.63
11017111F	6 aspect	Nr	6.00	229.27	–	3206.22	3435.49	4809.68	6870.97

11017121 Banner signals, to BR spec 1651 part 3:

11017121A	NRS ref 86/1260	Nr	10.40	397.40	–	2394.28	2791.68	3908.35	5583.36

Signalling

Rail 2003 Ref		Unit	Labour Hours	Labour Cost	Plant Cost	Materials Cost	Unit Rate	Unit Rate	Unit Rate
			£	£	£	Base Cost £	Green Zone £	Red Zone £
11	**NEW INSTALLATIONS – RESIGNALLING WORKS**								
110176	**Equipment generally**								
11017601	Miscellaneous indicators and equipment; supply and install:								
11017601A	CDSRA indicator	Nr	–	–	–	1418.40	1418.40	1985.76	2836.80
11017601B	TRTS/CD/RA plunger	Nr	–	–	–	320.10	320.10	448.14	640.20
11017601C	patrolman's L/O device	Nr	–	–	–	827.70	827.70	1158.78	1655.40
11017601D	off indicator	Nr	–	–	–	1645.50	1645.50	2303.70	3291.00
11017601E	indicator support	Nr	–	–	–	123.45	123.45	172.83	246.90
11017601F	TC interruptor	Nr	–	–	–	1125.00	1125.00	1575.00	2250.00
110181	**Extra over for miscellaneous fittings**								
11018101	Signal number plates, including mounting brackets:								
11018101A	for main signals or shunt signals, including mounting bracket	Nr	0.50	15.30	–	54.00	69.30	97.03	138.61
11018111	Security:								
11018111A	anti-climb device for signal post	Nr	2.00	61.20	–	409.29	470.49	658.67	940.96
11018121	Signal lamps; type:								
11018121A	SL35	Nr	0.50	15.30	–	9.92	25.22	35.30	50.43
11018121B	110V 35W	Nr	0.50	15.30	–	4.97	20.27	28.37	40.53
11018131	Mounting kits and brackets; extra over signal rate:								
11018131A	for roof/wall mounting kit	Nr	15.80	556.46	–	669.09	1225.55	1715.76	2451.09
11018131B	for off indicator mounting kit	Nr	12.00	422.63	–	165.30	587.93	823.10	1175.87
11018131C	to fix signal head to gantry	Nr	12.00	422.63	–	165.30	587.93	823.10	1175.87
11018131D	for RA/CD indicator mounting kit	Nr	12.00	422.63	–	658.99	1081.62	1514.27	2163.25
11018141	Extra over for additional height to post structure:								
11018141A	3300 – 3300mm high	Nr	–	–	–	553.13	553.13	774.38	1106.26
11018141B	3300 – 3600mm high	Nr	–	–	–	597.70	597.70	836.78	1195.40
11018141C	3600 – 3900mm high	Nr	–	–	–	643.73	643.73	901.22	1287.46
11018141D	3900 – 4200mm high	Nr	–	–	–	670.22	670.22	938.31	1340.44
11018141E	4200 – 4500mm high	Nr	–	–	–	717.19	717.19	1004.06	1434.37

Signalling

Rail 2003 Ref		Unit	Labour Hours	Labour Cost £	Plant Cost £	Materials Cost £	Unit Rate Base Cost £	Unit Rate Green Zone £	Unit Rate Red Zone £

11 NEW INSTALLATIONS – RESIGNALLING WORKS

1106 SIGNALLING STRUCTURES

110601 Gantries and cantilevers; (bases measured separately)

11060101 Spanning roads:

11060101A	1 road cantilever	Nr	80.00	2844.80	–	13555.32	16400.12	22960.16	32800.23
11060101B	2 road cantilever	Nr	112.00	3982.72	–	14671.64	18654.36	26116.11	37308.73
11060101C	2 road gantry	Nr	112.00	3982.72	–	16999.97	20982.69	29375.76	41965.37
11060101D	3 road gantry	Nr	172.00	6116.32	–	21560.93	27677.25	38748.16	55354.51
11060101E	4 road gantry	Nr	216.00	7680.96	–	26121.91	33802.87	47324.02	67605.74
11060101F	5 road gantry	Nr	284.00	10099.04	–	26329.21	36428.25	50999.55	72856.50
11060101G	6 road gantry	Nr	312.00	11094.72	–	35243.83	46338.55	64873.98	92677.11

1111 BASES FOR STRUCTURES

111101 Precast reinforced concrete bases; including holding down bolts and fixings (excavation, back filling and surfacing measured separately)

11110101 For signal posts, size:

11110101A	single, including 100mm duct	Nr	18.28	219.13	24.95	209.79	453.87	635.41	907.73
11110101B	double	Nr	27.45	329.07	37.38	291.34	657.79	920.90	1315.58
11110101C	triple	Nr	36.50	437.57	49.20	379.45	866.22	1212.69	1732.42
11110101D	quadruple	Nr	54.88	657.90	74.89	546.63	1279.42	1791.19	2558.84

11110121 For cantilevers, span:

11110121A	1 road	Nr	72.83	873.01	99.83	646.74	1619.58	2267.40	3239.15
11110121B	2 roads	Nr	99.09	1187.79	144.30	861.84	2193.93	3071.49	4387.86
11110121C	3 roads	Nr	135.56	1624.98	194.21	1532.51	3351.70	4692.38	6703.41

11110141 For gantries, span:

11110141A	2 roads	Nr	36.57	438.44	49.91	378.35	866.70	1213.38	1733.41
11110141B	3 roads	Nr	36.64	439.31	49.91	391.94	881.16	1233.63	1762.34
11110141C	4 roads	Nr	36.72	440.31	49.91	406.33	896.55	1255.18	1793.11
11110141D	5 roads	Nr	36.81	441.43	49.91	429.09	920.43	1288.61	1840.88
11110141E	6 roads	Nr	36.91	442.68	49.91	452.65	945.24	1323.34	1890.49

Signalling

Rail 2003 Ref		Unit	Labour Hours	Labour Cost £	Plant Cost £	Materials Cost £	Unit Rate Base Cost £	Unit Rate Green Zone £	Unit Rate Red Zone £
11	**NEW INSTALLATIONS – RESIGNALLING WORKS**								
1116	**CABLING**								
111601	Power and general cable and ancillaries								
11160101	External cable; lineside multicore, type B2; 0.75 mm; specification:								
11160101A	$0.75mm^2$; 1 core	m	0.09	3.17	–	0.14	3.31	4.64	6.62
11160101B	$0.75mm^2$; 2 core	m	0.10	3.52	–	0.27	3.79	5.31	7.58
11160101C	$0.75mm^2$; 4 core	m	0.13	4.58	–	0.56	5.14	7.19	10.28
11160101D	$0.75mm^2$; 7 core	m	0.15	5.28	–	1.41	6.69	9.38	13.40
11160101E	$0.75mm^2$; 10 core	m	0.18	6.34	–	2.11	8.45	11.83	16.91
11160101F	$0.75mm^2$; 12 core	m	0.20	7.04	–	1.97	9.01	12.62	18.03
11160101G	$0.75mm^2$; 19 core	m	0.24	8.45	–	2.83	11.28	15.80	22.58
11160101H	$0.75mm^2$; 27 core	m	0.30	10.57	–	3.71	14.28	19.99	28.56
11160101I	$0.75mm^2$; 37 core	m	0.45	15.85	–	4.72	20.57	28.80	41.15
11160101J	$0.75mm^2$; 48 core	m	0.50	17.61	–	5.87	23.48	32.87	46.96
11160105	External cable; lineside multicore, type B2; 1.5 mm; specification:								
11160105A	$1.5mm^2$; 1 core	m	0.10	3.52	–	0.17	3.69	5.16	7.37
11160105B	$1.5mm^2$; 2 core	m	0.11	3.87	–	0.32	4.19	5.87	8.40
11160105C	$1.5mm^2$; 4 core	m	0.14	4.93	–	0.84	5.77	8.07	11.54
11160105D	$1.5mm^2$; 7 core	m	0.17	5.99	–	1.78	7.77	10.87	15.53
11160105E	$1.5mm^2$; 10 core	m	0.20	7.04	–	2.41	9.45	13.23	18.91
11160105F	$1.5mm^2$; 12 core	m	0.22	7.75	–	2.81	10.56	14.78	21.12
11160105G	$1.5mm^2$; 19 core	m	0.28	9.86	–	3.55	13.41	18.78	26.82
11160105H	$1.5mm^2$; 27 core	m	0.35	12.33	–	5.12	17.45	24.42	34.88
11160105I	$1.5mm^2$; 37 core	m	0.50	17.61	–	6.63	24.24	33.93	48.48
11160105J	$1.5mm^2$; 48 core	m	0.66	23.25	–	8.38	31.63	44.27	63.25
11160110	External cable; lineside multicore, type B2; 2.5 mm; specification:								
11160110A	$2.5mm^2$; 1 core	m	0.12	4.23	–	0.88	5.11	7.15	10.21
11160110B	$2.5mm^2$; 2 core	m	0.13	4.58	–	1.45	6.03	8.44	12.06
11160110C	$2.5mm^2$; 4 core	m	0.16	5.64	–	2.13	7.77	10.88	15.54
11160110D	$2.5mm^2$; 7 core	m	0.18	6.34	–	2.02	8.36	11.70	16.73
11160110E	$2.5mm^2$; 10 core	m	0.22	7.75	–	2.70	10.45	14.63	20.90
11160110F	$2.5mm^2$; 12 core	m	0.25	8.80	–	3.32	12.12	16.98	24.25
11160110G	$2.5mm^2$; 19 core	m	0.30	10.57	–	4.63	15.20	21.27	30.39
11160110H	$2.5mm^2$; 27 core	m	0.36	12.68	–	6.28	18.96	26.54	37.91
11160110I	$2.5mm^2$; 37 core	m	0.52	18.31	–	4.72	23.03	32.25	46.08
11160110J	$2.5mm^2$; 48 core	m	0.68	23.95	–	9.19	33.14	46.39	66.27

Signalling

Rail 2003 Ref		Unit	Labour Hours	Labour Cost	Plant Cost	Materials Cost	Unit Rate	Unit Rate	Unit Rate
				£	£	£	Base Cost £	Green Zone £	Red Zone £
11	**NEW INSTALLATIONS – RESIGNALLING WORKS**								
111601	Power and general cable and ancillaries								
11160121	Power cable; aluminium; to BS6346; specification:								
11160121B	25mm^2, 2 core	m	0.36	12.68	–	4.11	16.79	23.50	33.58
11160121C	35mm^2, 2 core	m	0.40	14.09	–	5.35	19.44	27.21	38.88
11160121D	50mm^2, 2 core	m	0.45	15.85	–	6.28	22.13	30.98	44.26
11160121E	70mm^2, 2 core	m	0.52	18.31	–	8.53	26.84	37.59	53.70
11160121F	95mm^2, 2 core	m	0.65	22.89	–	12.19	35.08	49.11	70.16
11160141	Power cable; copper; type C1/C2; specification:								
11160141A	2.5mm^2, 1 core	m	0.12	4.23	–	0.99	5.22	7.31	10.44
11160141B	10mm^2, 1 core	m	0.22	7.75	–	1.81	9.56	13.39	19.12
11160141C	35mm^2, 1 core	m	0.30	10.57	–	2.52	13.09	18.32	26.17
11160141D	2.5mm^2, 2 core	m	0.13	4.58	–	1.43	6.01	8.41	12.02
11160141E	10mm^2, 2 core	m	0.22	7.75	–	2.33	10.08	14.11	20.16
11160141H	25mm^2, 2 core	m	0.30	10.57	–	4.23	14.80	20.72	29.60
11160141I	35mm^2, 2 core	m	0.42	14.79	–	5.51	20.30	28.43	40.60
11160141J	50mm^2, 2 core	m	0.48	16.91	–	6.60	23.51	32.90	47.00
11160141K	70mm^2, 2 core	m	0.56	19.72	–	8.96	28.68	40.16	57.37
11160141L	95mm^2, 2 core	m	0.65	22.89	–	12.80	35.69	49.97	71.38
11160151	Data cable:								
11160151A	0.9mm^2	m	0.12	4.36	–	2.10	6.46	9.04	12.91
11160156	Telephone cable:								
11160156A	2 pr.	m	0.10	3.63	–	1.85	5.48	7.67	10.96
11160156B	3 pr.	m	0.13	4.72	–	2.26	6.98	9.77	13.95
11160156C	4 pr.	m	0.18	6.53	–	2.79	9.32	13.06	18.66
11160156D	6 pr.	m	0.24	8.71	–	4.04	12.75	17.86	25.51
11160156E	10 pr.	m	0.30	10.89	–	5.65	16.54	23.16	33.08
11160156H	16 pr.	m	0.35	12.71	–	7.62	20.33	28.46	40.66
11160156I	20 pr.	m	0.40	14.52	–	8.35	22.87	32.02	45.74
11160156J	25 pr.	m	0.45	16.34	–	8.66	25.00	35.00	49.99
11160156K	30 pr.	m	0.52	18.88	–	9.52	28.40	39.76	56.80
11160156M	64 pr.	m	0.65	23.60	–	12.77	36.37	50.91	72.73
11160156N	128 pr.	m	0.75	27.23	–	14.96	42.19	59.07	84.38
11160156O	201 pr.	m	0.75	27.23	–	14.96	42.19	59.07	84.38

Signalling

Rail 2003 Ref		Unit	Labour Hours	Labour Cost £	Plant Cost £	Materials Cost £	Unit Rate Base Cost £	Unit Rate Green Zone £	Unit Rate Red Zone £
11	**NEW INSTALLATIONS – RESIGNALLING WORKS**								
111601	**Power and general cable and ancillaries**								
11160158	**Fibre optic cable; size:**								
11160158A	12 fibre	m	0.30	10.57	–	3.56	14.13	19.77	28.25
11160158C	20 fibre	m	0.32	11.27	–	4.40	15.67	21.94	31.34
11160158E	32 fibre	m	0.35	12.33	–	5.36	17.69	24.76	35.36
11160161	**Tail cables; type C1/C2; specification:**								
11160161A	2.5mm^2; 1 core	m	0.08	2.82	–	0.88	3.70	5.18	7.39
11160161B	2.5mm^2; 2 core	m	0.08	2.82	–	1.66	4.48	6.27	8.95
11160161C	2.5mm^2; 4 core	m	0.08	2.82	–	1.92	4.74	6.64	9.48
11160161D	2.5mm^2; 7 core	m	0.08	2.82	–	2.76	5.58	7.81	11.15
11160161E	2.5mm^2; 10 core	m	0.08	2.82	–	3.76	6.58	9.22	13.16
11160161F	2.5mm^2; 12 core	m	0.08	2.82	–	4.19	7.01	9.81	14.01
11160161G	2.5mm^2; 19 core	m	0.10	3.52	–	4.54	8.06	11.29	16.12
11160161H	2.5mm^2; 27 core	m	0.11	3.87	–	6.16	10.03	14.04	20.07
11160161I	2.5mm^2; 37 core	m	0.12	4.23	–	7.52	11.75	16.45	23.49
11160161J	2.5mm^2; 48 core	m	0.13	4.58	–	9.01	13.59	19.03	27.19
11160171	**Jointing; through joints; insulated copper cable; size:**								
11160171A	0.75 mm^2; 1 core	Nr	0.30	10.57	–	8.28	18.85	26.38	37.69
11160171B	0.75 mm^2; 2 core	Nr	0.33	11.62	–	8.28	19.90	27.86	39.80
11160171C	0.75 mm^2; 4 core	Nr	0.48	16.91	–	8.28	25.19	35.26	50.37
11160171D	0.75 mm^2; 7 core	Nr	0.78	27.47	–	8.28	35.75	50.05	71.50
11160171E	0.75 mm^2; 10 core	Nr	1.00	35.22	–	8.28	43.50	60.90	87.00
11160171F	0.75 mm^2; 12 core	Nr	1.20	42.26	–	8.28	50.54	70.76	101.09
11160171G	0.75 mm^2; 19 core	Nr	1.48	52.12	–	13.35	65.47	91.66	130.95
11160171H	0.75 mm^2; 27 core	Nr	2.00	70.44	–	17.40	87.84	122.98	175.69
11160171I	0.75 mm^2; 37 core	Nr	2.40	84.53	–	26.27	110.80	155.11	221.58
11160171J	0.75 mm^2; 48 core	Nr	3.00	105.66	–	26.27	131.93	184.69	263.84
11160173	**Jointing; through joints; insulated copper cable; size:**								
11160173A	2.5 mm^2; 1 core	Nr	0.32	11.27	–	8.28	19.55	27.37	39.10
11160173B	2.5 mm^2; 2 core	Nr	0.39	13.73	–	8.28	22.01	30.82	44.03
11160173C	2.5 mm^2; 4 core	Nr	0.58	20.43	–	8.28	28.71	40.19	57.41
11160173D	2.5 mm^2; 7 core	Nr	0.75	26.41	–	8.28	34.69	48.57	69.39
11160173E	2.5 mm^2; 10 core	Nr	1.04	36.63	–	8.28	44.91	62.87	89.82
11160173F	2.5 mm^2; 12 core	Nr	1.25	44.02	–	8.28	52.30	73.22	104.61
11160173G	2.5 mm^2; 19 core	Nr	1.52	53.53	–	13.35	66.88	93.64	133.77
11160173H	2.5 mm^2; 27 core	Nr	2.05	72.20	–	17.40	89.60	125.45	179.21

Signalling

Rail 2003 Ref		Unit	Labour Hours	Labour Cost	Plant Cost	Materials Cost	Unit Rate	Unit Rate	Unit Rate
				£	£	£	Base Cost £	Green Zone £	Red Zone £
11	**NEW INSTALLATIONS – RESIGNALLING WORKS**								
111601	Power and general cable and ancillaries								
11160173	Jointing; through joints; insulated copper cable; size:								
11160173I	2.5 mm^2; 37 core	Nr	2.55	89.81	–	38.92	128.73	180.22	257.46
11160173J	2.5 mm^2; 48 core	Nr	3.15	110.94	–	38.92	149.86	209.81	299.72
11160177	Jointing; through joints; telephone cable; size:								
11160177A	2 pr	Nr	0.10	3.52	–	4.70	8.22	11.51	16.44
11160177B	10 pr	Nr	0.15	5.28	–	4.70	9.98	13.98	19.97
11160177D	30 pr	Nr	0.20	7.04	–	6.85	13.89	19.45	27.79
11160177H	50 pr	Nr	0.35	12.33	–	9.35	21.68	30.35	43.35
11160177I	75 pr	Nr	0.65	22.89	–	9.35	32.24	45.14	64.48
11160181	Terminations; cable size 0.75 mm^2:								
11160181A	0.75 mm^2; 1 core	Nr	0.45	15.85	–	28.56	44.41	62.17	88.81
11160181B	0.75 mm^2; 2 core	Nr	0.49	17.26	–	28.56	45.82	64.14	91.62
11160181C	0.75 mm^2; 4 core	Nr	0.72	25.36	–	28.56	53.92	75.48	107.83
11160181D	0.75 mm^2; 7 core	Nr	1.15	40.50	–	28.56	69.06	96.68	138.11
11160181E	0.75 mm^2; 10 core	Nr	1.45	51.07	–	28.56	79.63	111.48	159.25
11160181F	0.75 mm^2; 12 core	Nr	1.62	57.06	–	28.56	85.62	119.86	171.22
11160181G	0.75 mm^2; 19 core	Nr	2.15	75.72	–	28.56	104.28	145.99	208.55
11160181H	0.75 mm^2; 27 core	Nr	2.70	95.09	–	28.56	123.65	173.11	247.29
11160181I	0.75 mm^2; 37 core	Nr	3.42	120.45	–	28.56	149.01	208.61	298.01
11160181J	0.75 mm^2; 48 core	Nr	3.75	132.07	–	28.56	160.63	224.88	321.25
11160183	Terminations; cable size 2.5 mm^2:								
11160183A	2.5 mm^2; 1 core	Nr	0.47	16.55	–	28.56	45.11	63.15	90.22
11160183B	2.5 mm^2; 2 core	Nr	0.50	17.61	–	28.56	46.17	64.63	92.33
11160183C	2.5 mm^2; 4 core	Nr	0.74	26.06	–	28.56	54.62	76.47	109.23
11160183D	2.5 mm^2; 7 core	Nr	1.20	42.26	–	28.56	70.82	99.15	141.64
11160183E	2.5 mm^2; 10 core	Nr	1.49	52.48	–	28.56	81.04	113.45	162.06
11160183F	2.5 mm^2; 12 core	Nr	1.68	59.17	–	28.56	87.73	122.82	175.45
11160183G	2.5 mm^2; 19 core	Nr	2.35	82.77	–	28.56	111.33	155.85	222.64
11160183H	2.5 mm^2; 27 core	Nr	2.90	102.13	–	28.56	130.69	182.97	261.38
11160183I	2.5 mm^2; 37 core	Nr	3.65	128.55	–	28.56	157.11	219.95	314.21
11160183J	2.5 mm^2; 48 core	Nr	3.99	140.52	–	28.56	169.08	236.71	338.16

Signalling

Rail 2003 Ref		Unit	Labour Hours	Labour Cost	Plant Cost	Materials Cost	Unit Rate	Unit Rate	Unit Rate
				£	£	£	Base Cost £	Green Zone £	Red Zone £

11 NEW INSTALLATIONS – RESIGNALLING WORKS

111601 Power and general cable and ancillaries

11160185 Terminations; 2 core cable; size:

11160185A	2.5 mm^2	Nr	0.50	17.61	–	28.56	46.17	64.63	92.33
11160185B	10 mm^2	Nr	0.59	20.78	–	28.56	49.34	69.07	98.67
11160185C	12 mm^2	Nr	0.68	23.95	–	28.56	52.51	73.51	105.01
11160185D	16 mm^2	Nr	1.15	40.50	–	34.59	75.09	105.13	150.19
11160185E	25 mm^2	Nr	1.35	47.55	–	34.59	82.14	114.99	164.28
11160185F	35 mm^2	Nr	1.95	68.68	–	68.13	136.81	191.53	273.61
11160185G	50 mm^2	Nr	2.35	82.77	–	68.13	150.90	211.25	301.79
11160185H	70 mm^2	Nr	3.25	114.46	–	68.13	182.59	255.63	365.18
11160185I	95 mm^2	Nr	5.40	190.18	–	99.54	289.72	405.62	579.45

11160187 Terminations; telephone wire; size:

11160187A	2 pr	Nr	0.22	7.75	–	4.95	12.70	17.78	25.40
11160187B	10 pr	Nr	0.28	9.86	–	4.95	14.81	20.74	29.62
11160187D	30 pr	Nr	0.36	12.68	–	7.30	19.98	27.97	39.96
11160187H	50 pr	Nr	0.56	19.72	–	9.35	29.07	40.70	58.15
11160187I	75 pr	Nr	0.86	30.29	–	9.35	39.64	55.49	79.28

11160191 Miscellaneous activities:

11160191A	draw 1 Nr cable through 90mm duct	m	0.50	15.30	–	–	15.30	21.42	30.60
11160191B	draw 2 Nr cables through 90mm duct	m	0.75	22.95	–	–	22.95	32.13	45.90

1121 CABLE ROUTES

112101 Trough; reinforced concrete, prefabricated, complete with lid

11210101 Straight; 1.0m long, with locating nib and recess and 2 Nr knockout panels; trough internal dimensions:

11210101A	130mm wide, 130mm deep	m	1.00	30.60	–	7.63	38.23	53.52	76.45
11210101B	190mm wide, 130mm deep	m	1.00	30.60	–	8.68	39.28	54.99	78.56
11210101C	350mm wide, 300mm deep	m	1.50	45.90	–	26.74	72.64	101.69	145.27
11210101D	350mm wide, 130mm deep	m	1.25	38.25	–	14.32	52.57	73.60	105.14

Signalling

Rail 2003 Ref		Unit	Labour Hours	Labour Cost £	Plant Cost £	Materials Cost £	Unit Rate Base Cost £	Unit Rate Green Zone £	Unit Rate Red Zone £
11	**NEW INSTALLATIONS – RESIGNALLING WORKS**								
112101	Trough; reinforced concrete, prefabricated, complete with lid								
11210151	Tee junction; with locating recess at the end of each arm; trough internal dimensions:								
11210151A	130mm wide, 130mm deep	Nr	1.50	45.90	–	38.85	84.75	118.64	169.48
11210151B	190mm wide, 130mm deep	Nr	1.50	45.90	–	44.39	90.29	126.41	180.58
11210151C	350mm wide, 300mm deep	Nr	1.88	57.52	–	52.78	110.30	154.42	220.61
11210151D	350mm wide, 130mm deep	Nr	2.25	68.85	–	53.07	121.92	170.67	243.82
11210191	Miscellaneous activities:								
11210191A	apply anti-vermin foam sealant at junction of location case and cable troughing	Nr	0.30	9.18	–	2.50	11.68	16.35	23.36
11210191B	apply anti-vermin foam sealant to joints generally	m	0.50	15.30	–	2.50	17.80	24.92	35.60
11210191C	carefully lift and realign existing cable trough, including realignment of existing cables within	m	1.50	45.90	–	1.36	47.26	66.17	94.52
112121	Pipe; PVC, orange, for cabling								
11212101	90 mm diameter; laid:								
11212101A	laid in 6m length under track, including pipe jointing and clips; excavating in ballast and drawing cables measured separately	m	0.40	12.24	–	3.94	16.18	22.65	32.37

Signalling

Rail 2003 Ref		Unit	Labour Hours	Labour Cost £	Plant Cost £	Materials Cost £	Unit Rate Base Cost £	Unit Rate Green Zone £	Unit Rate Red Zone £
11	NEW INSTALLATIONS – RESIGNALLING WORKS								
1126	TRACKSIDE CIRCUITS								
112601	AC traction								
11260101	TI21 equipment; frequencies A-H:								
11260101A	track transmitter; jointless	Nr	4.60	175.77	–	630.49	806.26	1128.76	1612.52
11260101B	receiver; jointless	Nr	4.60	175.77	–	810.62	986.39	1380.95	1972.80
11260101C	tuning unit; jointless	Nr	4.60	175.77	–	645.50	821.27	1149.77	1642.54
11260101D	termination unit; end; jointless	Nr	4.60	175.77	–	870.67	1046.44	1465.02	2092.89
11260101E	mounting stake; angle; galvanised 1370mm long	Nr	1.50	45.90	–	15.14	61.04	85.45	122.06
112671	DC traction, AC circuits								
11267101	Impedance bonds, including connection to track:								
11267101A	tuned; frequencies A-H	Nr	4.60	175.77	–	279.92	455.69	637.97	911.39
11267101B	B3 4000a	Nr	4.60	175.77	–	1570.26	1746.03	2444.44	3492.07
11267101C	non-tuned, Type 3, BR863, suitable for 600-1500V DC, 3000A per band, complete with variable capacitor, 20 micro farad	Nr	4.60	175.77	–	1976.79	2152.56	3013.59	4305.14
112681	Miscellaneous equipment								
11268101	Accessories to track circuits:								
11268101A	power supply unit with mounting plate	Nr	4.00	152.85	–	320.29	473.14	662.38	946.27
11268101B	surge arrestor, 26A, striking voltage 150-350V, 3 electrode/2 wire type complete with mounting	Nr	4.00	152.85	–	44.91	197.76	276.85	395.51
11268121	High voltage impulse:								
11268121A	fuse holder, complete with 6A fuse, and install	Nr	1.00	35.22	–	11.01	46.23	64.73	92.46

Signalling

Rail 2003 Ref		Unit	Labour Hours	Labour Cost £	Plant Cost £	Materials Cost £	Unit Rate Base Cost £	Unit Rate Green Zone £	Unit Rate Red Zone £
11	**NEW INSTALLATIONS – RESIGNALLING WORKS**								
1131	**LEVEL CROSSINGS**								
113101	Lifting barriers; to BRS 843, right or left hand, complete with machine (bases measured separately); type:								
11310101	Without skirt or support member; length pivot to tip:								
11310101A	3600mm	Nr	28.00	995.68	–	7550.89	8546.57	11965.20	17093.15
11310101B	4100mm	Nr	28.00	995.68	–	7614.08	8609.76	12053.66	17219.52
11310101C	4600mm	Nr	28.00	995.68	–	7662.57	8658.25	12121.55	17316.50
11310101D	5100mm	Nr	28.00	995.68	–	7746.76	8742.44	12239.42	17484.89
11310101E	5600mm	Nr	28.00	995.68	–	7813.11	8808.79	12332.30	17617.57
11310101F	6100mm	Nr	28.00	995.68	–	7997.60	8993.28	12590.59	17986.57
11310101G	6600mm	Nr	28.00	995.68	–	8107.62	9103.30	12744.62	18206.60
11310101H	7100mm (right hand only)	Nr	34.00	1209.04	–	10536.00	11745.04	16443.06	23490.08
11310101I	7600mm (right hand only)	Nr	34.00	1209.04	–	10703.59	11912.63	16677.69	23825.26
11310101J	8100mm (right hand only)	Nr	34.00	1209.04	–	10853.49	12062.53	16887.55	24125.07
11310101K	8600mm (right hand only)	Nr	34.00	1209.04	–	11081.51	12290.55	17206.77	24581.09
11310101L	9100mm (right hand only)	Nr	34.00	1209.04	–	11300.52	12509.56	17513.38	25019.11
11310111	With support member but without skirt; length pivot to tip:								
11310111A	7100mm	Nr	39.20	1393.95	–	9969.48	11363.43	15908.80	22726.86
11310111B	7600mm	Nr	39.20	1393.95	–	10071.44	11465.39	16051.55	22930.78
11310111C	8100mm	Nr	39.20	1393.95	–	10224.34	11618.29	16265.61	23236.59
11310111D	8600mm	Nr	39.20	1393.95	–	10399.44	11793.39	16510.75	23586.78
11310111E	9100mm	Nr	39.20	1393.95	–	10547.69	11941.64	16718.29	23883.27
11310121	With skirt but without support member; length pivot to tip:								
11310121A	3600mm	Nr	33.60	1194.82	–	7822.82	9017.64	12624.69	18035.27
11310121B	4100mm	Nr	33.60	1194.82	–	7960.32	9155.14	12817.19	18310.28
11310121C	4600mm	Nr	33.60	1194.82	–	8038.04	9232.86	12925.99	18465.71
11310121D	5100mm	Nr	33.60	1194.82	–	8172.30	9367.12	13113.97	18734.24
11310121E	5600mm	Nr	33.60	1194.82	–	8269.45	9464.27	13249.97	18928.53
11310121F	6100mm	Nr	33.60	1194.82	–	8591.53	9786.35	13700.88	19572.68
11310121G	6600mm	Nr	33.60	1194.82	–	8657.87	9852.69	13793.76	19705.37

Signalling

Rail 2003 Ref		Unit	Labour Hours	Labour Cost £	Plant Cost £	Materials Cost £	Unit Rate Base Cost £	Unit Rate Green Zone £	Unit Rate Red Zone £

11   **NEW INSTALLATIONS – RESIGNALLING WORKS**

113101   Lifting barriers; to BRS 843, right or left hand, complete with machine (bases measured separately); type:

11310131   With skirt and support member; length pivot to tip:

11310131A	7100mm (left hand only)............	Nr	44.80	1593.09	–	10536.00	12129.09	16980.72	24258.18
11310131B	7600mm (left hand only)............	Nr	44.80	1593.09	–	10687.79	12280.88	17193.23	24561.77
11310131C	8100mm (left hand only)............	Nr	44.80	1593.09	–	10853.49	12446.58	17425.21	24893.17
11310131D	8600mm (left hand only)............	Nr	44.80	1593.09	–	11081.51	12674.60	17744.43	25349.19
11310131E	9100mm (left hand only)............	Nr	44.80	1593.09	–	11300.52	12893.61	18051.04	25787.21

113121   Lights

11312101   Flashing road warning light:

11312101A	lamp and backboard assembly, comprising arm, backboard, lamps, conduit, brackets and cover, and fix to structure	Nr	16.00	611.38	–	1583.35	2194.73	3072.63	4389.48
11312101B	border, red/white retro reflective material, and fix to lamp and backboard assembly	Nr	8.00	305.69	–	144.81	450.50	630.70	900.99

11312111   Driver's crossing indicator:

11312111A	signal unit, 2 aspect, (white upper, red lower), BR spec 1970, and fix to structure	Nr	10.60	405.04	–	434.33	839.37	1175.12	1678.73

11312121   Miniature warning light:

11312121A	unit, miniature red and green light, BRS SE82 (issue H), and fix to structure ...	Nr	10.60	405.04	–	935.51	1340.55	1876.77	2681.10
11312121B	hood, aluminium, 140mm long, for miniature warning light, and fix to light	Nr	3.00	98.73	–	24.20	122.93	172.09	245.84

11312181   Posts with bases:

11312181A	for road signal, 4875mm high, steel cylindrical hollow section 115m diameter. Bolt to precast reinforced concrete plinth BRS-SC43 and fix to plinth to ground level surfacing	Nr	60.00	2133.60	–	504.25	2637.85	3692.99	5275.70
11312181B	for miniature red and green warning light and associated warning notices, tubular steel. Bolt to precast reinforced concrete base to BRS-SM91 and fix base to ground level surfacing	Nr	40.00	1422.40	–	242.59	1664.99	2330.99	3329.99

Signalling

Rail 2003 Ref		Unit	Labour Hours	Labour Cost £	Plant Cost £	Materials Cost £	Unit Rate Base Cost £	Unit Rate Green Zone £	Unit Rate Red Zone £

11 NEW INSTALLATIONS – RESIGNALLING WORKS

113141 Cable

11314101 Communications cable:

11314101A	coaxial cable type A, 1/1.829 copper, polythene insulated, taped and sheathed, including sundry ties, markers and labels as required (jointing, glands and sleeves measured separately)................	Nr	0.08	2.82	–	9.80	12.62	17.67	25.24
11314101B	extra over to form joint to coaxial cable type A with splicing kit.............	Nr	0.50	17.61	–	55.10	72.71	101.79	145.42

113161 Treadles

11316101 Non-directional:

11316101A	'cantor' single arm, with 2 Nr c/o contacts and mounting bracket for fixing bolts for flat bottomed rail......	Nr	16.00	611.38	–	958.53	1569.91	2197.88	3139.83

11316151 Direction sensitive; A-B and/or B-A:

11316151A	'forfex' double arm, with 2 Nr c/o contacts and mounting bracket and fixing bolts for flat bottomed rail......	Nr	16.00	611.38	–	1631.71	2243.09	3140.33	4486.18

Rail 2003 Ref		Unit		Composite Cost £	Unit Rate Base Cost £	Unit Rate Green Zone £	Unit Rate Red Zone £

113171 CCTV (closed circuit television)

11317110 Monitors:

11317110A	monitor complete...................	Nr	–	–	–	4638.00	4638.00	6493.20	9276.00
11317110B	PECA model 5143.................	Nr	–	–	–	1855.00	1855.00	2597.00	3710.00
11317110C	monitor housing...................	Nr	–	–	–	2319.00	2319.00	3246.60	4638.00
11317110D	double monitor housing	Nr	–	–	–	1391.00	1391.00	1947.40	2782.00
11317110E	two tier double monitor housing	Nr	–	–	–	1623.00	1623.00	2272.20	3246.00
11317110F	hydraulic pole; 2-section............	Nr	–	–	–	2783.00	2783.00	3896.20	5566.00
11317110G	monitor pole; 1.5m long	Nr	–	–	–	2165.00	2165.00	3031.00	4330.00
11317110H	core drill precast concrete platform; 50mm dia	Nr	–	–	–	371.00	371.00	519.40	742.00
11317110I	core drill precast concrete platform; 75mm dia	Nr	–	–	–	618.00	618.00	865.20	1236.00
11317110J	barrier fence to monitor pole.........	Nr	–	–	–	1546.00	1546.00	2164.40	3092.00
11317110K	video enhancer; PECA type 6040	Nr	–	–	–	928.00	928.00	1299.20	1856.00
11317110L	train mass detector	Nr	–	–	–	1160.00	1160.00	1624.00	2320.00

Signalling

Rail 2003 Ref		Unit	Composite Cost £	Unit Rate Base Cost £	Unit Rate Green Zone £	Unit Rate Red Zone £
11	**NEW INSTALLATIONS – RESIGNALLING WORKS**					
113171	**CCTV (closed circuit television)**					
11317115	**Cameras:**					
11317115A	camera complete	Nr	– – – 2938.00	2938.00	4113.20	5876.00
11317115B	Burle ref. TC652 BTX	Nr	– – – 541.00	541.00	757.40	1082.00
11317115C	auto-iris lens; 16mm	Nr	– – – 124.00	124.00	173.60	248.00
11317115D	auto-iris lens; 25mm	Nr	– – – 140.00	140.00	196.00	280.00
11317115E	camera housing	Nr	– – – 348.00	348.00	487.20	696.00
11317115F	hinged camera pole	Nr	– – – 1855.00	1855.00	2597.00	3710.00
11317115G	core drill precast concrete platform; 75mm dia	Nr	– – – 541.00	541.00	757.40	1082.00
11317115H	camera pole extension arm; 500mm long	Nr	– – – 309.00	309.00	432.60	618.00
11317115I	camera mounting bracket	Nr	– – – 1005.00	1005.00	1407.00	2010.00
11317115J	camera wall mounting bracket	Nr	– – – 1044.00	1044.00	1461.60	2088.00
11317120	**Power supply:**					
11317120A	distribution box	Nr	– – – 657.00	657.00	919.80	1314.00
11317120B	low voltage step-down transformer; 0.5 kVA 240/110V; complete	Nr	– – – 348.00	348.00	487.20	696.00
11317125	**Heated mirrors:**					
11317125A	complete	Nr	– – – 5411.00	5411.00	7575.40	10822.00
11317125B	landscape hooded membrane; 5 × 4	Nr	– – – 4870.00	4870.00	6818.00	9740.00
11317125C	portrait hooded membrane; 4 × 5	Nr	– – – 5643.00	5643.00	7900.20	11286.00
11317125D	landscape hooded membrane; 4 × 3	Nr	– – – 4020.00	4020.00	5628.00	8040.00
11317125E	portrait hooded membrane; 3 × 4	Nr	– – – 4329.00	4329.00	6060.60	8658.00
11317125F	mirror pole; 3.5m high	Nr	– – – 696.00	696.00	974.40	1392.00
11317125G	mirror pole; 4.5m high	Nr	– – – 773.00	773.00	1082.20	1546.00
11317125H	flange plate and fixing bolts	Nr	– – – 618.00	618.00	865.20	1236.00
11317125I	mirror yoke and safety collar	Nr	– – – 232.00	232.00	324.80	464.00
11317135	**Ancillaries:**					
11317135A	LLPA interface	Nr	– – – 25124.00	25124.00	35173.60	50248.00
11317135B	spares; allowance	Nr	– – – 1623.00	1623.00	2272.20	3246.00
11317135C	system integration	Nr	– – – 7731.00	7731.00	10823.40	15462.00
11317135D	optical patch cords	Nr	– – – 155.00	155.00	217.00	310.00
11317135E	GDT	Nr	– – – 39.00	39.00	54.60	78.00
11317135F	TMS adapters	Nr	– – – 2474.00	2474.00	3463.60	4948.00
11317135G	socket strip	Nr	– – – 309.00	309.00	432.60	618.00
11317135H	service terminals	Nr	– – – 696.00	696.00	974.40	1392.00

Signalling

Rail 2003 Ref		Unit	Labour Hours	Labour Cost £	Plant Cost £	Materials Cost £	Unit Rate Base Cost £	Unit Rate Green Zone £	Unit Rate Red Zone £
11	**NEW INSTALLATIONS – RESIGNALLING WORKS**								
1136	**LOCATION CASES AND RELOCATABLE EQUIPMENT BUILDINGS**								
113601	Bases and platforms; precast reinforced concrete, including holding down bolts and fixings (excavation, back filling and surfacing measured separately)								
11360101	For large apparatus location case:								
11360101A	base unit, precast reinforced concrete to BRS SC31/1, complete with 4 Nr corner units	Nr	9.00	296.18	–	72.33	368.51	515.91	737.01
11360101B	block filler for large steel apparatus case base	Nr	3.00	98.73	–	3.05	101.78	142.49	203.56
11360111	For small apparatus location case:								
11360111A	base unit, precast reinforced concrete to BRS SC32	Nr	3.00	98.73	–	56.20	154.93	216.90	309.85
11360120	For Relocatable Equipment Buildings; to suit REB size:								
11360120A	2.4 × 2.4 m long	Nr	13.04	169.13	6.59	109.68	285.40	399.55	570.79
11360120B	2.4 × 3.6 m long	Nr	17.81	229.54	15.02	168.58	413.14	578.40	826.30
11360120C	2.4 × 4.8 m long	Nr	22.50	288.99	17.70	226.74	533.43	746.79	1066.84
11360120D	2.4 × 6.0 m long	Nr	27.19	348.45	24.96	277.39	650.80	911.13	1301.62
11360120E	2.4 × 7.2 m long	Nr	31.89	408.02	29.93	327.48	765.43	1071.60	1530.86
11360120F	2.4 × 8.4 m long	Nr	36.59	467.61	34.89	385.66	888.16	1243.43	1776.31
11360120G	2.4 × 9.6 m long	Nr	41.35	527.90	39.99	437.00	1004.89	1406.84	2009.77
11360120H	2.4 × 10.8 m long	Nr	46.04	587.35	44.95	487.63	1119.93	1567.91	2239.85
11360120I	2.4 × 12.0 m long	Nr	50.75	647.05	49.91	545.30	1242.26	1739.18	2484.54
113611	Apparatus location cases; steel, including fixing to bases. Internal fittings measured separately. Size:								
11361101	Large:								
11361101A	case type A1; to BRS-SM431	Nr	13.00	496.75	–	625.81	1122.56	1571.58	2245.11
11361151	Small:								
11361151B	case type B1; to BRS-SM431	Nr	12.00	458.54	–	517.22	975.76	1366.05	1951.51

Signalling

Rail 2003 Ref		Unit	Labour Hours	Labour Cost £	Plant Cost £	Materials Cost £	Unit Rate Base Cost £	Unit Rate Green Zone £	Unit Rate Red Zone £
11	**NEW INSTALLATIONS – RESIGNALLING WORKS**								
113621	**Relocatable equipment buildings; 2.4m wide, air conditioned; excluding delivery**								
11362101	Equipment room only; size:								
11362101A	Code 1; 2.4m long	Nr	96.00	3413.76	–	6289.53	9703.29	13584.60	19406.57
11362101b	Code 2; 3.6m long	Nr	104.00	3698.24	–	7271.33	10969.57	15357.40	21939.14
11362101c	Code 3; 4.8m long	Nr	112.00	3982.72	–	8221.95	12204.67	17086.54	24409.35
11362101d	Code 4; 6.0m long	Nr	120.00	4267.20	–	9370.55	13637.75	19092.85	27275.50
11362101e	Code 5; 7.2m long	Nr	128.00	4551.68	–	10380.83	14932.51	20905.51	29865.02
11362101f	Code 6; 8.4m long	Nr	136.00	4836.16	–	11496.89	16333.05	22866.27	32666.11
11362101g	Code 7; 9.6m long	Nr	144.00	5120.64	–	12532.94	17653.58	24715.02	35307.17
11362101h	Code 8; 10.8m long	Nr	152.00	5405.12	–	13600.18	19005.30	26607.42	38010.59
11362101i	Code 9; 12.0m long	Nr	160.00	5689.60	–	14629.45	20319.05	28446.67	40638.10
11362111	Extra over for equipment room and battery room; size:								
11362111A	Code 2 to Code 9 inclusive	Nr	14.00	497.84	–	878.74	1376.58	1927.22	2753.17
11362121	Extra over for equipment room and operators room; size:								
11362121A	Code 2 to Code 9 inclusive	Nr	14.00	497.84	–	869.26	1367.10	1913.94	2734.19
11362131	Extra over for equipment room, operators room and battery room; size:								
11362131A	Code 4 to Code 9 inclusive	Nr	22.00	782.32	–	1478.13	2260.45	3164.64	4520.91
11362181	Extra over for computer sub floor, installed; size:								
11362181A	Code 1 to Code 9 inclusive	m²	4.00	142.24	–	146.43	288.67	404.14	577.34
11362186	Extra over for operator's window; installed:								
11362186A	fitting out	Nr	–	–	–	142.40	142.40	199.35	284.79

Signalling

Rail 2003 Ref		Unit	Labour Hours	Labour Cost £	Plant Cost £	Materials Cost £	Unit Rate Base Cost £	Unit Rate Green Zone £	Unit Rate Red Zone £
11	**NEW INSTALLATIONS – RESIGNALLING WORKS**								
113626	Relocatable Equipment Buildings; extra over for delivery, off loading and levelling (bases and drainage measured separately)								
11362601	Size code 1; radius of depot:								
11362601A	50 miles	Nr	–	–	–	88.30	88.30	123.62	176.61
11362601B	100 miles	Nr	–	–	–	126.15	126.15	176.61	252.30
11362601C	150 miles	Nr	–	–	–	151.38	151.38	211.93	302.75
11362601D	200 miles	Nr	–	–	–	181.65	181.65	254.31	363.30
11362601E	250 miles	Nr	–	–	–	227.07	227.07	317.89	454.13
11362601F	300 miles	Nr	–	–	–	317.89	317.89	445.05	635.78
11362601G	400 miles	Nr	–	–	–	489.45	489.45	685.23	978.91
11362602	Size code 2; radius of depot:								
11362602A	50 miles	Nr	–	–	–	103.44	103.44	144.82	206.88
11362602B	100 miles	Nr	–	–	–	126.15	126.15	176.61	252.30
11362602C	150 miles	Nr	–	–	–	151.38	151.38	211.93	302.75
11362602D	200 miles	Nr	–	–	–	181.65	181.65	254.31	363.30
11362602E	250 miles	Nr	–	–	–	227.07	227.07	317.89	454.13
11362602F	300 miles	Nr	–	–	–	317.89	317.89	445.05	635.78
11362602G	400 miles	Nr	–	–	–	489.45	489.45	685.23	978.91
11362603	Size code 3; radius of depot:								
11362603A	50 miles	Nr	–	–	–	113.53	113.53	158.95	227.07
11362603B	100 miles	Nr	–	–	–	151.38	151.38	211.93	302.75
11362603C	150 miles	Nr	–	–	–	181.65	181.65	254.31	363.30
11362603D	200 miles	Nr	–	–	–	242.20	242.20	339.09	484.41
11362603E	250 miles	Nr	–	–	–	302.75	302.75	423.86	605.51
11362603F	300 miles	Nr	–	–	–	423.86	423.86	593.40	847.71
11362603G	400 miles	Nr	–	–	–	567.66	567.66	794.73	1135.33
11362604	Size code 4; radius of depot:								
11362604A	50 miles	Nr	–	–	–	126.15	126.15	176.61	252.30
11362604B	100 miles	Nr	–	–	–	176.61	176.61	247.25	353.21
11362604C	150 miles	Nr	–	–	–	227.07	227.07	317.89	454.13
11362604D	200 miles	Nr	–	–	–	302.75	302.75	423.86	605.51
11362604E	250 miles	Nr	–	–	–	454.13	454.13	635.78	908.26
11362604F	300 miles	Nr	–	–	–	529.82	529.82	741.75	1059.64
11362604G	400 miles	Nr	–	–	–	708.95	708.95	992.53	1417.90

Signalling

Rail 2003 Ref		Unit	Labour Hours	Labour Cost	Plant Cost	Materials Cost	Unit Rate Base Cost	Unit Rate Green Zone	Unit Rate Red Zone
			£	£	£	£	£	£	£
11	**NEW INSTALLATIONS – RESIGNALLING WORKS**								
113626	Relocatable Equipment Buildings; extra over for delivery, off loading and levelling (bases and drainage measured separately)								
11362605	Size code 5, radius of depot:								
11362605A	50 miles	Nr	–	–	–	138.76	138.76	194.27	277.52
11362605B	100 miles	Nr	–	–	–	201.84	201.84	282.57	403.67
11362605C	150 miles	Nr	–	–	–	272.48	272.48	381.47	544.96
11362605D	200 miles	Nr	–	–	–	363.31	363.31	508.63	726.61
11362605E	250 miles	Nr	–	–	–	454.13	454.13	635.78	908.26
11362605F	300 miles	Nr	–	–	–	635.78	635.78	890.10	1271.57
11362605G	400 miles	Nr	–	–	–	851.50	851.50	1192.09	1702.99
11362606	Size code 6; radius of depot:								
11362606A	50 miles	Nr	–	–	–	151.38	151.38	211.93	302.75
11362606B	100 miles	Nr	–	–	–	227.07	227.07	317.89	454.13
11362606C	150 miles	Nr	–	–	–	302.75	302.75	423.86	605.51
11362606D	200 miles	Nr	–	–	–	403.67	403.67	565.14	807.34
11362606E	250 miles	Nr	–	–	–	504.59	504.59	706.43	1009.18
11362606F	300 miles	Nr	–	–	–	706.43	706.43	989.00	1412.85
11362606G	400 miles	Nr	–	–	–	946.11	946.11	1324.55	1892.21
11362607	Size code 7; radius of depot:								
11362607A	50 miles	Nr	–	–	–	189.22	189.22	264.91	378.44
11362607B	100 miles	Nr	–	–	–	271.22	271.22	379.70	542.43
11362607C	150 miles	Nr	–	–	–	340.60	340.60	476.84	681.20
11362607D	200 miles	Nr	–	–	–	428.90	428.90	600.46	857.80
11362607E	250 miles	Nr	–	–	–	536.13	536.13	750.58	1072.25
11362607F	300 miles	Nr	–	–	–	750.58	750.58	1050.81	1501.15
11362607G	400 miles	Nr	–	–	–	1009.18	1009.18	1412.85	2018.36

Signalling

Rail 2003 Ref		Unit	Labour Hours	Labour Cost £	Plant Cost £	Materials Cost £	Unit Rate Base Cost £	Unit Rate Green Zone £	Unit Rate Red Zone £

11 NEW INSTALLATIONS – RESIGNALLING WORKS

113626 Relocatable Equipment Buildings; extra over for delivery, off loading and levelling (bases and drainage measured separately)

11362608 Size code 8; radius of depot:

11362608A	50 miles	Nr	–	–	–	220.76	220.76	309.06	441.51
11362608B	100 miles	Nr	–	–	–	315.37	315.37	441.52	630.74
11362608C	150 miles	Nr	–	–	–	378.44	378.44	529.82	756.89
11362608D	200 miles	Nr	–	–	–	454.13	454.13	635.78	908.26
11362608E	250 miles	Nr	–	–	–	567.66	567.66	794.73	1135.33
11362608F	300 miles	Nr	–	–	–	794.73	794.73	1112.62	1589.46
11362608G	400 miles	Nr	–	–	–	1072.25	1072.25	1501.15	2144.51

11362609 Size code 9; radius of depot:

11362609A	50 miles	Nr	–	–	–	220.76	220.76	309.06	441.51
11362609B	100 miles	Nr	–	–	–	315.37	315.37	441.52	630.74
11362609C	150 miles	Nr	–	–	–	378.44	378.44	529.82	756.89
11362609D	200 miles	Nr	–	–	–	454.13	454.13	635.78	908.26
11362609E	250 miles	Nr	–	–	–	567.66	567.66	794.73	1135.33
11362609F	300 miles	Nr	–	–	–	794.73	794.73	1112.62	1589.46
11362609G	400 miles	Nr	–	–	–	1072.25	1072.25	1501.15	2144.51

113631 Internal fittings for location cases and relocatable equipment buildings, installed additionally (cabling, equipment and connections measured separately)

11363101 Channel:

11363101A	slotted upright channel for REBs	Nr	0.50	17.61	–	20.07	37.68	52.74	75.35
11363101B	plain horizontal channel for REBs	Nr	0.50	17.61	–	14.00	31.61	44.25	63.22

11363106 Mounting plates:

11363106A	for AC track feed	Nr	1.00	35.22	–	8.04	43.26	60.56	86.51
11363106B	for AC track relay	Nr	1.00	35.22	–	11.20	46.42	64.99	92.84
11363106C	for SSI fuses	Nr	1.00	35.22	–	19.62	54.84	76.77	109.67
11363106D	for fuseholder and busbar	Nr	1.00	35.22	–	4.61	39.83	55.76	79.65
11363106E	for track capacitor	Nr	1.00	35.22	–	9.83	45.05	63.07	90.09

Signalling

Rail 2003 Ref		Unit	Labour Hours	Labour Cost £	Plant Cost £	Materials Cost £	Unit Rate Base Cost £	Unit Rate Green Zone £	Unit Rate Red Zone £

11   NEW INSTALLATIONS – RESIGNALLING WORKS

113631   Internal fittings for location cases and relocatable equipment buildings, installed additionally (cabling, equipment and connections measured separately)

11363111   Equipment rails:

11363111A	large	Nr	0.50	17.61	–	13.54	31.15	43.60	62.30
11363111B	small	Nr	0.35	12.33	–	9.92	22.25	31.15	44.49
11363111C	adjustable clamp	Nr	0.25	8.80	–	4.82	13.62	19.08	27.25

11363116   Shelf rails:

11363116A	for battery	Nr	0.50	17.61	–	16.03	33.64	47.09	67.27
11363116B	for equipment	Nr	0.50	17.61	–	8.29	25.90	36.26	51.80

11363121   Terminal rails:

11363121A	14 2BA links	Nr	0.35	12.33	–	5.51	17.84	24.98	35.67
11363121B	30 2BA links	Nr	0.40	14.09	–	6.98	21.07	29.49	42.14
11363121C	48 2BA links	Nr	0.45	15.85	–	12.79	28.64	40.10	57.29

11363126   Relay rails:

11363126A	top or bottom, large	Nr	0.50	17.61	–	6.58	24.19	33.87	48.39
11363126B	intermediate, large	Nr	0.50	17.61	–	14.78	32.39	45.34	64.78
11363126C	top or bottom, small	Nr	0.35	12.33	–	9.51	21.84	30.57	43.67
11363126D	intermediate, small	Nr	0.35	12.33	–	9.99	22.32	31.24	44.63

11363131   SSI rails:

11363131A	large	Nr	0.50	17.61	–	14.24	31.85	44.58	63.70
11363131B	small	Nr	0.35	12.33	–	11.68	24.01	33.61	48.01

11363136   Label rails:

11363136A	large	Nr	0.50	17.61	–	8.02	25.63	35.87	51.25
11363136B	small	Nr	0.35	12.33	–	6.82	19.15	26.81	38.29

Signalling

Rail 2003 Ref		Unit	Labour Hours	Labour Cost £	Plant Cost £	Materials Cost £	Unit Rate Base Cost £	Unit Rate Green Zone £	Unit Rate Red Zone £
11	**NEW INSTALLATIONS – RESIGNALLING WORKS**								
113631	Internal fittings for location cases and relocatable equipment buildings, installed additionally (cabling, equipment and connections measured separately)								
11363141	Top hat mounting rails:								
11363141A	large	Nr	0.50	17.61	–	15.72	33.33	46.66	66.67
11363141B	small	Nr	0.35	12.33	–	11.66	23.99	33.58	47.96
11363141C	vertical	Nr	0.30	10.57	–	6.58	17.15	24.01	34.30
11363146	Fuseholder rails:								
11363146A	large 29 × BS714	Nr	0.50	17.61	–	15.96	33.57	47.00	67.14
11363146B	small 14 × BS714	Nr	0.35	12.33	–	11.79	24.12	33.76	48.23
11363146C	large 20 × BS88	Nr	0.50	17.61	–	11.91	29.52	41.32	59.03
11363151	Cable bars:								
11363151A	large	Nr	0.50	17.61	–	16.21	33.82	47.35	67.64
11363151B	small	Nr	0.35	12.33	–	12.05	24.38	34.14	48.76
11363156	Battery shelves:								
11363156A	large; for apparatus case	Nr	3.00	98.73	–	79.12	177.85	248.98	355.68
11363156B	small; for apparatus case	Nr	2.50	82.27	–	40.90	123.17	172.44	246.35
11363156C	large; for REB	Nr	6.00	197.45	–	358.98	556.43	779.00	1112.86
1141	**POINTS OPERATING EQUIPMENT**								
114101	MK2 rail clamp point locks; assemblies supplied complete								
11410101	One point end of 113A FB; vertical plain lead; tandem lead or single slip:								
11410101A	bar 890mm long; thrust bracket complete and drive lug	Nr	48.00	1834.15	–	3676.92	5511.07	7715.50	11022.14

Signalling

Rail 2003 Ref		Unit	Labour Hours	Labour Cost £	Plant Cost £	Materials Cost £	Unit Rate Base Cost £	Unit Rate Green Zone £	Unit Rate Red Zone £

11   **NEW INSTALLATIONS – RESIGNALLING WORKS**

114101   MK2 rail clamp point locks; assemblies supplied complete

11410111   One point end of 110A and 113A FB inclined plain lead; tandem lead or single slip:

11410111A	palletised package assembly, comprising: lock and detector assembly; adaptor block complete; switch rail bracket assembly complete; taper packing; tie bar 890mm long; thrust bracket complete and drive lug .	Nr	48.00	1834.15	–	3676.92	5511.07	7715.50	11022.14

11410121   One point end of UIC 54B/113A plain lead on timber:

11410121A	palletised package layout, comprising lock and detector assembly; adaptor block assembly; switch rail bracket assembly; tie bar and centre thrust bracket............................	Nr	48.00	1834.15	–	3806.79	5640.94	7897.32	11281.88

11410131   One point end of UIC 54B/113A plain lead on concrete:

11410131A	palletised package layout, comprising: lock and detector assembly; adaptor block assembly; switch rail bracket assembly; tie bar and centre thrust bracket............................	Nr	48.00	1834.15	–	5041.69	6875.84	9626.17	13751.67

114111   MK2 rail clamp point locks; accessories

11411101   Power pack, hydro-electric, hydraulic, 4 Nr outlets; type:

11411101A	HV110, 110V DC working (2.5 seconds)	Nr	20.00	658.17	–	1051.98	1710.15	2394.21	3420.30
11411101B	LV2, 50V DC working (4.0 seconds) ..	Nr	20.00	658.17	–	1079.48	1737.65	2432.71	3475.30

Signalling

Rail 2003 Ref		Unit	Labour Hours	Labour Cost	Plant Cost	Materials Cost	Unit Rate	Unit Rate	Unit Rate
				£	£	£	Base Cost £	Green Zone £	Red Zone £
11	**NEW INSTALLATIONS – RESIGNALLING WORKS**								
114111	**MK2 rail clamp point locks; accessories**								
11411111	Displacement ram; pair; application:								
11411111A	clamp locks, 24mm effective diameter 210mm stroke.	Nr	10.00	329.09	–	150.44	479.53	671.34	959.06
11411111B	switch diamonds, 32mm effective diameter × 210mm stroke	Nr	10.00	329.09	–	795.81	1124.90	1574.86	2249.80
11411116	Hose; hydraulic; pair; for clamp lock:								
11411116A	3.0m supplied length	Nr	3.00	98.73	–	68.65	167.38	234.33	334.75
11411121	Sundries for clamp locks:								
11411121A	base; precast reinforced concrete block.	Nr	8.00	263.27	–	8.81	272.08	380.91	544.15
11411121B	hydraulic oil for clamp lock; drain and top up clamp lock	Nr	1.00	30.60	–	13.17	43.77	61.28	87.54
114121	**Points machines**								
11412101	HW2000 range (GEC); model/voltage:								
11412101A	HW1121, 30/110V split field	Nr	96.00	3302.86	–	7392.24	10695.10	14973.14	21390.20
11412101B	HW2121, 110V, AC immune	Nr	96.00	3302.86	–	7495.68	10798.54	15117.96	21597.08
11412111	Style 63 range (WSL); model/voltage:								
11412111A	30V DC split field	Nr	96.00	3302.86	–	7568.85	10871.71	15220.39	21743.41
11412111B	110V AC immune	Nr	96.00	3302.86	–	8451.88	11754.74	16456.63	23509.47
11412121	Point telephones:								
11412121A	supply and install	Nr	3.00	103.21	–	18.75	121.96	170.75	243.92

Signalling

Rail 2003 Ref		Unit	Labour Hours	Labour Cost £	Plant Cost £	Materials Cost £	Unit Rate Base Cost £	Unit Rate Green Zone £	Unit Rate Red Zone £
11	**NEW INSTALLATIONS – RESIGNALLING WORKS**								
114181	**Miscellaneous additional works**								
11418101	**Point identification plates, drilled c.i.:**								
11418101A	letter, number or arrow plate: approximate size 100 × 68 × 20mm; including fixings as required.........	Nr	0.25	7.65	–	4.97	12.62	17.67	25.24
1146	**AUTOMATIC WARNING SYSTEMS (AWS)**								
114601	**Inductors; including bolting to mounting plate and rubber pad; (mounting plate and packing plate measured separately); type:**								
11460101	**Electro inductor:**								
11460101A	standard strength, 12/24V coil.......	Nr	15.00	493.63	–	516.11	1009.74	1413.64	2019.48
11460101B	extra strength, 60V coil.............	Nr	15.00	493.63	–	698.83	1192.46	1669.44	2384.90
11460111	**Permanent inductor:**								
11460111A	standard strength..................	Nr	15.00	493.63	–	317.58	811.21	1135.70	1622.42
11460111B	extra strength.....................	Nr	15.00	493.63	–	454.56	948.19	1327.46	1896.36
11460121	**Suppressor:**								
11460121A	standard strength, 24V coil..........	Nr	15.00	493.63	–	1808.18	2301.81	3222.53	4603.60
11460121B	extra strength, 110V coil............	Nr	15.00	493.63	–	3306.28	3799.91	5319.87	7599.81
114611	**Ramp; including bolting in place; (packing plate measured separately)**								
11461101	**Standard:**								
11461101A	pattern track indicator..............	Nr	8.00	263.27	–	84.19	347.46	486.44	694.92
11461151	**Extra over for ramp extension:**								
11461151A	in electro permanent electro track inductor installations...............	Nr	8.00	263.27	–	35.79	299.06	418.68	598.11

Signalling

Rail 2003 Ref		Unit	Labour Hours	Labour Cost £	Plant Cost £	Materials Cost £	Unit Rate Base Cost £	Unit Rate Green Zone £	Unit Rate Red Zone £

11 NEW INSTALLATIONS – RESIGNALLING WORKS

114671 Fixing plates and brackets; including bolting in place; (packing plates measured separately)

11467101 Standard single:

Ref	Description	Unit	Hours	Labour	Plant	Materials	Base	Green	Red
11467101A	non-adjustable; for track ramp	Nr	4.00	131.63	–	40.74	172.37	241.33	344.75
11467101B	adjustable; for track inductors and ramp	Nr	4.00	131.63	–	58.10	189.73	265.63	379.47

11467121 Standard double:

| 11467121A | for track inductors | Nr | 4.00 | 131.63 | – | 50.92 | 182.55 | 255.58 | 365.11 |

11467141 Special double:

| 11467141A | for inductor; suppressor; extra strength | Nr | 6.00 | 197.45 | – | 54.05 | 251.50 | 352.11 | 503.01 |
| 11467141B | for electro permanent electro track inductor installations | Nr | 6.00 | 197.45 | – | 61.13 | 258.58 | 362.01 | 517.16 |

11467161 Bridge:

| 11467161A | for gap between electro and permanent inductors; in electro permanent electro track inductor installations | Nr | 4.00 | 131.63 | – | 22.36 | 153.99 | 215.59 | 307.99 |

Rail 2003 Ref		Unit			Specialist Cost £	Unit Rate Base Cost £	Unit Rate Green Zone £	Unit Rate Red Zone £

114676 TOWS (Train Operated Warning Systems)

11467601 Yodalarm; type:

| 11467601A | Y05 | Nr | – | – | – | 163.00 | 163.00 | 228.20 | 326.00 |

11467611 Switches; type:

| 11467611A | TOWS | Nr | – | – | – | 118.25 | 118.25 | 165.55 | 236.50 |

11467616 Brackets; type:

| 11467616A | switch mounting | Nr | – | – | – | 37.75 | 37.75 | 52.85 | 75.50 |

Signalling

Rail 2003 Ref		Unit			Specialist Cost £	Unit Rate Base Cost £	Unit Rate Green Zone £	Unit Rate Red Zone £
11	**NEW INSTALLATIONS – RESIGNALLING WORKS**							
114676	**TOWS (Train Operated Warning Systems)**							
11467621	Posts and concrete bases:							
11467621A	generally	Nr	–	–	–	83.25	83.25	116.55	166.50
11467626	Identification plates and notices:							
11467626A	label	Nr	–	–	–	19.00	19.00	26.60	38.00
11467626B	standard notice board	Nr	–	–	–	220.00	220.00	308.00	440.00
11467631	SPAD indicators:							
11467631A	generally	Nr	–	–	–	6690.00	6690.00	9366.00	13380.00
11467636	DIS box:							
11467636A	complete with stake	Nr	–	–	–	307.00	307.00	429.80	614.00

Rail 2003 Ref		Unit	Labour Hours	Labour Cost £	Plant Cost £	Materials Cost £	Unit Rate Base Cost £	Unit Rate Green Zone £	Unit Rate Red Zone £
114681	**Miscellaneous activities**								
11468101	Extra over for packing plates; type:								
11468101A	single, for 1Nr bolt	Nr	0.05	1.76	–	1.91	3.67	5.14	7.33
11468101B	combined; for track inductors and ramp	Nr	1.00	35.22	–	8.83	44.05	61.68	88.11
11468111	Signage; including fixing to post set in ground; (making good surfacing measured separately); sign:								
11468111A	start of AWS Gap	Nr	4.50	137.69	–	211.87	349.56	489.38	699.11
11468111B	end of AWS Gap	Nr	4.50	137.69	–	211.87	349.56	489.38	699.11
11468111C	cancelling indicator	Nr	4.50	137.69	–	211.87	349.56	489.38	699.11

Signalling

Rail 2003 Ref		Unit	Labour Hours	Labour Cost £	Plant Cost £	Materials Cost £	Unit Rate Base Cost £	Unit Rate Green Zone £	Unit Rate Red Zone £
11	**NEW INSTALLATIONS – RESIGNALLING WORKS**								
1148	**TRAIN PROTECTION WARNING SYSTEM (TPWS)**								
114810	Equipment supply generally								
11481010	Prewired enclosures – standard; type:								
11481010A	SB+FIB	Nr	–	–	–	5075.00	5075.00	7105.00	10150.00
11481010B	SB+FIL	Nr	–	–	–	5225.00	5225.00	7315.00	10450.00
11481010C	SB+AB	Nr	–	–	–	5175.00	5175.00	7245.00	10350.00
11481020	Prewired enclosures – special; type:								
11481020A	SB+FIL+AB	Nr	–	–	–	5515.00	5515.00	7721.00	11030.00
11481030	Miscellaneous equipment:								
11481030A	podium	Nr	–	–	–	242.50	242.50	339.50	485.00
11481030B	beacon 6m cable	Nr	–	–	–	1181.50	1181.50	1654.10	2363.00
11481030C	little bracket	Nr	–	–	–	392.00	392.00	548.80	784.00
11481030D	TDA box	Nr	–	–	–	1898.75	1898.75	2658.25	3797.50
11481040	Standard encoder; SSI input; type:								
11481040A	SB+	Nr	–	–	–	8612.75	8612.75	12057.85	17225.50
11481040B	SB+FIB	Nr	–	–	–	10890.00	10890.00	15246.00	21780.00
11481040C	SB+FIL	Nr	–	–	–	11224.00	11224.00	15713.60	22448.00
11481040D	SB+AB	Nr	–	–	–	11255.00	11255.00	15757.00	22510.00
11481040E	SB+ExtPCB	Nr	–	–	–	12320.00	12320.00	17248.00	24640.00
11481040F	Data plug	Nr	–	–	–	284.65	284.65	398.51	569.30
114820	Equipment installation								
11482010	Generally:								
11482010A	per encoder	Nr	225.00	8179.50	–	–	8179.50	11451.30	16359.00
11482010B	enclosure	Nr	101.00	3668.00	–	–	3668.00	5135.20	7336.00
11482010C	podium	Nr	6.00	212.75	–	–	212.75	297.85	425.50
11482010D	beacon	Nr	12.85	466.80	–	–	466.80	653.52	933.60
11482010E	little bracket	Nr	3.80	138.75	–	–	138.75	194.25	277.50
11482010F	TDA box	Nr	25.75	935.50	–	–	935.50	1309.70	1871.00

Signalling

Rail 2003 Ref		Unit	Labour Hours	Labour Cost	Plant Cost	Materials Cost	Unit Rate	Unit Rate	Unit Rate
				£	£	£	Base Cost £	Green Zone £	Red Zone £

11   NEW INSTALLATIONS – RESIGNALLING WORKS

1151   SIGNALLING POWER SUPPLY EQUIPMENT

115101   Transformers; to BRS 924

11510101   Mains; isolating; open; 250/250V; rating:

Ref	Rating	Unit	L.Hrs	L.Cost	Plant	Mat.Cost	Base	Green	Red
11510101A	0.5KVA	Nr	12.00	458.54	–	91.84	550.38	770.52	1100.75
11510101B	1.0KVA	Nr	12.00	458.54	–	149.75	608.29	851.60	1216.58
11510101C	2.5KVA	Nr	12.00	458.54	–	278.52	737.06	1031.88	1474.12

11510111   Track feed (capacitor); 110/110V; rating:

11510111A	150VA	Nr	12.00	458.54	–	61.91	520.45	728.63	1040.91

11510121   Mains; closed; 650/110V; rating:

11510121A	0.5KVA	Nr	12.00	458.54	–	114.60	573.14	802.40	1146.29
11510121B	1.0KVA	Nr	12.00	458.54	–	279.82	738.36	1033.70	1476.72
11510121C	2.5KVA	Nr	12.00	458.54	–	303.39	761.93	1066.69	1523.85
11510121D	5.0KVA	Nr	12.00	458.54	–	491.98	950.52	1330.72	1901.03
11510121E	10.0KVA	Nr	12.00	458.54	–	1160.56	1619.10	2266.73	3238.19

11510131   Mains; open; 250/110V; rating:

11510131A	0.5KVA	Nr	12.00	458.54	–	91.13	549.67	769.53	1099.34
11510131B	1.0KVA	Nr	12.00	458.54	–	152.20	610.74	855.03	1221.47
11510131C	2.5KVA	Nr	12.00	458.54	–	278.52	737.06	1031.88	1474.12
11510131D	5.0KVA	Nr	12.00	458.54	–	474.79	933.33	1306.66	1866.67

11510141   Open; for illuminated diagrams or track feeds; 110/12V; rating:

11510141A	250VA	Nr	12.00	458.54	–	88.30	546.84	765.57	1093.69
11510141B	0.5KVA	Nr	12.00	458.54	–	123.98	582.52	815.52	1165.04

Signalling

Rail 2003 Ref		Unit	Labour Hours	Labour Cost £	Plant Cost £	Materials Cost £	Unit Rate Base Cost £	Unit Rate Green Zone £	Unit Rate Red Zone £
11	NEW INSTALLATIONS – RESIGNALLING WORKS								
115111	Transformer rectifiers; to BRS 865								
11511101	For TI 21 track circuits; rating:								
11511101A	95-120/24V DC 4.4A	Nr	12.00	458.54	–	100.92	559.46	783.24	1118.92
11511111	For locations and AWS inductors or suppressors; rating:								
11511111A	110/24V DC 1A	Nr	12.00	458.54	–	63.07	521.61	730.25	1043.23
11511121	For locations and relay rooms; rating:								
11511121A	110/50-60V DC 1A	Nr	12.00	458.54	–	61.31	519.85	727.78	1039.70
11511121B	110/50-60V DC 3A	Nr	12.00	458.54	–	81.00	539.54	755.35	1079.08
11511121C	110/50-60V DC 10A	Nr	12.00	458.54	–	202.23	660.77	925.07	1321.54
11511121D	110/50-60V DC 20A	Nr	12.00	458.54	–	294.87	753.41	1054.77	1506.82
11511121E	110/50-60V DC 30A	Nr	12.00	458.54	–	812.73	1271.27	1779.77	2542.54
11511121F	110/50-60V DC 40A	Nr	12.00	458.54	–	1204.71	1663.25	2328.54	3326.50
11511131	For clamp locks or points machines; rating:								
11511131A	650/120-130V DC, Type A 2.5A, continuous	Nr	12.00	458.54	–	176.88	635.42	889.59	1270.85
11511131B	650/120-130V DC, Type B 5A, continuous	Nr	12.00	458.54	–	351.50	810.04	1134.05	1620.08
11511131C	650/120-130V DC, Type C 10A, continuous	Nr	12.00	458.54	–	667.19	1125.73	1576.02	2251.47
115121	Battery standby power supplies								
11512101	For points batteries, 50V batteries and 6V track feeds:								
11512101A	nickel cadmium cell RE16-16AH in 3 or 5 cell crates	Nr	3.00	105.66	–	433.37	539.03	754.63	1078.04

Signalling

Rail 2003 Ref		Unit	Specialist Cost £	Unit Rate Base Cost £	Unit Rate Green Zone £	Unit Rate Red Zone £
11	**NEW INSTALLATIONS – RESIGNALLING WORKS**					
115181	**Bases**					
11518101	Bases for transformers; concrete; transformer size:					
11518101A	not exceeding 5 Kv	Nr	–	145.00	203.00	290.00
11518101B	exceeding 5 Kv	Nr	–	197.50	276.50	395.00
1156	**INTERLOCKING SYSTEMS**					
115610	Solid State Interlocking (SSI) equipment					
11561010	SSI equipment generally; supply and install:					
11561010A	signal module Mk III	Nr	–	8117.00	11363.80	16234.00
11561010B	points module Mk III	Nr	–	9470.00	13258.00	18940.00
11561010C	data link module	Nr	–	4750.00	6650.00	9500.00
11561010D	long distance terminal	Nr	–	4750.00	6650.00	9500.00
11561010E	fused supply module	Nr	–	1082.25	1515.15	2164.50
11561010F	plug coupler 50-way	Nr	–	338.20	473.48	676.40
11561010G	plug coupler 75-way	Nr	–	377.73	528.82	755.46
11561010M	module power supply; signal 650/110/110	Nr	–	457.25	640.15	914.50
1176	**STRIPPING OUT AND REMOVALS**					
117610	Equipment and signals					
11761012	Location cases:					
11761012A	full size	Nr	–	550.00	770.00	1100.00
11761012B	half size	Nr	–	775.00	1085.00	1550.00
11761015	General equipment; disposal off-site:					
11761015C	relays	Nr	–	191.20	267.68	382.40
11761017	Points machines:					
11761017A	type 5A	Nr	–	195.00	273.00	390.00

Signalling

Rail 2003 Ref		Unit	Specialist Cost £	Unit Rate Base Cost £	Unit Rate Green Zone £	Unit Rate Red Zone £
11	**NEW INSTALLATIONS – RESIGNALLING WORKS**					
117610	**Equipment and signals**					
11761022	**Heater installations:**					
11761022A	cartridge heater.....................	Nr	– – –	135.00	135.00	189.00	270.00
11761022B	strip heater	Nr	– – –	147.50	147.50	206.50	295.00
11761025	**Track circuit and ancillary equipment:**					
11761025A	generally	Nr	– – –	150.00	150.00	210.00	300.00
11761032	**Signal heads, post and ancillary equipment:**					
11761032A	single	Nr	– – –	300.00	300.00	420.00	600.00
11761032B	multiple	Nr	– – –	425.00	425.00	595.00	850.00
11761035	**Miscellaneous equipment:**					
11761035A	signage	Nr	– – –	42.00	42.00	58.80	84.00
11761040	**Control centres; signal box constructions:**					
11761040A	brick construction	m^{2}GFA	– – –	100.50	100.50	140.69	200.99
11761040B	concrete construction	m^{2}GFA	– – –	110.55	110.55	154.76	221.09
11761040C	timber construction	m^{2}GFA	– – –	80.40	80.40	112.55	160.79
11761042	**Control centres; signal boxes; strip out equipment and dispose:**					
11761042A	generally	Nr	– – –	4500.00	4500.00	6300.00	9000.00
11761045	**Relay rooms; strip out equipment and dispose:**					
11761045A	generally	Nr	– – –	1950.00	1950.00	2730.00	3900.00

Signalling

Rail 2003 Ref		Unit	Labour Hours	Labour Cost £	Plant Cost £	Materials Cost £	Unit Rate Base Cost £	Unit Rate Green Zone £	Unit Rate Red Zone £
11	**NEW INSTALLATIONS – RESIGNALLING WORKS**								
117630	**Cable routes**								
11763010	General route clearance; site vegetation and scrub; removal of debris off-site:								
11763010A	light vegetation	m^2	0.06	0.72	0.42	–	1.14	1.59	2.28
11763010B	heavy vegetation	m^2	0.12	1.43	0.85	–	2.28	3.19	4.56
11763010C	light scrub	m^2	0.16	1.91	1.13	–	3.04	4.26	6.08
11763010D	heavy scrub	m^2	0.22	2.63	1.55	–	4.18	5.85	8.36
11763015	General route clearance; trees; backfilling with imported materials; removal of debris off-site; girth:								
11763015A	not exceeding 600 mm	Nr	4.50	53.78	5.29	4.30	63.37	88.71	126.73
11763015B	600 mm – 1.00 m	Nr	6.00	71.70	17.63	14.33	103.66	145.12	207.32
11763015C	over 1.00 m	Nr	9.00	107.55	28.21	22.92	158.68	222.15	317.37
11763020	General route clearance; stumps; backfilling with imported materials; removal of debris off-site; diameter:								
11763020A	not exceeding 600 mm	Nr	9.00	107.55	10.58	8.60	126.73	177.42	253.45
11763020B	600 mm – 1.00 m	Nr	13.50	161.33	17.63	14.33	193.29	270.60	386.57
11763020C	over 1.00 m	Nr	18.00	215.10	28.21	22.92	266.23	372.72	532.47
11763025	General route clearance; excavation of top soil by hand; removal of debris off-site; depth:								
11763025A	not exceeding 150 mm	m^2	0.40	4.78	2.35	–	7.13	9.99	14.27
11763025B	150 – 300 mm	m^2	0.75	8.96	4.71	–	13.67	19.14	27.34
11763025C	300 – 450 mm	m^2	1.15	13.74	7.06	–	20.80	29.13	41.61
11763030	Cable trough removal and disposal; complete with lid; type:								
11763030A	concrete; cross sectional area n.e. 0.03 m^2	m	0.80	9.56	0.50	–	10.06	14.08	20.13
11763030B	concrete; cross sectional area 0.03–0.11 m^2	m	1.00	11.95	0.73	–	12.68	17.75	25.36
11763030C	concrete; cross sectional area over 0.11 m^2	m	1.33	15.89	1.81	–	17.70	24.78	35.41
11763030D	timber; generally	m	0.90	10.76	0.35	–	11.11	15.55	22.22
11763030E	GRP; generally	m	1.00	11.95	1.71	–	13.66	19.12	27.32

Signalling

Rail 2003 Ref		Unit	Labour Hours	Labour Cost £	Plant Cost £	Materials Cost £	Unit Rate Base Cost £	Unit Rate Green Zone £	Unit Rate Red Zone £
11	**NEW INSTALLATIONS – RESIGNALLING WORKS**								
117630	Cable routes								
11763030	Cable trough removal and disposal; complete with lid; type:								
11763030F	asbestos; generally; including approved disposal	m	3.00	35.85	12.65	–	48.50	67.90	97.00
11763035	Cable trough removal and setting aside for re-use; type:								
11763035A	C1/6, 7, 9 and 10	m	0.75	30.15	–	–	30.15	42.21	60.30
11763035B	C1/8 and 29	m	0.80	32.16	–	–	32.16	45.02	64.32
11763035C	C1/43	m	0.90	36.18	–	–	36.18	50.65	72.36
11763035D	lid to C1/6, 7, 9 and 10	m	0.12	4.82	–	–	4.82	6.75	9.65
11763035E	lid C1/8 and 29	m	0.15	6.03	–	–	6.03	8.44	12.06
11763035F	lid to C1/43	m	0.20	8.04	–	–	8.04	11.26	16.08
11763035G	tee piece and lid to C1/6, 7, 9, and 10.	Nr	0.75	30.15	–	–	30.15	42.21	60.30
11763035H	tee piece and lid to C1/8 and 29	Nr	0.85	34.17	–	–	34.17	47.84	68.34
11763035I	tee piecec and lid to C1/43	Nr	0.95	38.19	–	–	38.19	53.46	76.38
11763040	Cable trough realignment; type:								
11763040A	C1/6, 7, 9 and 10	m	0.75	30.15	–	–	30.15	42.21	60.30
11763040B	C1/8 and 29	m	0.85	34.17	–	–	34.17	47.84	68.34
11763040C	C1/43	m	1.05	42.21	–	–	42.21	59.09	84.42
11763045	Cable trough cleaning; size:								
11763045A	cross sectional area n.e. 0.03 m^2; single cable	m	0.05	2.01	–	–	2.01	2.81	4.02
11763045B	cross sectional area 0.03 – 0.11 m^2; single cable	m	0.06	2.41	–	–	2.41	3.38	4.82
11763045C	cross sectional area over 0.11 m^2; single cable	m	0.08	3.02	–	–	3.02	4.22	6.03
11763045D	cross sectional area n.e. 0.03 m^2; multiple cable	m	0.06	2.41	–	–	2.41	3.38	4.82
11763045E	cross sectional area 0.03 – 0.11 m^2; multiple cable	m	0.07	2.89	–	–	2.89	4.05	5.79
11763045F	cross sectional area over 0.11 m^2; multiple cable	m	0.09	3.62	–	–	3.62	5.07	7.24

Signalling

Rail 2003 Ref		Unit	Labour Hours	Labour Cost £	Plant Cost £	Materials Cost £	Unit Rate Base Cost £	Unit Rate Green Zone £	Unit Rate Red Zone £
11	**NEW INSTALLATIONS – RESIGNALLING WORKS**								
117630	Cable routes								
11763050	Cabling; taking up and realigning in trough units; type:								
11763050A	power cable	m	0.06	2.41	–	–	2.41	3.38	4.82
11763050B	signal cable	m	0.05	2.01	–	–	2.01	2.81	4.02
11763050C	communication	m	0.05	1.81	–	–	1.81	2.53	3.62
11763050D	data link	m	0.05	1.81	–	–	1.81	2.53	3.62
117640	Cabling								
11764010	Removal and disposal of redundant cabling; type:								
11764010A	power cable	m	0.13	5.23	–	–	5.23	7.32	10.45
11764010B	signal cable	m	0.10	4.02	–	–	4.02	5.63	8.04
11764010C	communication	m	0.09	3.62	–	–	3.62	5.07	7.24
11764010D	data link	m	0.09	3.62	–	–	3.62	5.07	7.24
1181	**RECOVERIES**								
118101	Remove material and clear off site; (breaking out bases and surfacing, charges for disposal off site and electrical power supply isolations measured separately)								
11810101	Cable, including sundries;								
11810101A	laid on trackside	m	0.10	3.06	–	–	3.06	4.28	6.12
11810101B	clipped to structures, up to 3.5m above ground level	m	0.20	6.12	–	–	6.12	8.57	12.24
11810101C	clipped to structures, over 3.5m above ground level	m	0.30	9.18	–	–	9.18	12.85	18.36
11810101D	in trough, lifting and replacing lid	m	0.25	7.65	–	0.33	7.98	11.17	15.95
11810106	Troughing; excavation, back filling and surfacing and cable removal measured separately;								
11810106A	lid only	m	0.20	6.12	–	–	6.12	8.57	12.24
11810106B	lid and trough complete, single 1.0m section	Nr	0.50	15.30	–	–	15.30	21.42	30.60
11810106C	lid and trough complete, consecutive 1.0m sections	m	0.40	12.24	–	–	12.24	17.13	24.48

Signalling

Rail 2003 Ref		Unit	Labour Hours	Labour Cost £	Plant Cost £	Materials Cost £	Unit Rate Base Cost £	Unit Rate Green Zone £	Unit Rate Red Zone £
11	**NEW INSTALLATIONS – RESIGNALLING WORKS**								
118101	Remove material and clear off site; (breaking out bases and surfacing, charges for disposal off site and electrical power supply isolations measured separately)								
11810106	Troughing; excavation, back filling and surfacing and cable removal measured separately;								
11810106D	lid and trough complete, single 1.0m section; prepare for insertion of tee junction trough and supply and fit	Nr	2.38	72.82	–	44.39	117.21	164.10	234.44
11810111	Ducting; cable removal measured separately:								
11810111A	orange pipe, 90mm dia plastic, above ground level	m	0.15	4.59	–	–	4.59	6.43	9.18
11810111B	rod through and prove existing duct, clearing out debris as required, leaving duct in place	m	0.33	10.10	–	–	10.10	14.14	20.19
11810116	Structures; including sundries, mounting brackets, identification plates and signage and the like:								
11810116A	colour light signal post only	Nr	40.25	1345.70	–	–	1345.70	1883.97	2691.39
11810116B	netting from colour light signal post, single section	Nr	3.00	91.79	–	–	91.79	128.51	183.59
11810116C	signal or indicator head, elevated mounting, up to 3.5m above ground level	Nr	17.50	585.09	–	–	585.09	819.12	1170.17
11810116D	signal or indicator head, elevated mounting, over 3.5m above ground level	Nr	24.50	819.12	–	–	819.12	1146.77	1638.24
11810116E	signal or indicator head, ground mounting	Nr	14.00	468.07	–	–	468.07	655.30	936.14
11810116F	1 road cantilever	Nr	40.25	1345.70	–	–	1345.70	1883.97	2691.39
11810116G	2 road cantilever	Nr	80.50	2691.39	–	–	2691.39	3767.95	5382.78
11810116H	2 road gantry	Nr	105.00	3510.51	–	–	3510.51	4914.71	7021.02

Signalling

Rail 2003 Ref		Unit	Labour Hours	Labour Cost £	Plant Cost £	Materials Cost £	Unit Rate Base Cost £	Unit Rate Green Zone £	Unit Rate Red Zone £

11   **NEW INSTALLATIONS – RESIGNALLING WORKS**

118101   Remove material and clear off site; (breaking out bases and surfacing, charges for disposal off site and electrical power supply isolations measured separately)

11810121   AWS and associated ramps:

| 11810121A | complete | Nr | 12.00 | 412.86 | – | – | 412.86 | 578.00 | 825.71 |

11810126   Track circuits; cabling measured separately:

| 11810126A | transmitter | Nr | 20.00 | 658.17 | – | – | 658.17 | 921.44 | 1316.34 |

11810131   Level crossings; cabling measured separately:

11810131A	lifting barrier and machine	Nr	60.00	2064.29	–	–	2064.29	2890.00	4128.57
11810131B	extra over for skirt for lifting barrier	Nr	4.00	122.39	–	–	122.39	171.35	244.78
11810131C	extra over for support member for lifting barrier	Nr	8.00	244.78	–	–	244.78	342.70	489.57

11810136   Points; cabling measured separately:

| 11810136A | clamp lock, complete | Nr | 21.00 | 674.91 | – | – | 674.91 | 944.87 | 1349.81 |
| 11810136B | points machine, complete | Nr | 21.00 | 674.91 | – | – | 674.91 | 944.87 | 1349.81 |

11810141   Location cases; (cabling measured separately); size:

| 11810141A | large, complete | Nr | 14.00 | 460.72 | – | – | 460.72 | 645.01 | 921.44 |
| 11810141B | small, complete | Nr | 10.00 | 329.09 | – | – | 329.09 | 460.72 | 658.17 |

11810191   Miscellaneous items; (cabling measured separately):

11810191A	transformer	Nr	10.00	329.09	–	–	329.09	460.72	658.17
11810191B	signal post telephone	Nr	1.50	52.83	–	–	52.83	73.96	105.66
11810191C	road sign and post	Nr	3.00	91.79	–	–	91.79	128.51	183.59
11810191D	sign, fixed to structure, over 1.00m^2, up to 3.5m above ground level	Nr	1.50	45.90	–	–	45.90	64.26	91.79
11810191E	sign, fixed to structure, over 1.00m^2, over 3.5m above ground level	Nr	2.50	76.50	–	–	76.50	107.09	152.99
11810191F	extra over for signal post anti climb device	Nr	2.00	61.20	–	–	61.20	85.67	122.39

Signalling

Rail 2003 Ref		Unit	Labour Hours	Labour Cost £	Plant Cost £	Materials Cost £	Unit Rate Base Cost £	Unit Rate Green Zone £	Unit Rate Red Zone £
11	**NEW INSTALLATIONS – RESIGNALLING WORKS**								
118101	Remove material and clear off site; (breaking out bases and surfacing, charges for disposal off site and electrical power supply isolations measured separately)								
11810191	Miscellaneous items; (cabling measured separately):								
11810191G	battery power supply unit, per battery of 5 cells, excluding racking	Nr	3.00	98.73	–	–	98.73	138.22	197.45
11810191H	clear away light vegetation from around track side installations	m²	0.04	1.22	–	–	1.22	1.71	2.45
1191	**MISCELLANEOUS EQUIPMENT AND ACTIVITIES**								
119101	Signal post telephones; to BS 4800; including weatherproof housing and fixing to structure; (power cabling measured separately); type:								
11910101	CB, loop calling,								
11910101A	for general use, colour grey, complete with moulded polycarbonate handset and armoured or extensible cord	Nr	2.00	70.44	–	224.57	295.01	413.01	590.02
11910101B	for use at level crossings, colour yellow, self illuminated, complete with moulded polycarbonate handset and armoured cord	Nr	2.00	70.44	–	421.06	491.50	688.09	982.99
11910111	Magneto calling:								
11910111A	complete with ring button vandal resistant hand set and steel armoured cord	Nr	2.00	70.44	–	291.09	361.53	506.13	723.05
11910121	Auto-720:								
11910121A	with switchable decadic/DTMF signalling, complete with switchable earth loop timed break recall, 14 button keypad and moulded polycarbonate handset	Nr	2.00	70.44	–	280.70	351.14	491.60	702.29

Signalling

Rail 2003 Ref		Unit	Labour Hours	Labour Cost £	Plant Cost £	Materials Cost £	Unit Rate Base Cost £	Unit Rate Green Zone £	Unit Rate Red Zone £
11	**NEW INSTALLATIONS – RESIGNALLING WORKS**								
119101	Signal post telephones; to BS 4800; including weatherproof housing and fixing to structure; (power cabling measured separately); type:								
11910131	Single button Autodialler, Auteldac-721:								
11910131A	for general use, colour grey or yellow with armoured handset cord	Nr	2.00	70.44	–	308.77	379.21	530.89	758.42
11910131B	for use at level crossings, colour yellow, self illuminated, with armoured handset cord	Nr	2.00	70.44	–	517.56	588.00	823.19	1176.00
1196	**TESTING AND COMMISSIONING**								
119601	CONTRACTOR'S TESTING AND COMMISSIONING								
11960110	Signalling system:								
11960110A	generally	hour	3.00	142.61	–	–	142.61	199.66	285.23

AC Electrification

Rail 2003 Ref		Unit	Composite Cost £	Unit Rate Base Cost £	Unit Rate Green Zone £	Unit Rate Red Zone £
2	**AC ELECTRIFICATION**					
241	**NEW WORKS – INNER CITY**					
24102	**OVERHEAD LINE WORKS – 2 TRACK WORK**					
2410201	Electrification works generally					
241020101	Bases for structures; in situ reinforced concrete bases, including holding down bolts and fixings:					
241020101A	masts or the like	Nr		979.06	979.06	1370.68	1958.11
241020111	Headspans:					
241020111A	supply	Nr	–	–	–	889.00	889.00	1244.60	1778.00
241020111B	erection	Nr	–	–	–	177.80	177.80	248.92	355.60
241020121	Cantilevers:					
241020121A	supply	Nr	–	–	–	711.21	711.21	995.69	1422.42
241020121B	erection	Nr	–	–	–	533.40	533.40	746.76	1066.80
241020131	Gantries:					
241020131A	steel supply	Nr	–	–	–	3556.03	3556.03	4978.44	7112.06
241020131B	steel erection	Nr	–	–	–	2667.02	2667.02	3733.83	5334.04
241020141	Brackets:					
241020141A	single track cantilever – supply	Nr	–	–	–	177.80	177.80	248.92	355.60
241020141B	single track cantilever – install	Nr	–	–	–	266.70	266.70	373.38	533.40
241020151	Metalwork; miscellaneous:					
241020151A	general metalwork supply	kg	–	–	–	0.95	0.95	1.33	1.90
241020151B	general metalwork erection	kg	–	–	–	0.54	0.54	0.76	1.08
241020161	Wiring:					
241020161A	headspan – supply	Nr	–	–	–	17780.15	17780.15	24892.21	35560.30
241020161B	headspan – install	Nr	–	–	–	1778.02	1778.02	2489.23	3556.04
241020161C	earth/return – supply	m	–	–	–	3.56	3.56	4.98	7.12
241020161D	earth/return – install	m	–	–	–	0.89	0.89	1.25	1.78
241020161E	conductor wire – supply	m	–	–	–	8.89	8.89	12.45	17.78
241020161F	conductor wire – install	m	–	–	–	5.33	5.33	7.46	10.66
241020161G	registration – supply	Nr	–	–	–	35.56	35.56	49.78	71.12
241020161H	registration – install	Nr	–	–	–	44.45	44.45	62.23	88.90
241020161I	paning – install	Nr	–	–	–	26.67	26.67	37.34	53.34

AC Electrification

Rail 2003 Ref		Unit	Composite Cost £	Unit Rate Base Cost £	Unit Rate Green Zone £	Unit Rate Red Zone £
241	**NEW WORKS – INNER CITY**					
2410201	Electrification works generally					
241020171	Transformers; booster					
241020171A	supply	Nr	– – – 26670.23	26670.23	37338.32	53340.46
241020171B	install	Nr	– – – 8890.08	8890.08	12446.11	17780.16
241020181	Bonding					
241020181A	supply	Nr	– – – 44.45	44.45	62.23	88.90
241020181B	install	Nr	– – – 88.90	88.90	124.46	177.80
24104	**OVERHEAD LINE WORKS – 4 TRACK WORK**					
2410401	Electrification works generally					
241040101	Bases for structures; in situ concrete bases, including holding down bolts and fixings:					
241040101A	masts or the like	Nr	– – – 1064.22	1064.22	1489.91	2128.44
241040111	Headspans:					
241040111A	steel supply	Nr	– – – 966.31	966.31	1352.83	1932.62
241040111B	steel erection	Nr	– – – 193.26	193.26	270.56	386.52
241040121	Cantilevers:					
241040121A	steel supply	Nr	– – – 773.05	773.05	1082.27	1546.10
241040121B	steel erection	Nr	– – – 579.79	579.79	811.71	1159.58
241040131	Gantries:					
241040131A	steel supply	Nr	– – – 3865.25	3865.25	5411.35	7730.50
241040131B	steel erection	Nr	– – – 2898.94	2898.94	4058.52	5797.88

AC Electrification

Rail 2003 Ref		Unit	Composite Cost £	Unit Rate Base Cost £	Unit Rate Green Zone £	Unit Rate Red Zone £
241	**NEW WORKS – INNER CITY**					
2410401	**Electrification works generally**					
241040141	**Brackets:**					
241040141A	single track cantilever – supply	Nr	–	193.26	270.56	386.52
241040141B	single track cantilever – install	Nr	–	289.89	405.85	579.78
241040151	**Metalwork:**					
241040151A	steel supply	kg	–	2.32	3.25	4.64
241040151B	steel erection	kg	–	0.59	0.83	1.18
241040161	**Wiring:**					
241040161A	headspan – supply	Nr	–	19326.25	27056.75	38652.50
241040161B	headspan – install	Nr	–	1932.63	2705.68	3865.26
241040161C	earth/return – supply	m	–	3.86	5.40	7.72
241040161D	earth/return – install	m	–	0.97	1.36	1.94
241040161E	conductor wire – supply	m	–	9.66	13.52	19.32
241040161F	conductor wire – install	m	–	5.80	8.12	11.60
241040161G	registration – supply	Nr	–	38.65	54.11	77.30
241040161H	registration – install	Nr	–	48.32	67.65	96.64
241040161I	paning – install	Nr	–	28.99	40.59	57.98
241040171	**Transformers:**					
241040171A	supply	Nr	–	28989.38	40585.13	57978.76
241040171B	install	Nr	–	9663.13	13528.38	19326.26
241040181	**Bonding:**					
241040181A	supply	Nr	–	48.32	67.65	96.64
241040181B	install	Nr	–	96.63	135.28	193.26

AC Electrification

Rail 2003 Ref		Unit	Composite Cost £	Unit Rate Base Cost £	Unit Rate Green Zone £	Unit Rate Red Zone £
241	**NEW WORKS – INNER CITY**					
24106	**OVERHEAD LINE WORKS – 6 TRACK WORK**					
2410601	**Electrification works generally**					
241060101	Bases for structures; in situ concrete bases, including holding down bolts and fixings:					
241060101A	masts or the like	Nr	– – –	1276.88	1787.63	2553.77
241060111	**Headspans:**					
241060111A	steel supply	Nr	– – –	1159.58	1623.41	2319.16
241060111B	steel erection	Nr	– – –	231.92	324.69	463.84
241060121	**Cantilevers:**					
241060121A	steel supply	Nr	– – –	927.66	1298.72	1855.32
241060121B	steel erection	Nr	– – –	695.75	974.05	1391.50
241060131	**Gantries:**					
241060131A	steel supply	Nr	– – –	4638.30	6493.62	9276.60
241060131B	steel erection	Nr	– – –	3478.73	4870.22	6957.46
241060141	**Brackets:**					
241060141A	single track cantilever – supply	Nr	– – –	231.92	324.69	463.84
241060141B	single track cantilever – install	Nr	– – –	347.87	487.02	695.74
241060151	**Metalwork:**					
241060151A	steel supply	kg	– – –	3.24	4.54	6.48
241060151B	steel erection	kg	– – –	0.69	0.97	1.38
241060161	**Wiring:**					
241060161A	headspan – supply	Nr	– – –	23191.50	32468.10	46383.00
241060161B	headspan – install	Nr	– – –	2319.15	3246.81	4638.30
241060161C	earth/return – supply	m	– – –	4.64	6.50	9.28
241060161D	earth/return – install	m	– – –	1.16	1.62	2.32
241060161E	conductor wire – supply	m	– – –	11.60	16.24	23.20
241060161F	conductor wire – install	m	– – –	6.96	9.74	13.92
241060161G	registration – supply	Nr	– – –	46.58	65.21	93.16
241060161H	registration – install	Nr	– – –	57.98	81.17	115.96
241060161I	paning – install	Nr	– – –	34.79	48.71	69.58

Note: The Composite Cost column shows 1276.88, 1159.58, 231.92, 927.66, 695.75, 4638.30, 3478.73, 231.92, 347.87, 3.24, 0.69, 23191.50, 2319.15, 4.64, 1.16, 11.60, 6.96, 46.58, 57.98, 34.79 respectively for each row (same as Base Cost).

AC Electrification

Rail 2003 Ref		Unit	Composite Cost £	Unit Rate Base Cost £	Unit Rate Green Zone £	Unit Rate Red Zone £
241	**NEW WORKS – INNER CITY**					
2410601	**Electrification works generally**					
241060171	Transformers:					
241060171A	supply	Nr	– – –	34787.25	34787.25	48702.15	69574.50
241060171B	install	Nr	– – –	11595.75	11595.75	16234.05	23191.50
241060181	Bonding:					
241060181A	supply	Nr	– – –	57.98	57.98	81.17	115.96
241060181B	install	Nr	– – –	115.96	115.96	162.34	231.92
24108	**OVERHEAD LINE WORKS – 8 TRACK WORK**					
2410801	**Electrification works generally**					
241080101	Bases for structures; in situ concrete bases, including holding down bolts and fixings:					
241080101A	masts or the like	Nr	– – –	1702.34	2383.27	3404.66	4766.54
241080111	Headspans:					
241080111A	steel supply	Nr	– – –	1546.10	1546.10	2164.54	3092.20
241080111B	steel erection	Nr	– – –	309.22	309.22	432.91	618.44
241080121	Cantilevers:					
241080121A	steel supply	Nr	– – –	1236.88	1236.88	1731.63	2473.76
241080121B	steel erection	Nr	– – –	927.66	927.66	1298.72	1855.32
241080131	Gantries:					
241080131A	steel supply	Nr	– – –	6184.40	6184.40	8658.16	12368.80
241080131B	steel erection	Nr	– – –	4638.30	4638.30	6493.62	9276.60
241080141	Brackets:					
241080141A	single track cantilever – supply	Nr	– – –	309.22	309.22	432.91	618.44
241080141B	single track cantilever – install	Nr	– – –	463.83	463.83	649.36	927.66

AC Electrification

Rail 2003 Ref		Unit	Composite Cost £	Unit Rate Base Cost £	Unit Rate Green Zone £	Unit Rate Red Zone £
241	**NEW WORKS – INNER CITY**					
2410801	**Electrification works generally**					
241080151	**Metalwork:**					
241080151A	steel supply	kg	–	3.71	5.19	7.42
241080151B	steel erection	kg	–	0.92	1.29	1.84
241080161	**Wiring:**					
241080161A	headspan – supply	Nr	–	30922.00	43290.80	61844.00
241080161B	headspan – install	Nr	–	3092.20	4329.08	6184.40
241080161C	earth/return – supply	m	–	6.18	8.65	12.36
241080161D	earth/return – install	m	–	1.55	2.17	3.10
241080161E	conductor wire – supply	m	–	15.46	21.64	30.92
241080161F	conductor wire – install	m	–	9.28	12.99	18.56
241080161G	registration – supply	Nr	–	61.84	86.58	123.68
241080161H	registration – install	Nr	–	77.31	108.23	154.62
241080161I	paning – install	Nr	–	46.38	64.93	92.76
241080171	**Transformers:**					
241080171A	supply	Nr	–	46383.00	64936.20	92766.00
241080171B	install	Nr	–	15461.00	21645.40	30922.00
241080181	**Bonding:**					
241080181A	supply	Nr	–	77.30	108.22	154.60
241080181B	install	Nr	–	154.61	216.45	309.22
24111	**DISTRIBUTION EQUIPMENT**					
2411101	**Feeder stations**					
241110101	**Busbars:**					
241110101A	ref	m	–	42.52	59.53	85.04
241110111	**Cable:**					
241110111A	ref	m	–	34.01	47.61	68.02

AC Electrification

Rail 2003 Ref		Unit	Composite Cost £	Unit Rate Base Cost £	Unit Rate Green Zone £	Unit Rate Red Zone £
241	**NEW WORKS – INNER CITY**					
2411116	**Switchgear**					
241111601	Double pole disconnector:					
241111601A	ref	Nr	–	17007.10	23809.94	34014.20
			– – 17007.10			
241111611	Circuit breakers:					
241111611A	25Kv SMOS	Nr	– – – 20408.52	20408.52	28571.93	40817.04
241111616	Remote monitoring equipment:					
241111616A	2 circuit breaker location	Nr	– – – 6500.00	6500.00	9100.00	13000.00
241111616B	4 circuit breaker location	Nr	– – – 8750.00	8750.00	12250.00	17500.00
241111616C	6 circuit breaker location	Nr	– – – 11575.00	11575.00	16205.00	23150.00
241111616D	8 circuit breaker location	Nr	– – – 14575.00	14575.00	20405.00	29150.00
241111616F	12 circuit breaker location	Nr	– – – 21250.00	21250.00	29750.00	42500.00
241111616G	PC workstation	Nr	– – – 2575.00	2575.00	3605.00	5150.00
241111621	Voltage transformers:					
241111621A	25Kv auxilliary supply transformer	Nr	– – – 1700.71	1700.71	2380.99	3401.42
241111631	Ancillary equipment cubicle:					
241111631A	ref	Nr	– – – 1275.53	1275.53	1785.74	2551.06
241111641	Protection panels:					
241111641A	ref	Nr	– – – 1700.71	1700.71	2380.99	3401.42
241111651	Motorised switch mechanism:					
241111651A	supply	Nr	– – – 3401.42	3401.42	4761.99	6802.84
241111651B	install	Nr	– – – 1700.71	1700.71	2380.99	3401.42
241111661	Overhead line SPS/operating mechanism:					
241111661A	ref	Nr	– – – 34014.20	34014.20	47619.88	68028.40

Rail 2003 Ref		Unit				Composite Cost	Unit Rate	Unit Rate	Unit Rate
						£	Base Cost £	Green Zone £	Red Zone £
241	**NEW WORKS – INNER CITY**								
2411131	**Cable**								
241113101	**Polymeric concentric, 300 mm^2:**								
241113101A	supply	m	–	–	–	51.02	51.02	71.43	102.04
241113101B	install	m	–	–	–	8.50	8.50	11.90	17.00
241113121	**Track feeder:**								
241113121A	supply	m	–	–	–	34.01	34.01	47.61	68.02
241113121B	install	m	–	–	–	17.00	17.00	23.80	34.00
241113141	**Along track bare feeder:**								
241113141A	supply	m	–	–	–	34.01	34.01	47.61	68.02
241113141B	install	m	–	–	–	17.00	17.00	23.80	34.00
241113161	**Trenching:**								
241113161A	trench; hand excavation, fill, reinstate	m	–	–	–	21.25	21.25	29.75	42.50

AC Electrification

Rail 2003 Ref		Unit	Composite Cost £	Unit Rate Base Cost £	Unit Rate Green Zone £	Unit Rate Red Zone £
246	**NEW WORKS – SUBURBAN**					
24601	**OVERHEAD LINE WORKS – 1 TRACK WORK**					
2460101	**Electrification works generally**					
246010101	Bases for structures; in situ reinforced concrete bases, including holding down bolts and fixings:					
246010101A	masts or the like	Nr	– – –	851.16	851.16	1191.63	1702.34
246010111	Headspans:					
246010111A	steel supply	Nr	– – –	773.05	773.05	1082.27	1546.10
246010111B	steel erection	Nr	– – –	154.61	154.61	216.45	309.22
246010121	Cantilevers:					
246010121A	steel supply	Nr	– – –	618.44	618.44	865.82	1236.88
246010121B	steel erection	Nr	– – –	463.83	463.83	649.36	927.66
246010131	Gantries:					
246010131A	steel supply	Nr	– – –	3092.20	3092.20	4329.08	6184.40
246010131B	steel erection	Nr	– – –	2319.15	2319.15	3246.81	4638.30
246010141	Brackets:					
246010141A	single track cantilever – supply	Nr	– – –	154.61	154.61	216.45	309.22
246010141B	single track cantilever – install	Nr	– – –	231.92	231.92	324.69	463.84
246010151	Metalwork:					
246010151A	steel supply	kg	– – –	1.85	1.85	2.59	3.70
246010151B	steel erection	kg	– – –	0.46	0.46	0.64	0.92
246010161	Wiring:					
246010161A	headspan – supply	Nr	– – –	15461.00	15461.00	21645.40	30922.00
246010161B	headspan – install	Nr	– – –	1546.10	1546.10	2164.54	3092.20
246010161C	earth/return – supply	m	– – –	3.09	3.09	4.33	6.18
246010161D	earth/return – install	m	– – –	0.77	0.77	1.08	1.54
246010161E	conductor wire – supply	m	– – –	7.73	7.73	10.82	15.46
246010161F	conductor wire – install	m	– – –	4.64	4.64	6.50	9.28
246010161G	registration – supply	Nr	– – –	30.92	30.92	43.29	61.84
246010161H	registration – install	Nr	– – –	38.65	38.65	54.11	77.30
246010161I	paning – install	Nr	– – –	23.19	23.19	32.47	46.38

AC Electrification

Rail 2003 Ref		Unit	Composite Cost £	Unit Rate Base Cost £	Unit Rate Green Zone £	Unit Rate Red Zone £
246	**NEW WORKS – SUBURBAN**					
2460101	**Electrification works generally**					
246010171	Transformers:					
246010171A	supply	Nr	–	–	–	23191.50	23191.50	32468.10	46383.00
246010171B	install	Nr	–	–	–	7730.50	7730.50	10822.70	15461.00
246010181	Bonding:					
246010181A	supply	Nr	–	–	–	38.65	38.65	54.11	77.30
246010181B	install	Nr	–	–	–	77.31	77.31	108.23	154.62
24602	**OVERHEAD LINE WORKS – 2 TRACK WORK**					
2460201	**Electrification works generally**					
246020101	Bases for structures; in situ reinforced concrete bases, including holding down bolts and fixings:					
246020101A	masts or the like	Nr	–	–	–	893.75	893.75	1251.23	1787.47
246020111	Headspans:					
246020111A	steel supply	Nr	–	–	–	811.70	811.70	1136.38	1623.40
246020111B	steel erection	Nr	–	–	–	162.34	162.34	227.28	324.68
246020121	Cantilevers:					
246020121A	steel supply	Nr	–	–	–	649.36	649.36	909.10	1298.72
246020121B	steel erection	Nr	–	–	–	487.02	487.02	681.83	974.04
246020131	Gantries:					
246020131A	steel supply	Nr	–	–	–	3246.81	3246.81	4545.53	6493.62
246020131B	steel erection	Nr	–	–	–	2435.11	2435.11	3409.15	4870.22
246020141	Brackets:					
246020141A	single track cantilever – supply	Nr	–	–	–	162.34	162.34	227.28	324.68
246020141B	single track cantilever – install	Nr	–	–	–	243.51	243.51	340.91	487.02

AC Electrification

Rail 2003 Ref		Unit	Composite Cost £	Unit Rate Base Cost £	Unit Rate Green Zone £	Unit Rate Red Zone £
246	**NEW WORKS – SUBURBAN**					
2460201	Electrification works generally					
246020151	Metalwork:					
246020151A	steel supply	kg	– – –	1.95	1.95	2.73	3.90
246020151B	steel erection	kg	– – –	0.49	0.49	0.69	0.98
246020161	Wiring:					
246020161A	headspan – supply	Nr	– – –	16234.05	16234.05	22727.67	32468.10
246020161B	headspan – install	Nr	– – –	1623.41	1623.41	2272.77	3246.82
246020161C	earth/return – supply	m	– – –	3.25	3.25	4.55	6.50
246020161D	earth/return – install	m	– – –	0.82	0.82	1.15	1.64
246020161E	conductor wire – supply	m	– – –	8.12	8.12	11.37	16.24
246020161F	conductor wire – install	m	– – –	4.87	4.87	6.82	9.74
246020161G	registration – supply	Nr	– – –	32.47	32.47	45.46	64.94
246020161H	registration – install	Nr	– – –	40.59	40.59	56.83	81.18
246020161I	paning – install	Nr	– – –	24.35	24.35	34.09	48.70
246020171	Transformers:					
246020171A	supply	Nr	– – –	24351.08	24351.08	34091.51	48702.16
246020171B	install	Nr	– – –	8117.03	8117.03	11363.84	16234.06
246020181	Bonding:					
246020181A	supply	Nr	– – –	40.59	40.59	56.83	81.18
246020181B	install	Nr	– – –	81.17	81.17	113.64	162.34
24604	**OVERHEAD LINE WORKS – 4 TRACK WORK**					
2460401	Electrification works generally					
246040101	Bases for structures; in situ reinforced concrete bases, including holding down bolts and fixings:					
246040101A	masts or the like	Nr	– – –	936.35	936.35	1310.86	1872.67

AC Electrification

Rail 2003 Ref	Unit	Composite Cost £	Unit Rate Base Cost £	Unit Rate Green Zone £	Unit Rate Red Zone £
246	**NEW WORKS – SUBURBAN**				
2460401	**Electrification works generally**				
246040111	**Headspans:**				
246040111A	steel supply .... Nr	–	–	–	850.36	850.36	1190.50	1700.72
246040111B	steel erection .... Nr	–	–	–	170.07	170.07	238.10	340.14
246040121	**Cantilevers:**				
246040121A	steel supply .... Nr	–	–	–	680.28	680.28	952.39	1360.56
246040121B	steel erection .... Nr	–	–	–	510.21	510.21	714.29	1020.42
246040131	**Gantries:**				
246040131A	steel supply .... Nr	–	–	–	3401.42	3401.42	4761.99	6802.84
246040131B	steel erection .... Nr	–	–	–	2551.07	2551.07	3571.50	5102.14
246040141	**Brackets:**				
246040141A	single track cantilever – supply .... Nr	–	–	–	170.07	170.07	238.10	340.14
246040141B	single track cnatilever – install .... Nr	–	–	–	255.11	255.11	357.15	510.22
246040151	**Metalwork:**				
246040151A	steel supply .... kg	–	–	–	2.04	2.04	2.86	4.08
246040151B	steel erection .... kg	–	–	–	0.51	0.51	0.71	1.02
246040161	**Wiring:**				
246040161A	headspan – supply .... Nr	–	–	–	17007.10	17007.10	23809.94	34014.20
246040161B	headspan – install .... Nr	–	–	–	1700.71	1700.71	2380.99	3401.42
246040161C	earth/return – supply .... m	–	–	–	3.40	3.40	4.76	6.80
246040161D	earth/return – install .... m	–	–	–	0.85	0.85	1.19	1.70
246040161E	conductor wire – supply .... m	–	–	–	8.50	8.50	11.90	17.00
246040161F	conductor wire – install .... m	–	–	–	5.10	5.10	7.14	10.20
246040161G	registration – supply .... Nr	–	–	–	34.01	34.01	47.61	68.02
246040161H	registration – install .... Nr	–	–	–	42.52	42.52	59.53	85.04
246040161I	paning – install .... Nr	–	–	–	25.51	25.51	35.71	51.02

AC Electrification

Rail 2003 Ref		Unit	Composite Cost £	Unit Rate Base Cost £	Unit Rate Green Zone £	Unit Rate Red Zone £

246 NEW WORKS – SUBURBAN

2460401 Electrification works generally

246040171 Transformers:

246040171A	supply	Nr	–	–	–	25510.65	25510.65	35714.91	51021.30
246040171B	install	Nr	–	–	–	8503.55	8503.55	11904.97	17007.10

246040181 Bonding:

246040181A	supply	Nr	–	–	–	42.52	42.52	59.53	85.04
246040181B	install	Nr	–	–	–	85.04	85.04	119.06	170.08

24606 OVERHEAD LINE WORKS – 6 TRACK WORK

2460601 Electrification works generally

246060101 Bases for structures; in situ reinforced concrete bases; including holding down bolts and fixings:

246060101A	masts or the like	Nr	–	–	–	979.06	979.06	1370.68	1958.11

246060111 Headspans:

246060111A	steel supply	Nr	–	–	–	889.00	889.00	1244.60	1778.00
246060111B	steel erection	Nr	–	–	–	177.80	177.80	248.92	355.60

246060121 Cantilevers:

246060121A	steel supply	Nr	–	–	–	711.20	711.20	995.68	1422.40
246060121B	steel erection	Nr	–	–	–	533.40	533.40	746.76	1066.80

246060131 Gantries:

246060131A	steel supply	Nr	–	–	–	3556.03	3556.03	4978.44	7112.06
246060131B	steel erection	Nr	–	–	–	2667.02	2667.02	3733.83	5334.04

246060141 Brackets:

246060141A	single track cantilever – supply	Nr	–	–	–	177.80	177.80	248.92	355.60
246060141B	single track cantilever – install	Nr	–	–	–	266.70	266.70	373.38	533.40

AC Electrification

Rail 2003 Ref	Unit	Composite Cost £	Unit Rate Base Cost £	Unit Rate Green Zone £	Unit Rate Red Zone £
246	**NEW WORKS – SUBURBAN**				
2460601	**Electrification works generally**				
246060151	**Metalwork:**				
246060151A	steel supply .... kg	– – – 2.13	2.13	2.98	4.26
246060151B	steel erection .... kg	– – – 0.54	0.54	0.76	1.08
246060161	**Wiring:**				
246060161A	headspan – supply .... Nr	– – – 17780.15	17780.15	24892.21	35560.30
246060161B	headspan – install .... Nr	– – – 1778.02	1778.02	2489.23	3556.04
246060161C	earth/return – supply .... m	– – – 3.56	3.56	4.98	7.12
246060161D	earth/return – install .... m	– – – 0.89	0.89	1.25	1.78
246060161E	conductor wire – supply .... m	– – – 8.89	8.89	12.45	17.78
246060161F	conductor wire – install .... m	– – – 5.33	5.33	7.46	10.66
246060161G	registration – supply .... Nr	– – – 35.56	35.56	49.78	71.12
246060161H	registration – install .... Nr	– – – 44.45	44.45	62.23	88.90
246060161I	paning – install .... Nr	– – – 26.67	26.67	37.34	53.34
246060171	**Transformers:**				
246060171A	supply .... Nr	– – – 26670.23	26670.23	37338.32	53340.46
246060171B	install .... Nr	– – – 8890.08	8890.08	12446.11	17780.16
246060181	**Bonding:**				
246060181A	supply .... Nr	– – – 44.45	44.45	62.23	88.90
246060181B	install .... Nr	– – – 88.90	88.90	124.46	177.80
24611	**DISTRIBUTION EQUIPMENT**				
2461101	**Feeder stations**				
246110101	**Busbars:**				
246110101A	generally .... m	– – – 38.65	38.65	54.11	77.30
246110111	**Cable:**				
246110111A	generally .... m	– – – 30.92	30.92	43.29	61.84

AC Electrification

Rail 2003 Ref		Unit			Composite Cost £	Unit Rate Base Cost £	Unit Rate Green Zone £	Unit Rate Red Zone £
246	NEW WORKS – SUBURBAN							
2461116	Switchgear:							
246111601	Double pole disconnector							
246111611	Circuit breakers:							
246111611A	generally	Nr	–	–	–	18553.20	18553.20	25974.48	37106.40
246111621	Voltage transformer:							
246111621A	generally	Nr	–	–	–	1546.10	1546.10	2164.54	3092.20
246111631	Ancillary equipment cubicle:							
246111631A	generally	Nr	–	–	–	1159.58	1159.58	1623.41	2319.16
246111641	Protection panels:							
246111641A	generally	Nr	–	–	–	1546.10	1546.10	2164.54	3092.20
246111651	Motorised switch mechanism:							
246111651A	supply	Nr	–	–	–	3092.20	3092.20	4329.08	6184.40
246111651B	install	Nr	–	–	–	1546.10	1546.10	2164.54	3092.20
246111661	Overhead line SPS/operating mechanism:							
246111661A	generally	Nr	–	–	–	30922.00	30922.00	43290.80	61844.00
2461131	Cable							
246113101	Polymetric concentric, 300 mm^2:							
246113101A	supply	m	–	–	–	46.38	46.38	64.93	92.76
246113101B	install	m	–	–	–	7.73	7.73	10.82	15.46
246113121	Track feeder:							
246113121A	supply	m	–	–	–	30.92	30.92	43.29	61.84
246113121B	install	m	–	–	–	15.46	15.46	21.64	30.92

AC Electrification

Rail 2003 Ref		Unit	Composite Cost £	Unit Rate Base Cost £	Unit Rate Green Zone £	Unit Rate Red Zone £
246	**NEW WORKS – SUBURBAN**					
2461131	**Cable**					
246113141	**Along bare track feeder:**					
246113141A	supply	m	– – –	30.92	43.29	61.84
246113141B	install	m	– – –	15.46	21.64	30.92
246113161	**Trenching:**					
246113161A	trench; hand excavation, fill, reinstate	m	– – –	19.33	27.06	38.66

AC Electrification

Rail 2003 Ref		Unit	Composite Cost £	Unit Rate Base Cost £	Unit Rate Green Zone £	Unit Rate Red Zone £
251	**NEW WORKS – RURAL**					
25101	**OVERHEAD LINE WORKS – SINGLE TRACK**					
2510101	Electrification works generally					
251010101	Bases for structures; in situ reinforced concrete bases, including holding down bolts and fixings:					
251010101A	masts or the like	Nr	– – –	851.16	851.16	1191.63	1702.34
251010111	Headspans:					
251010111A	supply	Nr	– – –	734.40	734.40	1028.16	1468.80
251010111B	erection	Nr	– – –	146.88	146.88	205.63	293.76
251010121	Cantilevers:					
251010121A	steel supply	Nr	– – –	587.52	587.52	822.53	1175.04
251010121B	steel erection	Nr	– – –	440.64	440.64	616.90	881.28
251010131	Gantries:					
251010131A	steel supply	Nr	– – –	2937.59	2937.59	4112.63	5875.18
251010131B	steel erection	Nr	– – –	2203.19	2203.19	3084.47	4406.38
251010141	Brackets:					
251010141A	single track cantilever – supply	Nr	– – –	146.88	146.88	205.63	293.76
251010141B	single track cantilever – install	Nr	– – –	220.32	220.32	308.45	440.64
251010151	Metalwork:					
251010151A	steel supply	kg	– – –	1.76	1.76	2.46	3.52
251010151B	steel erection	kg	– – –	0.45	0.45	0.63	0.90
251010161	Wiring:					
251010161A	headspan – supply	Nr	– – –	14687.95	14687.95	20563.13	29375.90
251010161B	headspan – install	Nr	– – –	1468.80	1468.80	2056.32	2937.60
251010161C	earth/return – supply	m	– – –	2.94	2.94	4.12	5.88
251010161D	earth/return – install	m	– – –	0.74	0.74	1.04	1.48
251010161E	conductor wire – supply	m	– – –	7.34	7.34	10.28	14.68
251010161F	conductor wire – install	m	– – –	7.73	7.73	10.82	15.46
251010161G	registration – supply	Nr	– – –	29.38	29.38	41.13	58.76
251010161H	registration – install	Nr	– – –	36.72	36.72	51.41	73.44
251010161I	paning – install	Nr	– – –	22.03	22.03	30.84	44.06

Rail 2003 Ref		Unit				Composite Cost	Unit Rate	Unit Rate	Unit Rate
						£	Base Cost £	Green Zone £	Red Zone £
251	**NEW WORKS – RURAL**								
2510101	**Electrification works generally**								
251010171	**Transformers:**								
251010171A	supply	Nr	–	–	–	22031.93	22031.93	30844.70	44063.86
251010171B	install	Nr	–	–	–	7343.95	7343.95	10281.53	14687.90
251010181	**Bonding:**								
251010181A	supply	Nr	–	–	–	36.72	36.72	51.41	73.44
251010181B	install	Nr	–	–	–	73.44	73.44	102.82	146.88
25102	**OVERHEAD LINE WORKS – TWO TRACK**								
2510201	**Electrification works generally**								
251020101	**Bases for structures; in situ reinforced concrete bases; including holding down bolts and fixings:**								
251020101A	mast foundation; 1.5 × 1.2 × 3.5 m	Nr	–	–	–	6875.00	6875.00	9625.00	13750.00
251020101B	mast foundation; 0.8 × 0.8 × 2.5 m	Nr	–	–	–	5450.00	5450.00	7630.00	10900.00
251020101C	mast foundation; 0.6 × 0.6 × 2.5 m	Nr	–	–	–	5200.00	5200.00	7280.00	10400.00
251020101F	tie foundation; 0.8 × 0.8 × 2.5 m	Nr	–	–	–	6850.00	6850.00	9590.00	13700.00
251020101K	portal foundation; 1.1 × 1.5 × 3.0 m	Nr	–	–	–	7000.00	7000.00	9800.00	14000.00
251020111	**Headspans:**								
251020111A	steel supply	Nr	–	–	–	773.05	773.05	1082.27	1546.10
251020111B	steel erection	Nr	–	–	–	154.61	154.61	216.45	309.22
251020121	**Cantilevers:**								
251020121A	steel supply	Nr	–	–	–	618.44	618.44	865.82	1236.88
251020121B	steel erection	Nr	–	–	–	463.83	463.83	649.36	927.66

AC Electrification

Rail 2003 Ref		Unit	Composite Cost £	Unit Rate Base Cost £	Unit Rate Green Zone £	Unit Rate Red Zone £
251	**NEW WORKS – RURAL**					
2510201	**Electrification works generally**					
251020131	Gantries:					
251020131A	steel supply	Nr	– – –	3092.20	4329.08	6184.40
251020131B	steel erection	Nr	– – –	2319.15	3246.81	4638.30
251020141	Brackets:					
251020141A	two track cantilever – supply	Nr	– – –	154.61	216.45	309.22
251020141B	two track cantilever – install	Nr	– – –	231.92	324.69	463.84
251020151	Metalwork:					
251020151A	steel supply	kg	– – –	1.85	2.59	3.70
251020151B	steel erection	kg	– – –	0.46	0.64	0.92
251020156	Anchor assemblies; complete installation:					
251020156A	tie wire anchor assembly	Nr	– – –	4575.00	6405.00	9150.00
251020156B	tie rod assembly	Nr	– – –	3100.00	4340.00	6200.00
251020156C	fixed end anchor assembly	Nr	– – –	4675.00	6545.00	9350.00
251020156D	balance weight anchor assembly	Nr	– – –	7750.00	10850.00	15500.00
251020161	Wiring:					
251020161A	headspan – supply	Nr	– – –	15461.00	21645.40	30922.00
251020161B	headspan – install	Nr	– – –	1546.10	2164.54	3092.20
251020161C	earth/return – supply	m	– – –	3.09	4.33	6.18
251020161D	earth/return – install	m	– – –	0.77	1.08	1.54
251020161E	conductor wire – supply	m	– – –	7.73	10.82	15.46
251020161F	conductor wire – install	m	– – –	4.63	6.48	9.26
251020161G	registration – supply	Nr	– – –	30.92	43.29	61.84
251020161H	registration – install	Nr	– – –	38.65	54.11	77.30
251020161I	paning – install	Nr	– – –	23.19	32.47	46.38
251020166	Along track generally:					
251020166A	contact wire; 107 mm^2; supply and install	m	– – 15.45	15.45	21.63	30.90
251020166B	catenary wire (AWAL); supply and install	m	– – 15.75	15.75	22.05	31.50
251020166C	earth wire; supply and install	m	– – 10.75	10.75	15.05	21.50
251020166D	bonding cable 19/3.25 mm^2; supply and install	m	– – 12.40	12.40	17.36	24.80
251020166E	spark gap device; supply and install	Nr	– – 7730.00	7730.00	10822.00	15460.00

AC Electrification

Rail 2003 Ref		Unit	Composite Cost £	Unit Rate Base Cost £	Unit Rate Green Zone £	Unit Rate Red Zone £
251	**NEW WORKS – RURAL**					
2510201	Electrification works generally					
251020171	Transformers:					
251020171A	supply	Nr	–	23191.50	32468.10	46383.00
251020171B	install	Nr	–	7730.50	10822.70	15461.00
251020181	Bonding:					
251020181A	supply	Nr	–	38.65	54.11	77.30
251020181B	install	Nr	–	77.31	108.23	154.62
251020191	Miscellaneous works; complete installation:					
251020191A	cantilever frame and registration assembly	Nr	–	1500.00	2100.00	3000.00
251020191B	single pole isolator assembly	Nr	–	5400.00	7560.00	10800.00
25103	**OVERHEAD LINE WORKS – THREE TRACK**					
2510301	Electrification works generally					
251030101	Bases for structures; in situ reinforced concrete bases; including holding down bolts and fixings:					
251030101A	mast foundation; 1.5 × 1.2 × 3.5 m	Nr	–	6875.00	9625.00	13750.00
251030101B	mast foundations; 0.8 × 0.8 × 2.5 m	Nr	–	5450.00	7630.00	10900.00
251030101C	mast foundation; 0.6 × 0.6 × 2.5 m	Nr	–	5200.00	7280.00	10400.00
251030101F	tie foundation; 0.8 × 0.8 × 2.5 m	Nr	–	6850.00	9590.00	13700.00
251030101K	portal foundation; 1.1 × 1.5 × 3.0 m	Nr	–	7000.00	9800.00	14000.00
251030111	Headspans:					
251030111A	steel supply	Nr	–	773.05	1082.27	1546.10
251030111B	steel erection	Nr	–	154.61	216.45	309.22
251030121	Cantilevers:					
251030121A	steel erection	Nr	–	618.44	865.82	1236.88
251030121B	steel erection	Nr	–	463.83	649.36	927.66

AC Electrification

Rail 2003 Ref		Unit	Composite Cost £	Unit Rate Base Cost £	Unit Rate Green Zone £	Unit Rate Red Zone £
251	**NEW WORKS – RURAL**					
2510301	**Electrification works generally**					
251030131	Gantries:					
251030131A	steel supply	Nr	– – –	3092.20	3092.20	4329.08	6184.40
251030131B	steel erection	Nr	– – –	2319.15	2319.15	3246.81	4638.30
251030141	Brackets:					
251030141A	three track cantilever – supply	Nr	– – –	154.61	154.61	216.45	309.22
251030141B	three track cantilever – install	Nr	– – –	231.92	231.92	324.69	463.84
251030151	Metalwork:					
251030151A	steel supply	kg	– – –	1.85	1.85	2.59	3.70
251030151B	steel erection	kg	– – –	0.46	0.46	0.64	0.92
251030156	Anchor assemblies; complete installation:					
251030156A	tie wire anchor assembly	Nr	– – –	4575.00	4575.00	6405.00	9150.00
251030156B	tie wire anchor assembly	Nr	– – –	3100.00	3100.00	4340.00	6200.00
251030156C	fixed end anchor assembly	Nr	– – –	4675.00	4675.00	6545.00	9350.00
251030156D	balance weight anchor assembly	Nr	– – –	7750.00	7750.00	10850.00	15500.00
251030161	Wiring:					
251030161A	headspan – supply	Nr	– – –	15461.00	15461.00	21645.40	30922.00
251030161B	headspan – install	Nr	– – –	1546.10	1546.10	2164.54	3092.20
251030161C	earth/return – supply	m	– – –	3.09	3.09	4.33	6.18
251030161D	earth/return – install	m	– – –	0.77	0.77	1.08	1.54
251030161E	conductor wire – supply	m	– – –	7.73	7.73	10.82	15.46
251030161G	registration – supply	Nr	– – –	30.92	30.92	43.29	61.84
251030161H	registration – install	Nr	– – –	38.65	38.65	54.11	77.30
251030161I	paning – install	Nr	– – –	23.19	23.19	32.47	46.38
251030166	Along track generally:					
251030166A	contact wire; 107 mm^2; supply and install	m	– – –	15.45	15.45	21.63	30.90
251030166B	catenary wire (AWAL); supply and install	m	– – –	15.75	15.75	22.05	31.50
251030166C	earth wire; supply and install	m	– – –	10.75	10.75	15.05	21.50
251030166D	bonding cable 19/3.25 mm^2; supply and install	m	– – –	12.40	12.40	17.36	24.80
251030166E	spark gap device; supply and install	Nr	– – –	7730.00	7730.00	10822.00	15460.00

AC Electrification

Rail 2003 Ref		Unit	Composite Cost £	Unit Rate Base Cost £	Unit Rate Green Zone £	Unit Rate Red Zone £
251	**NEW WORKS – RURAL**					
2510301	**Electrification works generally**					
251030171	Transformers:					
251030171A	supply	Nr	–	23191.50	32468.10	46383.00
251030171B	install	Nr	–	7730.50	10822.70	15461.00
251030181	Bonding:					
251030181A	supply	Nr	–	38.65	54.11	77.30
251030181B	install	Nr	–	77.31	108.23	154.62
251030191	Miscellaneous works; complete installation:					
251030191A	cantilever frame and registration assembly	Nr	–	1500.00	2100.00	3000.00
251030191B	single pole isolator assembly	Nr	–	5400.00	7560.00	10800.00
25104	**OVERHEAD LINE WORKS – FOUR TRACK**					
2510401	**Electrification works generally**					
251040101	Bases for structures; in situ reinforced concrete bases; including holding down bolts and fixings:					
251040101A	mast foundation; 1.5 × 1.2 × 3.5 m	Nr	–	6875.00	9625.00	13750.00
251040101B	mast foundation; 0.8 × 0.8 × 2.5 m	Nr	–	5450.00	7630.00	10900.00
251040101C	mast foundation; 0.6 × 0.6 × 2.5 m	Nr	–	5200.00	7280.00	10400.00
251040101F	tie foundation; 0.8 × 0.8 × 2.5 m	Nr	–	6850.00	9590.00	13700.00
251040101K	portal foundation; 1.1 × 1.5 × 3.0 m	Nr	–	7000.00	9800.00	14000.00
251040111	Headspans:					
251040111A	steel supply	Nr	–	773.05	1082.27	1546.10
251040111B	steel erection	Nr	–	154.61	216.45	309.22
251040121	Cantilevers:					
251040121A	steel erection	Nr	–	618.44	865.82	1236.88
251040121B	steel erection	Nr	–	463.83	649.36	927.66

AC Electrification

Rail 2003 Ref		Unit	Composite Cost £	Unit Rate Base Cost £	Unit Rate Green Zone £	Unit Rate Red Zone £

251 NEW WORKS – RURAL

2510401 Electrification works generally

251040131 Gantries:

251040131A	steel supply	Nr	–	–	–	3092.20	3092.20	4329.08	6184.40
251040131B	steel erection	Nr	–	–	–	2319.15	2319.15	3246.81	4638.30

251040141 Brackets:

251040141A	four track cantilever – supply	Nr	–	–	–	154.61	154.61	216.45	309.22
251040141B	four track cantilever – install	Nr	–	–	–	231.92	231.92	324.69	463.84

251040151 Metalwork:

251040151A	steel supply	kg	–	–	–	1.85	1.85	2.59	3.70
251040151B	steel erection	kg	–	–	–	0.46	0.46	0.64	0.92

251040156 Anchor assemblies; complete installation:

251040156A	tie wire anchor assembly	Nr	–	–	–	4575.00	4575.00	6405.00	9150.00
251040156B	tie wire anchor assembly	Nr	–	–	–	3100.00	3100.00	4340.00	6200.00
251040156C	fixed end anchor assembly	Nr	–	–	–	4675.00	4675.00	6545.00	9350.00
251040156D	balance weight anchor assembly	Nr	–	–	–	7750.00	7750.00	10850.00	15500.00

251040161 Wiring:

251040161A	headspan – supply	Nr	–	–	–	15461.00	15461.00	21645.40	30922.00
251040161B	headspan – install	Nr	–	–	–	1546.10	1546.10	2164.54	3092.20
251040161C	earth/return – supply	m	–	–	–	3.09	3.09	4.33	6.18
251040161D	earth/return – install	m	–	–	–	0.77	0.77	1.08	1.54
251040161E	conductor wire – supply	m	–	–	–	7.73	7.73	10.82	15.46
251040161G	registration – supply	Nr	–	–	–	30.92	30.92	43.29	61.84
251040161H	registration – install	Nr	–	–	–	38.65	38.65	54.11	77.30
251040161I	paning – install	Nr	–	–	–	23.19	23.19	32.47	46.38

251040166 Along track generally:

251040166A	contact wire; 107 mm^2; supply and install	m	–	–	–	15.45	15.45	21.63	30.90
251040166B	catenary wire (AWAL); supply and install	m	–	–	–	15.75	15.75	22.05	31.50
251040166C	earth wire; supply and install	m	–	–	–	10.75	10.75	15.05	21.50
251040166D	bonding cable 19/3.25 mm^2; supply and install	m	–	–	–	12.40	12.40	17.36	24.80
251040166E	spark gap device; supply and install	Nr	–	–	–	7730.00	7730.00	10822.00	15460.00

AC Electrification

Rail 2003 Ref		Unit	Composite Cost £	Unit Rate Base Cost £	Unit Rate Green Zone £	Unit Rate Red Zone £
251	**NEW WORKS – RURAL**					
2510401	**Electrification works generally**					
251040171	Transformers:					
251040171A	supply	Nr	– – –	23191.50	23191.50	32468.10	46383.00
251040171B	install	Nr	– – –	7730.50	7730.50	10822.70	15461.00
251040181	Bonding:					
251040181A	supply	Nr	– – –	38.65	38.65	54.11	77.30
251040181B	install	Nr	– – –	77.31	77.31	108.23	154.62
251040191	Miscellaneous works; complete installation:					
251040191A	cantilever frame and registration assembly	Nr	– – –	1500.00	1500.00	2100.00	3000.00
251040191B	single pole isolator assembly	Nr	– – –	5400.00	5400.00	7560.00	10800.00
25111	**DISTRIBUTION EQUIPMENT**					
2511101	**Feeder stations**					
251110101	Busbars:					
251110101A	generally	m	– – –	36.72	36.72	51.41	73.44
251110111	Cable:					
251110111A	generally	m	– – –	29.38	29.38	41.13	58.76
2511116	**Switchgear**					
251111601	Double pole disconnector:					
251111601A	generally	Nr	– – –	14687.95	14687.95	20563.13	29375.90
251111611	Circuit breakers:					
251111611A	generally	Nr	– – –	17625.54	17625.54	24675.76	35251.08

AC Electrification

Rail 2003 Ref	Description	Unit	Composite Cost £	Unit Rate Base Cost £	Unit Rate Green Zone £	Unit Rate Red Zone £
251	**NEW WORKS – RURAL**					
2511116	**Switchgear**					
251111621	Voltage transformer:					
251111621A	generally	Nr	– – –	1468.80	1468.80	2056.32	2937.60
251111631	Ancillary equipment cubicle:					
251111631A	generally	Nr	– – –	1101.60	1101.60	1542.24	2203.20
251111641	Protection panels:					
251111641A	generally	Nr	– – –	1468.80	1468.80	2056.32	2937.60
251111651	Motorised switch mechanism:					
251111651A	supply	Nr	– – –	2937.59	2937.59	4112.63	5875.18
251111651B	install	Nr	– – –	1468.80	1468.80	2056.32	2937.60
251111661	Overhead line SPS/operating mechanism:					
251111661A	generally	Nr	– – –	29375.90	29375.90	41126.26	58751.80
2511131	**Cable**					
251113101	Polymetric concentric; 300 mm^2:					
251113101A	supply	m	– – –	44.06	44.06	61.68	88.12
251113101B	install	m	– – –	7.34	7.34	10.28	14.68
251113121	Track feeder:					
251113121A	supply	m	– – –	29.38	29.38	41.13	58.76
251113121B	install	m	– – –	14.69	14.69	20.57	29.38
251113141	Along bare track feeder:					
251113141A	supply	m	– – –	29.38	29.38	41.13	58.76
251113141B	install	m	– – –	14.69	14.69	20.57	29.38

AC Electrification

Rail 2003 Ref		Unit	Composite Cost £	Unit Rate Base Cost £	Unit Rate Green Zone £	Unit Rate Red Zone £
251	**NEW WORKS – RURAL**					
2511131	**Cable**					
251113161	**Trenching:**					
251113161A	trench; hand excavation, fill, re-instate	m	– – – 18.37	18.37	25.72	36.74
271	**REPAIRS AND RENEWALS – INNER CITY**					
27101	**OVERHEAD LINE WORKS**					
2710101	**Single line installation**					
271010101	Removal of ceramic bead insulators in normal neutral section; replace with extended length of ceramic bead insulator:					
271010101A	1090 mm long (Type MK 3B STD)	Nr	– – – 11904.97	11904.97	16666.96	23809.94
271010101B	1260 mm long (Type MK 3B STD)	Nr	– – – 12755.33	12755.33	17857.46	25510.66
27111	**OLW 1**					
2711101	**Double line installation**					
271110101	Removal of ceramic bead insulators in normal neutral section; replace with extended length of ceramic bead insulator:					
271110101A	1090 mm long (Type MK 3B STD)	Nr	– – – 9353.91	9353.91	13095.47	18707.82
271110101B	1260 mm long (Type MK 3B STD)	Nr	– – – 10204.26	10204.26	14285.96	20408.52

AC Electrification

Rail 2003 Ref		Unit	Composite Cost £	Unit Rate Base Cost £	Unit Rate Green Zone £	Unit Rate Red Zone £
271	**REPAIRS AND RENEWALS – INNER CITY**					
27121	**OLW 2**					
2712101	**Contact wiring**					
271210101	Remove existing contract wiring; replace with new:					
271210101A	maximum length 4.00 m both sides of neutral section	Nr	–	–	–	
			340.14	340.14	476.20	680.28
271210101B	maximum length 8.00 m both sides of neutral section	Nr	–	–	–	
			510.21	510.21	714.29	1020.42
2712111	**Booster transformer**					
271211101	Supply only:					
271211101A	100 Amp 25 kV 1 phase	Nr	–	–	–	12755.33	12755.33	17857.46	25510.66
271211111	Remove existing transformer; install new Amp transformer to existing structure:					
271211111A	100/200 Amp transformer	Nr	–	–	–	3401.42	3401.42	4761.99	6802.84
27131	**DISTRIBUTION EQUIPMENT**					
2713101	**Switchgear**					
271310101	Optimho relays:					
271310101A	Relay	Nr	–	–	–	8503.55	8503.55	11904.97	17007.10
271310111	Remote monitoring equipment:					
271310111A	2 circuit breaker location	Nr	–	–	–	11904.97	11904.97	16666.96	23809.94
271310111B	4 circuit breaker location	Nr	–	–	–	17007.10	17007.10	23809.94	34014.20
271310111C	6 circuit breaker location	Nr	–	–	–	21258.88	21258.88	29762.43	42517.76
271310111D	8 circuit breaker location	Nr	–	–	–	27211.36	27211.36	38095.90	54422.72
271310111E	10 circuit breaker location	Nr	–	–	–	34014.20	34014.20	47619.88	68028.40
271310111F	12 circuit breaker location	Nr	–	–	–	42517.75	42517.75	59524.85	85035.50
271310121	Circuit breakers:					
271310121A	25 kV SMOS circuit breaker	Nr	–	–	–	17007.10	17007.10	23809.94	34014.20

AC Electrification

Rail 2003 Ref		Unit	Composite Cost £	Unit Rate Base Cost £	Unit Rate Green Zone £	Unit Rate Red Zone £
271	REPAIRS AND RENEWALS – INNER CITY					
2713101	Switchgear					
271310131	Transformers:					
271310131A	1000 kVA 25,000/850 v single phase..	Nr	– – – 34014.20	34014.20	47619.88	68028.40
276	REPAIRS AND RENEWALS – SUBURBAN					
27601	OVERHEAD LINE WORKS					
2760101	Single line installation					
276010101	Removal of ceramic bead insulators in normal neutral section; replace with extended length of ceramic bead insulator:					
276010101A	1090 mm long (Type MK 3B STD)	Nr	– – – 10822.70	10822.70	15151.78	21645.40
276010101B	1260 mm long (Type MK 3B STD)	Nr	– – – 11595.75	11595.75	16234.05	23191.50
27611	OLW 3					
2761101	Double line installation					
276110101	Removal of ceramic bead insulators in normal neutral section; replace with extended length of ceramic bead insulator:					
276110101A	1090 mm long (Type MK 3D STD)	Nr	– – – 8503.55	8503.55	11904.97	17007.10
276110101B	1260 mm long (Type MK 3B STD)	Nr	– – – 9276.60	9276.60	12987.24	18553.20

AC Electrification

Rail 2003 Ref		Unit	Composite Cost £	Unit Rate Base Cost £	Unit Rate Green Zone £	Unit Rate Red Zone £
276	**REPAIRS AND RENEWALS – SUBURBAN**					
27621	**OLW 4**					
2762101	**Contact wiring**					
276210101	Remove existing contact wiring; replace with new:					
276210101A	maximum length 4.00 m both sides of neutral section	Nr	–	–	–	309.22	309.22	432.91	618.44
276210101B	maximum length 8.00 m both sides of neutral section	Nr	–	–	–	463.83	463.83	649.36	927.66
2762111	**Booster transformer**					
276211101	Supply only:					
276211101A	100 Amp 25 kV 1 phase	Nr	–	–	–	11595.75	11595.75	16234.05	23191.50
276211111	Remove existing transformer; install new Amp transformer to existing structure:					
276211111A	100/200 Amp transformer	Nr	–	–	–	3092.20	3092.20	4329.08	6184.40
27631	**DISTRIBUTION EQUIPMENT**					
2763101	**Switchgear**					
276310101	Optimho relays:					
276310101A	relay	Nr	–	–	–	7730.50	7730.50	10822.70	15461.00
276310111	Remote monitoring equipment:					
276310111A	2 circuit breaker location	Nr	–	–	–	10822.70	10822.70	15151.78	21645.40
276310111B	4 circuit breaker location	Nr	–	–	–	15461.00	15461.00	21645.40	30922.00
276310111C	6 circuit breaker location	Nr	–	–	–	19326.25	19326.25	27056.75	38652.50
276310111D	8 circuit breaker location	Nr	–	–	–	24737.60	24737.60	34632.64	49475.20
276310111E	10 circuit breaker location	Nr	–	–	–	30922.00	30922.00	43290.80	61844.00
276310111F	12 circuit breaker location	Nr	–	–	–	38652.50	38652.50	54113.50	77305.00
276310121	Circuit breakers:					
276310121A	25 kV SMOS circuit breaker	Nr	–	–	–	15461.00	15461.00	21645.40	30922.00

AC Electrification

Rail 2003 Ref	Unit	Composite Cost £	Unit Rate Base Cost £	Unit Rate Green Zone £	Unit Rate Red Zone £
276	**REPAIRS AND RENEWALS – SUBURBAN**				
2763101	Switchgear				
276310131	Transformers:				
276310131A	1000 kVA 25,000/850 v single phase.. Nr	– – – 30922.00	30922.00	43290.80	61844.00
281	**REPAIRS AND RENEWALS – RURAL**				
28101	OVERHEAD LINE WORKS				
2810101	Single line installation				
281010101	Removal of ceramic bead insulators in normal neutral section; replace with extended length of ceramic bead insulator:				
281010101A	1090 mm long (Type MK 3B STD) .... Nr	– – – 10281.57	10281.57	14394.20	20563.14
281010101B	1260 mm long (Type MK 3B STD) .... Nr	– – – 11015.96	11015.96	15422.34	22031.92
28111	LINE INSTALLATION				
2811101	Double line installation				
281110101	Removal of ceramic bead insulators in normal neutral section; replace with extended length of ceramic bead insulator:				
281110101A	1090 mm long (Type MK 3B STD) .... Nr	– – – 8078.37	8078.37	11309.72	16156.74
281110101B	1260 mm long (Type MK 3B STD) .... Nr	– – – 8812.77	8812.77	12337.88	17625.54

AC Electrification

Rail 2003 Ref		Unit	Composite Cost £	Unit Rate Base Cost £	Unit Rate Green Zone £	Unit Rate Red Zone £
281	**REPAIRS AND RENEWALS – RURAL**					
28121	**WIRING**					
2812101	Contact wiring					
281210101	Remove existing contact wiring; replace with new:					
281210101A	maximum length 4.00 m both sides of neutral section	Nr	–	293.76	411.26	587.52
281210101B	maximum length 8.00 m both sides of neutral section	Nr	–	440.64	616.90	881.28
2812111	Booster transformers					
281211101	Supply only:					
281211101A	100 Amp 25 kV 1 phase	Nr	–	11015.96	15422.34	22031.92
281211111	Remove existing transformer; install new Amp transformer to existing structure:					
281211111A	100/200 Amp transformer	Nr	–	2937.59	4112.63	5875.18
28131	**DISTRIBUTION EQUIPMENT**					
2813101	Switchgear					
281310101	Optimho relays:					
281310101A	relays	Nr	–	7343.98	10281.57	14687.96
281310111	Remote monitoring equipment:					
281310111A	2 circuit breaker location	Nr	–	10281.57	14394.20	20563.14
281310111B	4 circuit breaker location	Nr	–	14687.95	20563.13	29375.90
281310111C	6 circuit breaker location	Nr	–	18359.94	25703.92	36719.88
281310111D	8 circuit breaker location	Nr	–	23500.72	32901.01	47001.44
281310111E	10 circuit breaker location	Nr	–	29375.90	41126.26	58751.80
281310111F	12 circuit breaker location	Nr	–	36719.88	51407.83	73439.76
281310121	Circuit breakers:					
281310121A	25 kV SMOS circuit breaker	Nr	–	14687.95	20563.13	29375.90

AC Electrification

Rail 2003 Ref		Unit				Composite Cost £	Unit Rate Base Cost £	Unit Rate Green Zone £	Unit Rate Red Zone £

281 REPAIRS AND RENEWALS – RURAL

2813101 Switchgear

281310131 Transformers:

| 281310131A | 1000 kVA 25,000/850 v single phase.. | Nr | – | – | – | 29375.90 | 29375.90 | 41126.26 | 58751.80 |

Rail 2003 Ref		Unit	Labour Hours	Labour Cost £	Plant Cost £	Materials Cost £	Unit Rate Base Cost £	Unit Rate Green Zone £	Unit Rate Red Zone £

291 TESTING AND COMMISSIONING

29101 CONTRACTOR'S TESTING AND COMMISSIONING

2910101 General testing and commissioning prior to handover

291010101 AC electrical installation:

| 291010101A | generally .... | hour | 3.00 | 142.61 | – | – | 142.61 | 199.66 | 285.23 |

DC Electrification

Rail 2003 Ref		Unit	Labour Hours	Labour Cost £	Plant Cost £	Materials Cost £	Unit Rate Base Cost £	Unit Rate Green Zone £	Unit Rate Red Zone £
3	DC ELECTRIFICATION								
341	NEW WORKS – INNER CITY								
34101	SUB-STATIONS								
3410101	11 kV								
341010181	Concrete bases, ducts and minor civils:								
341010181A	bases for structures; in situ concrete bases, including holding down bolts and fixings........................	Nr	40.73	488.12	22.95	199.65	710.72	995.02	1421.46
3410121	22 kV								
341012181	Concrete bases, ducts and minor civils:								
341012181A	bases for structures; in situ concrete bases, including holding down bolts and fixings........................	Nr	50.92	610.25	28.69	249.36	888.30	1243.60	1776.57
3410141	33 kV								
341014181	Concrete bases, ducts and minor civils:								
341014181A	bases for structures; in situ concrete bases, including holding down bolts and fixings........................	Nr	57.03	683.47	32.12	279.49	995.08	1393.12	1990.18
3410181	Miscellaneous equipment, supply								
341018101	Transformers:								
341018101A	40 kVA 3P	Nr	–	–	–	13335.11	13335.11	18669.15	26670.22
341018101B	60 kVA 1P	Nr	–	–	–	9779.08	9779.08	13690.71	19558.16
341018101C	60 kVA 3P	Nr	–	–	–	18669.16	18669.16	26136.82	37338.32
341018101D	80 kVA 3P	Nr	–	–	–	19558.17	19558.17	27381.44	39116.34
341018101E	120 kVA 3P	Nr	–	–	–	21336.18	21336.18	29870.65	42672.36
341018101F	150 kVA 1P	Nr	–	–	–	16002.14	16002.14	22403.00	32004.28
341018101G	200 kVA 3P	Nr	–	–	–	26670.23	26670.23	37338.32	53340.46

DC Electrification

Rail 2003 Ref		Unit	Labour Hours	Labour Cost £	Plant Cost £	Materials Cost £	Unit Rate Base Cost £	Unit Rate Green Zone £	Unit Rate Red Zone £

341 NEW WORKS – INNER CITY

3410181 Miscellaneous equipment, supply

341018121 2-panel boards:

341018121A	100 A	Nr	–	–	–	19558.17	19558.17	27381.44	39116.34
341018121B	200 A	Nr	–	–	–	23114.20	23114.20	32359.88	46228.40
341018121C	300 A	Nr	–	–	–	35560.30	35560.30	49784.42	71120.60

341018141 Modules:

| 341018141A | high voltage tie breaker | Nr | – | – | – | 222251.87 | 222251.87 | 311152.62 | 444503.74 |

34111 HIGH VOLTAGE CABLE

3411141 33 kV

341114101 Supply only, type:

| 341114101A | XLPL | m | – | – | – | 17.33 | 17.33 | 24.26 | 34.66 |

34121 TRACK FEEDER CABLE AND ETE

3412101 Traction works, supply only

341210101 Cable:

341210101A	240 mm^2 aluminium	m	–	–	–	7.11	7.11	9.95	14.22
341210101B	800 mm^2 aluminium	m	–	–	–	12.03	12.03	16.84	24.06
341210101C	1000 mm^2 aluminium	m	–	–	–	24.89	24.89	34.85	49.78

341210111 Bonds:

| 341210111A | copper/aluminium | Nr | – | – | – | 115.57 | 115.57 | 161.80 | 231.14 |

341210121 Hookswitches:

| 341210121A | Mk 7 | Nr | – | – | – | 400.05 | 400.05 | 560.07 | 800.10 |

341210131 Lugs:

341210131A	240 mm	Nr	–	–	–	17.78	17.78	24.89	35.56
341210131B	800 mm	Nr	–	–	–	35.56	35.56	49.78	71.12
341210131C	1000 mm	Nr	–	–	–	44.45	44.45	62.23	88.90

DC Electrification

Rail 2003 Ref		Unit	Labour Hours	Labour Cost £	Plant Cost £	Materials Cost £	Unit Rate Base Cost £	Unit Rate Green Zone £	Unit Rate Red Zone £
341	**NEW WORKS – INNER CITY**								
3412101	Traction works, supply only								
341210141	Fixings:								
341210141A	nut, bolt and washer set.............	Nr	–	–	–	8.00	8.00	11.20	16.00
3412116	Power, supply only								
341211601	Cable; 50 mm^2 SWA:								
341211601A	2 core............................	m	–	–	–	8.00	8.00	11.20	16.00
341211601B	3 core............................	m	–	–	–	11.56	11.56	16.18	23.12
341211601C	4 core............................	m	–	–	–	13.34	13.34	18.68	26.68
341211621	Cable; 70 mm^2 SWA:								
341211621A	2 core............................	m	–	–	–	10.22	10.22	14.31	20.44
341211621B	4 core............................	m	–	–	–	17.78	17.78	24.89	35.56
341211641	Cable; 185 mm^2 SWA:								
341211641A	2 core............................	m	–	–	–	24.89	24.89	34.85	49.78
341211641B	3 core............................	m	–	–	–	37.34	37.34	52.28	74.68
3412131	Points heating, supply only								
341213101	Cable:								
341213101A	6mm^2 SWA......................	m	–	–	–	1.83	1.83	2.56	3.66
341213101B	16mm^2 SWA.....................	m	–	–	–	2.95	2.95	4.14	5.91
341213101C	50mm^2 SWA.....................	m	–	–	–	5.02	5.02	7.03	10.04
341213101D	70mm^2 SWA.....................	m	–	–	–	6.19	6.19	8.66	12.37
341213101E	120mm^2 SWA....................	m	–	–	–	9.45	9.45	13.23	18.90
341213101F	8 core............................	m	–	–	–	1.88	1.88	2.63	3.75
341213111	Strip clip/connection block:								
341213111A	type Av...........................	Nr	–	–	–	396.30	396.30	554.82	792.60
341213111B	type Bv...........................	Nr	–	–	–	402.52	402.52	563.52	805.03
341213111C	type Cv...........................	Nr	–	–	–	402.52	402.52	563.52	805.03
341213111D	type Dv...........................	Nr	–	–	–	517.52	517.52	724.53	1035.04
341213111E	type Ev...........................	Nr	–	–	–	718.78	718.78	1006.29	1437.55
341213111F	type SGv	Nr	–	–	–	1023.38	1023.38	1432.73	2046.76

DC Electrification

Rail 2003 Ref		Unit	Labour Hours	Labour Cost £	Plant Cost £	Materials Cost £	Unit Rate Base Cost £	Unit Rate Green Zone £	Unit Rate Red Zone £
341	**NEW WORKS – INNER CITY**								
3412151	**Fittings**								
341215101	Strip clip/connection block:								
341215101A	AV	Nr	–	–	–	622.31	622.31	871.23	1244.62
3412171	**Labour rates, Basic**								
341217101	Standard day, Monday to Friday:								
341217101A	Assistant	hr	1.00	17.80	–	–	17.80	24.92	35.60
341217101B	Jointer	hr	1.00	22.28	–	–	22.28	31.19	44.56
341217101C	Electrician	hr	1.00	26.73	–	–	26.73	37.42	53.46
341217101D	Supervisor	hr	1.00	30.59	–	–	30.59	42.83	61.19
341217101E	Lookout	hr	1.00	16.55	–	–	16.55	23.17	33.09
3412175	**Labour rates, Overtime**								
341217501	17.00 – 2030, Monday – Friday (Rate × 1.2):								
341217501A	Assistant	hr	1.00	21.38	–	–	21.38	29.94	42.77
341217501B	Jointer	hr	1.00	26.74	–	–	26.74	37.43	53.47
341217501C	Electrician	hr	1.00	32.08	–	–	32.08	44.91	64.15
341217501D	Supervisor	hr	1.00	36.66	–	–	36.66	51.32	73.32
341217501E	Lookout	hr	1.00	19.86	–	–	19.86	27.80	39.71
341217511	0730 – 1200, Saturday (Rate × 1.2):								
341217511A	Assistant	hr	1.00	21.38	–	–	21.38	29.94	42.77
341217511B	Jointer	hr	1.00	26.74	–	–	26.74	37.43	53.47
341217511C	Electrician	hr	1.00	32.08	–	–	32.08	44.91	64.15
341217511D	Supervisor	hr	1.00	36.66	–	–	36.66	51.32	73.32
341217511E	Lookout	hr	1.00	19.86	–	–	19.86	27.80	39.71
341217521	2030 – 0730, Monday – Friday, (Rate × 1.3):								
341217521A	Assistant	hr	1.00	23.27	–	–	23.27	32.58	46.54
341217521B	Jointer	hr	1.00	28.96	–	–	28.96	40.55	57.93
341217521C	Electrician	hr	1.00	34.75	–	–	34.75	48.65	69.50
341217521D	Supervisor	hr	1.00	39.71	–	–	39.71	55.60	79.43
341217521E	Lookout	hr	1.00	21.51	–	–	21.51	30.12	43.02

DC Electrification

Rail 2003 Ref		Unit	Labour Hours	Labour Cost £	Plant Cost £	Materials Cost £	Unit Rate Base Cost £	Unit Rate Green Zone £	Unit Rate Red Zone £

341 NEW WORKS – INNER CITY

3412175 Labour rates, Overtime

341217531 1200 Saturday – 0730 Monday (Rate × 1.5):

Ref	Description	Unit	Hours	Cost	Plant	Mat	Base	Green	Red
341217531A	Assistant	hr	1.00	26.70	–	–	26.70	37.38	53.40
341217531B	Jointer	hr	1.00	33.42	–	–	33.42	46.79	66.84
341217531C	Electrician	hr	1.00	40.10	–	–	40.10	56.13	80.19
341217531D	Supervisor	hr	1.00	45.82	–	–	45.82	64.15	91.65
341217531E	Lookout	hr	1.00	24.82	–	–	24.82	34.75	49.64

Rail 2003 Ref		Unit	Composite Cost £	Unit Rate Base Cost £	Unit Rate Green Zone £	Unit Rate Red Zone £

34131 CONDUCTOR RAIL

3413101 Steel

341310101 Rail; supply and lay; mass:

341310101A	generally	m	– – –	129.59	129.59	181.43	259.18

3413151 Aluminium

341315101 Rail; supply and lay; mass:

341315101A	53 kg/m	m	– – –	145.80	145.80	204.12	291.60

Rail 2003 Ref		Unit	Labour Hours	Labour Cost £	Plant Cost £	Materials Cost £	Unit Rate Base Cost £	Unit Rate Green Zone £	Unit Rate Red Zone £

34141 POINTS HEATING

3414101 Points heating; supply

341410110 Constant wattage strip units; to FB rail including clips, fixings etc; type:

341410110A	Av	Nr	–	–	–	396.30	396.30	554.82	792.60
341410110B	Bv	Nr	–	–	–	402.52	402.52	563.52	805.03
341410110C	Cv	Nr	–	–	–	402.52	402.52	563.52	805.03
341410110D	Dv	Nr	–	–	–	517.52	517.52	724.53	1035.04
341410110E	Ev	Nr	–	–	–	718.78	718.78	1006.29	1437.55
341410110F	Fv	Nr	–	–	–	788.71	788.71	1104.20	1577.42
341410110G	Gv	Nr	–	–	–	1132.95	1132.95	1586.13	2265.89
341410110H	Hv	Nr	–	–	–	1529.25	1529.25	2140.94	3058.49
341410110I	SGv	Nr	–	–	–	1023.38	1023.38	1432.73	2046.76
341410110P	Extra for insulation	Nr	–	–	–	711.01	711.01	995.41	1422.01

DC Electrification

Rail 2003 Ref		Unit	Labour Hours £	Labour Cost £	Plant Cost £	Materials Cost	Unit Rate Base Cost £	Unit Rate Green Zone £	Unit Rate Red Zone £
341	**NEW WORKS – INNER CITY**								
3414101	**Points heating; supply**								
341410115	Constant wattage strip units; to BH rail including clips and fixings; type:								
341410115A	Av	Nr	–	–	–	396.30	396.30	554.82	792.60
341410115B	Bv	Nr	–	–	–	402.52	402.52	563.52	805.03
341410115C	Cv	Nr	–	–	–	402.52	402.52	563.52	805.03
341410115D	Dv	Nr	–	–	–	517.52	517.52	724.53	1035.04
341410115E	Ev	Nr	–	–	–	718.78	718.78	1006.29	1437.55
341410115F	Fv	Nr	–	–	–	788.71	788.71	1104.20	1577.42
341410115G	Gv	Nr	–	–	–	1132.95	1132.95	1586.13	2265.89
341410115H	Hv	Nr	–	–	–	1529.25	1529.25	2140.94	3058.49
341410115I	SGv	Nr	–	–	–	1023.38	1023.38	1432.73	2046.76
341410115P	Extra for insulation	Nr	–	–	–	711.01	711.01	995.41	1422.01
341410120	Transformers:								
341410120A	5kVa; 230/110V	Nr	–	–	–	437.09	437.09	611.93	874.19
341410120B	10kVa; 230/110V	Nr	–	–	–	960.17	960.17	1344.24	1920.34
341410120C	portable tool								
341410130	Control cubicle; 1.2x1.7x0.475m; with Icealert, sensors, MCCBs or fuse switch units:								
341410130A	single phase; galvanised	Nr	–	–	–	5405.19	5405.19	7567.27	10810.39
341410130B	single phase; stainless steel	Nr	–	–	–	5831.80	5831.80	8164.52	11663.60
341410130E	three phase; galvanised	Nr	–	–	–	6680.34	6680.34	9352.48	13360.68
341410130F	three phase; stainless steel	Nr	–	–	–	7106.95	7106.95	9949.72	14213.89
341410133	Supply cubicle; complete with backboard and meter tail; single/three phase; not exceeding 100A per phase:								
341410133A	galvanised	Nr	–	–	–	3065.48	3065.48	4291.68	6130.97
341410133B	stainless steel	Nr	–	–	–	3981.63	3981.63	5574.28	7963.26
341410135	Supply cubicle; complete with backboard and meter tail; single/three phase; over 100A per phase:								
341410135A	galvanised	Nr	–	–	–	3761.72	3761.72	5266.41	7523.45
341410135B	stainless steel	Nr	–	–	–	3981.63	3981.63	5574.28	7963.26

DC Electrification

Rail 2003 Ref		Unit	Labour Hours	Labour Cost £	Plant Cost £	Materials Cost £	Unit Rate Base Cost £	Unit Rate Green Zone £	Unit Rate Red Zone £
341	**NEW WORKS – INNER CITY**								
3414101	Points heating; supply								
341410138	Distribution cubicle; 315A TP&N:								
341410138A	galvanised	Nr	–	–	–	2564.28	2564.28	3589.99	5128.56
341410138B	stainless steel	Nr	–	–	–	2875.10	2875.10	4025.15	5750.21
3414151	Points heating; installation								
341415110	Constant wattage strip units; to FB rail including clips, fixings, welding etc; type:								
341415110A	Av	Nr	2.50	66.98	–	12.00	78.98	110.57	157.95
341415110B	Bv	Nr	2.50	66.98	–	12.00	78.98	110.57	157.95
341415110C	Cv	Nr	2.50	66.98	–	12.00	78.98	110.57	157.95
341415110D	Dv	Nr	3.25	87.07	–	12.00	99.07	138.69	198.13
341415110E	Ev	Nr	4.00	107.16	–	15.00	122.16	171.02	244.32
341415110F	Fv	Nr	4.00	107.16	–	15.00	122.16	171.02	244.32
341415110G	Gv	Nr	6.50	174.14	–	25.00	199.14	278.79	398.27
341415110H	Hv	Nr	6.50	174.14	–	25.00	199.14	278.79	398.27
341415110I	SGv	Nr	4.00	107.16	–	20.00	127.16	178.02	254.32
341415110P	Extra for insulation	Nr	0.50	13.40	–	5.00	18.40	25.75	36.79
341415115	Constant wattage strip units; to BH rail including clips, fixings, welding etc; type:								
341415115A	Av	Nr	10.00	267.90	–	12.00	279.90	391.86	559.80
341415115B	Bv	Nr	12.00	321.48	–	12.00	333.48	466.87	666.96
341415115C	Cv	Nr	12.00	321.48	–	12.00	333.48	466.87	666.96
341415115D	Dv	Nr	14.00	375.06	–	12.00	387.06	541.88	774.12
341415115E	Ev	Nr	20.00	535.80	–	15.00	550.80	771.12	1101.60
341415115F	Fv	Nr	23.75	636.26	–	15.00	651.26	911.77	1302.52
341415115G	Gv	Nr	33.50	897.47	–	25.00	922.47	1291.45	1844.93
341415115H	Hv	Nr	41.50	1111.79	–	25.00	1136.79	1591.50	2273.57
341415115I	SGv	Nr	27.75	743.42	–	20.00	763.42	1068.79	1526.85
341415115P	Extra for insulation	Nr	0.50	13.40	–	5.00	18.40	25.75	36.79

DC Electrification

Rail 2003 Ref		Unit	Labour Hours	Labour Cost £	Plant Cost £	Materials Cost £	Unit Rate Base Cost £	Unit Rate Green Zone £	Unit Rate Red Zone £
341	**NEW WORKS – INNER CITY**								
3414151	**Points heating; installation**								
341415120	Transformers; including concrete plinth base:								
341415120A	5kVa; 230/110V	Nr	3.25	87.07	–	20.00	107.07	149.89	214.13
341415120B	10ka; 230/110V	Nr	3.25	87.07	–	20.00	107.07	149.89	214.13
341415120C	portable tool	Nr	3.25	87.07	–	20.00	107.07	149.89	214.13
341415130	Control cubicle; 1.2x1.7x0.475m; with Icealert, sensors, MCCBs or fuse switch units:								
341415130A	single phase; galvanised	Nr	22.50	602.78	–	20.00	622.78	871.89	1245.55
341415130B	single phase; stainless steel	Nr	22.50	602.78	–	20.00	622.78	871.89	1245.55
341415130E	three phase; galvanised	Nr	32.00	857.28	–	40.00	897.28	1256.19	1794.56
341415130F	three phase; stainless steel	Nr	32.00	857.28	–	40.00	897.28	1256.19	1794.56
341415133	Supply cubicle; complete with backboard and meter tail; single/three phase; not exceeding 100A per phase:								
341415133A	galvanised	Nr	19.00	509.01	–	15.00	524.01	733.61	1048.02
341415133B	stainless steel	Nr	19.00	509.01	–	15.00	524.01	733.61	1048.02
341415135	Supply cubicle; complete with backboard and meter tail; single/three phase; over 100A per phase:								
341415135A	galvanised	Nr	22.00	589.38	–	15.00	604.38	846.13	1208.76
341415135B	stainless steel	Nr	22.00	589.38	–	15.00	604.38	846.13	1208.76
341415138	Distribution cubicle; 315A TP&N:								
341415138A	galvanised	Nr	42.00	1125.18	–	50.00	1175.18	1645.25	2350.36
341415138B	stainless steel	Nr	42.00	1125.18	–	50.00	1175.18	1645.25	2350.36

DC Electrification

Rail 2003 Ref		Unit	Labour Hours	Labour Cost £	Plant Cost £	Materials Cost £	Unit Rate Base Cost £	Unit Rate Green Zone £	Unit Rate Red Zone £
341	**NEW WORKS – INNER CITY**								
3414161	Ancillary items; supply and installation								
341416110	Cable; support, gland and termination measured separately:								
341416110B	6mm² SWA	m	0.20	7.26	–	1.83	9.09	12.73	18.18
341416110D	16mm² SWA	m	0.27	9.80	–	2.95	12.75	17.85	25.50
341416110G	50mm² SWA	m	0.48	17.42	–	5.02	22.44	31.42	44.88
341416110H	70mm² SWA	m	0.56	20.33	–	6.19	26.52	37.13	53.04
341416110J	120mm² SWA	m	0.93	33.88	–	9.45	43.33	60.66	86.66
341416110P	8 core	m	0.15	5.45	–	1.88	7.33	10.26	14.66
341416120	Gland and termination; including supply of gland and termination kit:								
341416120B	6mm² SWA	Nr	0.35	9.38	–	10.40	19.78	27.69	39.55
341416120D	16mm² SWA	Nr	0.85	22.77	–	14.98	37.75	52.85	75.49
341416120G	50mm² SWA	Nr	3.00	80.37	–	45.63	126.00	176.40	252.00
341416120H	70mm² SWA	Nr	6.00	160.74	–	74.27	235.01	329.02	470.02
341416120J	120mm² SWA	Nr	10.00	267.90	–	81.15	349.05	488.67	698.09
341416120P	8 core	Nr	1.20	32.15	–	8.65	40.80	57.13	81.61
3414181	Concrete bases; complete installation								
341418110	Control cubicles; type:								
341418110A	single phase; normal location	Nr	9.50	381.88	–	138.57	520.45	728.63	1040.90
341418110F	three phase; normal location								
341418120	Supply cubicles; type:								
341418120A	single phase; normal location	Nr	9.50	381.88	–	138.57	520.45	728.63	1040.90
341418120F	three phase; normal location								
341418130	Distribution cubicles; type:								
341418130A	315A TP normal location	Nr	9.50	381.88	–	138.57	520.45	728.63	1040.90

DC Electrification

Rail 2003 Ref		Unit	Labour Hours	Labour Cost £	Plant Cost £	Materials Cost £	Unit Rate Base Cost £	Unit Rate Green Zone £	Unit Rate Red Zone £
346	**NEW WORKS – SUBURBAN**								
34601	**SUB STATIONS**								
3460101	11 kV								
346010181	Concrete bases, ducts and minor civils:								
346010181A	bases for structures; in situ concrete bases, including holding down bolts and fixings........................	Nr	37.03	443.78	20.86	181.35	645.99	904.39	1291.99
3460121	22 kV								
346012181	Concrete bases, ducts and minor civils:								
346012181A	bases for structures; in situ concrete bases, including holding down bolts and fixings:	Nr	46.29	554.76	26.09	226.69	807.54	1130.54	1615.06
3460141	33 kV								
346014181	Concrete bases, ducts and minor civils:								
346014181A	bases for structures; in situ concrete bases, including holding down bolts and fixings:	Nr	51.84	621.27	29.21	253.72	904.20	1265.88	1808.40
3460181	Miscellaneous equipment; supply								
346018101	Transformers:								
346018101A	40 kVA 3P	Nr	–	–	–	11595.75	11595.75	16234.05	23191.50
346018101B	60 kVA 1P	Nr	–	–	–	8503.55	8503.55	11904.97	17007.10
346018101C	60 kVA 3P	Nr	–	–	–	16234.05	16234.05	22727.67	32468.10
346018101D	80 kVA 3P	Nr	–	–	–	17007.10	17007.10	23809.94	34014.20
346018101E	120 kVA 3P	Nr	–	–	–	18553.20	18553.20	25974.48	37106.40
346018101F	150 kVA 1P	Nr	–	–	–	13914.90	13914.90	19480.86	27829.80
346018101G	200 kVA 3P	Nr	–	–	–	23191.50	23191.50	32468.10	46383.00
346018121	2-panel boards:								
346018121A	100 A	Nr	–	–	–	17007.10	17007.10	23809.94	34014.20
346018121B	200 A	Nr	–	–	–	20099.30	20099.30	28139.02	40198.60
346018121C	300 A	Nr	–	–	–	30922.00	30922.00	43290.80	61844.00

DC Electrification

Rail 2003 Ref		Unit	Labour Hours	Labour Cost £	Plant Cost £	Materials Cost £	Unit Rate Base Cost £	Unit Rate Green Zone £	Unit Rate Red Zone £
346	**NEW WORKS – SUBURBAN**								
3460181	**Miscellaneous equipment; supply**								
346018141	**Modules:**								
346018141A	high voltage tie breaker	Nr	–	–	–	193262.50	193262.50	270567.50	386525.00
34611	**HIGH VOLTAGE CABLE**								
3461141	**33 kV**								
346114101	**Supply only; type:**								
346114101A	XLPL	m	–	–	–	15.07	15.07	21.10	30.14
34621	**TRACK FEEDER CABLE AND ETE**								
3462101	**Traction works; supply only**								
346210101	**Cable:**								
346210101A	240 mm^2 aluminium	m	–	–	–	6.18	6.18	8.65	12.36
346210101B	800 mm^2 aluminium	m	–	–	–	15.46	15.46	21.64	30.92
346210101C	1000 mm^2 aluminium	m	–	–	–	21.65	21.65	30.31	43.30
346210111	**Bonds:**								
346210111A	Copper/aluminium	Nr	–	–	–	100.50	100.50	140.70	201.00
346210121	**Hook switches:**								
346210121A	Mk 7	Nr	–	–	–	347.87	347.87	487.02	695.74
346210131	**Lugs:**								
346210131A	240 mm	Nr	–	–	–	15.46	15.46	21.64	30.92
346210131B	800 mm	Nr	–	–	–	30.92	30.92	43.29	61.84
346210131C	1000 mm	Nr	–	–	–	38.65	38.65	54.11	77.30
346210141	**Fixings:**								
346210141A	nut, bolt and washer set	Nr	–	–	–	6.96	6.96	9.74	13.92

DC Electrification

Rail 2003 Ref		Unit	Labour Hours	Labour Cost £	Plant Cost £	Materials Cost £	Unit Rate Base Cost £	Unit Rate Green Zone £	Unit Rate Red Zone £
346	**NEW WORKS – SUBURBAN**								
3462116	**Power, supply only**								
346211601	Cable; 50 mm² SWA:								
346211601A	2 core	m	–	–	–	6.96	6.96	9.74	13.92
346211601B	3 Core	m	–	–	–	10.05	10.05	14.07	20.10
346211601C	4 core	m	–	–	–	11.60	11.60	16.24	23.20
346211621	Cable; 70 mm² SWA:								
346211621A	2 core	m	–	–	–	8.89	8.89	12.45	17.78
346211621B	4 core	m	–	–	–	15.46	15.46	21.64	30.92
346211641	Cable; 185 mm² SWA:								
346211641A	2 core	m	–	–	–	21.65	21.65	30.31	43.30
346211641B	3 core	m	–	–	–	32.47	32.47	45.46	64.94
3462131	**Points heating, supply only**								
346213101	Cable:								
346213101A	6 mm² SWA	m	–	–	–	1.83	1.83	2.56	3.66
346213101B	16 mm² SWA	m	–	–	–	2.95	2.95	4.14	5.91
346213101C	50 mm² SWA	m	–	–	–	5.02	5.02	7.03	10.04
346213101D	70 mm² SWA	m	–	–	–	8.53	8.53	11.95	17.07
346213101E	120 mm² SWA	m	–	–	–	9.45	9.45	13.23	18.90
346213101F	8 core	m	–	–	–	1.88	1.88	2.63	3.75
3462151	**Fittings**								
346215101	Strip clip/connection block; FB; type:								
346215101A	Av	Nr	–	–	–	396.30	396.30	554.82	792.60
346215101B	Bv	Nr	–	–	–	402.52	402.52	563.52	805.03
346215101C	Cv	Nr	–	–	–	402.52	402.52	563.52	805.03
346215101D	Dv	Nr	–	–	–	517.52	517.52	724.53	1035.04
346215101E	Ev	Nr	–	–	–	718.78	718.78	1006.29	1437.55
346215101F	SGv	Nr	–	–	–	1023.38	1023.38	1432.73	2046.76

DC Electrification

Rail 2003 Ref		Unit	Labour Hours	Labour Cost	Plant Cost	Materials Cost	Unit Rate	Unit Rate	Unit Rate
				£	£	£	Base Cost £	Green Zone £	Red Zone £
346	**NEW WORKS – SUBURBAN**								
3462171	**Labour rates, Basic**								
346217101	**Standard day, Monday – Friday:**								
346217101A	Assistant	hr	1.00	17.80	–	–	17.80	24.92	35.60
346217101B	Jointer	hr	1.00	22.28	–	–	22.28	31.19	44.56
346217101C	Electrician	hr	1.00	26.73	–	–	26.73	37.42	53.46
346217101D	Supervisor	hr	1.00	30.55	–	–	30.55	42.77	61.10
346217101E	Lookout	hr	1.00	16.55	–	–	16.55	23.17	33.09
3462181	**Labour rates; Overtime**								
346218101	**1700 – 2030, Monday – Friday (Rate × 1.2):**								
346218101A	Assistant	hr	1.00	21.38	–	–	21.38	29.94	42.77
346218101B	Jointer	hr	1.00	26.74	–	–	26.74	37.43	53.47
346218101C	Electrician	hr	1.00	32.08	–	–	32.08	44.91	64.15
346218101D	Supervisor	hr	1.00	36.66	–	–	36.66	51.32	73.32
346218101E	Lookout	hr	1.00	19.86	–	–	19.86	27.80	39.71
346218111	**0730 – 1200 Saturday, (Rate × 1.2):**								
346218111A	Assistant	hr	1.00	21.38	–	–	21.38	29.94	42.77
346218111B	Jointer	hr	1.00	26.74	–	–	26.74	37.43	53.47
346218111C	Electrician	hr	1.00	32.08	–	–	32.08	44.91	64.15
346218111D	Supervisor	hr	1.00	36.66	–	–	36.66	51.32	73.32
346218111E	Lookout	hr	1.00	19.86	–	–	19.86	27.80	39.71
346218121	**2030 – 0730, Monday – Friday (Rate × 1.3):**								
346218121A	Assistant	hr	1.00	23.27	–	–	23.27	32.58	46.54
346218121B	Jointer	hr	1.00	28.96	–	–	28.96	40.55	57.93
346218121C	Electrician	hr	1.00	34.75	–	–	34.75	48.65	69.50
346218121D	Supervisor	hr	1.00	39.71	–	–	39.71	55.60	79.43
346218121E	Lookout	hr	1.00	21.51	–	–	21.51	30.12	43.02
346218131	**1200 Saturday – 0730 Monday (Rate × 1.5):**								
346218131A	Assistant	hr	1.00	26.70	–	–	26.70	37.38	53.40
346218131B	Jointer	hr	1.00	33.42	–	–	33.42	46.79	66.84
346218131C	Electrician	hr	1.00	40.10	–	–	40.10	56.13	80.19
346218131D	Supervisor	hr	1.00	45.82	–	–	45.82	64.15	91.65
346218131E	Lookout	hr	1.00	24.82	–	–	24.82	34.75	49.64

DC Electrification

Rail 2003 Ref	Unit	Composite Cost £	Unit Rate Base Cost £	Unit Rate Green Zone £	Unit Rate Red Zone £

346   **NEW WORKS – SUBURBAN**

34631   **CONDUCTOR RAIL**

3463101   Steel

346310101   Rail; supply and lay; mass:

346310101A	generally	m	–	–	–	129.59	129.59	181.43	259.18

3463151   Aluminium

346315101   Rail; supply and lay; mass:

346315101A	53 kg/m	m	–	–	–	145.80	145.80	204.12	291.60

DC Electrification

Rail 2003 Ref		Unit	Composite Cost £	Unit Rate Base Cost £	Unit Rate Green Zone £	Unit Rate Red Zone £
376	**REPAIRS AND RENEWALS – SUBURBAN**					
37601	**SUBSTATIONS**					
3760101	**Transformers**					
376010101	2 MW Class H transformer/rectifier:					
376010101A	supply	Nr	–	108227.00	151517.80	216454.00
376010101B	install	Nr	–	6184.40	8658.16	12368.80
376010101C	rail transport	Nr	–	30922.00	43290.80	61844.00
376010101D	civils base	Nr	–	7730.50	10822.70	15461.00
376010101E	33 kV cable installation/alterations	Nr	–	11595.75	16234.05	23191.50
376010101F	LV DC cable installation/alterations	Nr	–	11595.75	16234.05	23191.50
376010101G	protection equipment and control panels	Nr	–	2705.68	3787.95	5411.36
3760111	**Switch boards; non-packaged**					
376011101	Suppy and install:					
376011101A	3 panel	Nr	–	247376.00	346326.40	494752.00
376011101B	6 panel	Nr	–	425177.50	595248.50	850355.00
376011101C	8 panel	Nr	–	425177.50	595248.50	850355.00
376011101D	9 panel	Nr	–	463830.00	649362.00	927660.00
376011101E	11 panel	Nr	–	541135.00	757589.00	1082270.00

DC Electrification

Rail 2003 Ref		Unit	Labour Hours	Labour Cost	Plant Cost	Materials Cost	Unit Rate	Unit Rate	Unit Rate
				£	£	£	Base Cost £	Green Zone £	Red Zone £
376	**REPAIRS AND RENEWALS – SUBURBAN**								
37611	**TRACK FEEDER CABLE**								
3761101	**Cable**								
376110101	Take from store; refix including connections to existing terminations:								
376110101A	continuity cable, approx 14 m long	Nr	1.06	38.65	–	–	38.65	54.11	77.30
376110101B	main supply cables to aluminium rail	Nr	1.28	46.38	–	–	46.38	64.93	92.76
376110101C	2 × 500 mm² twin core cable as continuity cable	m	2.13	77.31	–	–	77.31	108.23	154.62
37621	**CONDUCTOR RAIL**								
3762101	**Conductor rail supply**								
376210101	ASC conductor rail:								
376210101A	sections over 100 m length	m	–	–	–	100.50	100.50	140.70	201.00
376210101B	bolted joints	Nr	–	–	–	69.57	69.57	97.40	139.14
376210101C	high speed ramp rail	Nr	–	–	–	695.75	695.75	974.05	1391.50
376210101D	ramp end support assembly	Nr	–	–	–	185.53	185.53	259.74	371.06
376210101E	mid-point anchor assembly	Nr	–	–	–	92.77	92.77	129.88	185.54
376210101F	insulator assembly; concrete sleeper	Nr	–	–	–	77.31	77.31	108.23	154.62
376210101G	insulator assembly; timber sleeper	Nr	–	–	–	77.31	77.31	108.23	154.62
376210101H	cable termination assembly	Nr	–	–	–	46.38	46.38	64.93	92.76
3762111	**Installation generally**								
376211101	ASC conductor rail:								
376211101A	sections over 100 m length	m	–	–	–	38.65	38.65	54.11	77.30
376211101B	low speed/high speed Ramp; 15 m length	m	–	–	–	541.14	541.14	757.60	1082.28
376211101C	ceramic insulator and cast iron base assembly; fixings and packings to concrete or timber sleeper	Nr	–	–	–	11.60	11.60	16.24	23.20
376211101D	ceramic insulator & cast iron mid-point anchor; fixings and packings to concrete or timber sleeper	Nr	–	–	–	13.14	13.14	18.40	26.28
376211101E	ceramic ramp end insulator & cast iron base assembly; fixings and packings to concrete or timber sleeper	Nr	–	–	–	10.05	10.05	14.07	20.10
376211101F	cable termination assemblies fixed to aluminium rails	Nr	–	–	–	19.33	19.33	27.06	38.66
376211111	**Ancillary items:**								
376211111A	conductor rail guard boards	m	0.25	10.05	–	17.00	27.05	37.87	54.10
376211111B	spot re-sleepering with timber sleepers for insulated assemblies	m	1.30	52.26	–	91.75	144.01	201.61	288.01
376211111C	spot re-sleepering with concrete sleepers for insulated assemblies	m	1.30	52.26	–	62.75	115.01	161.01	230.01

DC Electrification

Rail 2003 Ref		Unit	Labour Hours	Labour Cost £	Plant Cost £	Materials Cost £	Unit Rate Base Cost £	Unit Rate Green Zone £	Unit Rate Red Zone £
376	**REPAIRS AND RENEWALS – SUBURBAN**								
3762151	Supply and installation								
376215101	Ancillary items:								
376215101A	conductor rail guard boards	m	0.30	12.06	–	16.75	28.81	40.33	57.62
376215101B	spot re-sleepering with timber sleepers for insulated assemblies	Nr	–	–	–	69.57	69.57	97.40	139.14
376215101C	spot re-sleepering with concrete sleepers for insulated assemblies	Nr	–	–	–	85.04	85.04	119.06	170.08
37661	**STRIP OUT & RECOVERIES**								
3766101	Track feeder cable								
376610101	Disconnect; store for re-use:								
376610101A	continuity cable, approx 14 m long	Nr	1.06	38.65	–	–	38.65	54.11	77.30
376610111	Disconnect; dig access pit; insert draw wire; set aside for re-use:								
376610111A	continuity cable, approx 14 m long	Nr	1.70	61.84	–	–	61.84	86.58	123.68
376610121	Disconnect; set aside for re-use:								
376610121A	supply cable, approx 14 m long	Nr	1.28	46.38	–	–	46.38	64.93	92.76
3766111	Conductor rail								
376611101	Take up conductor rail; wooden guard boards:								
376611101A	disposal off site	m	0.32	3.87	–	–	3.87	5.42	7.74
376611111	Take up steel conductor rail and fittings; make good timber sleepers:								
376611111A	disposal off site	m	0.58	6.96	–	–	6.96	9.74	13.92

DC Electrification

Rail 2003 Ref		Unit	Labour Hours	Labour Cost	Plant Cost	Materials Cost	Unit Rate	Unit Rate	Unit Rate
				£	£	£	Base Cost £	Green Zone £	Red Zone £
391	TESTING AND COMMISSIONING								
39101	CONTRACTOR'S TESTING AND COMMISSIONING								
3910101	General testing and commissioning prior to handover								
391010101	DC electrical installation:								
391010101A	AC/DC interface	hour	3.00	142.61	–	–	142.61	199.66	285.23
391010101B	conductor rail	hour	3.00	142.61	–	–	142.61	199.66	285.23
391010101C	high voltage cable	hour	3.00	142.61	–	–	142.61	199.66	285.23
391010101D	track feeder cable	hour	3.00	142.61	–	–	142.61	199.66	285.23
391010101E	sub-stations	hour	3.00	142.61	–	–	142.61	199.66	285.23
391010101F	track paralleling huts	hour	3.00	142.61	–	–	142.61	199.66	285.23

Permanent Way

Rail 2003 Ref		Unit	Labour Hours	Labour Cost £	Plant Cost £	Materials Cost £	Unit Rate Base Cost £	Unit Rate Green Zone £	Unit Rate Red Zone £
4	**PERMANENT WAY**								
420	**TRACK FOUNDATIONS**								
42005	**REMOVALS**								
4200501	Take up and remove; including off-site disposal								
420050101	Track ballast; overall depth:								
420050101A	200 mm	m²	0.31	3.75	–	–	3.75	5.25	7.50
420050101B	300 mm	m²	0.44	5.25	–	–	5.25	7.35	10.50
420050101C	400 mm	m²	0.56	6.75	–	–	6.75	9.45	13.50
42010	**BASE CONSTRUCTION**								
4201010	Bottom ballast								
420101010	Crushed granite; graded 50-28mm; in layers:								
420101010A	bottom	m³	1.50	19.45	9.58	18.80	47.83	66.97	95.67
420101010B	top	m³	2.25	29.18	13.96	18.80	61.94	86.71	123.87
4201020	Blankets								
420102010	Sand; in layers; thickness:								
420102010A	300 mm in 100 mm and 200 mm	m²	0.20	2.70	1.83	0.96	5.49	7.68	10.98
420102010B	200 mm	m²	0.20	2.70	1.83	0.60	5.13	7.18	10.27
420102010C	100 mm	m²	0.16	2.16	1.47	0.30	3.93	5.49	7.84
4201030	Membranes								
420103010	Terram 4000; including side and end laps:								
420103010A	horizontal; over 300 mm wide	m²	0.05	0.60	–	1.49	2.09	2.93	4.18
420103010B	vertical; not exceeding 150 mm high	m	0.08	0.96	–	1.49	2.45	3.43	4.89
420103010C	vertical; 150 – 300 mm high	m	0.09	1.08	–	1.49	2.57	3.60	5.13
420103010D	vertical; over 300 mm high	m²	0.10	1.20	–	1.49	2.69	3.76	5.37

Permanent Way

Rail 2003 Ref		Unit	Labour Hours	Labour Cost £	Plant Cost £	Materials Cost £	Unit Rate Base Cost £	Unit Rate Green Zone £	Unit Rate Red Zone £
420	**TRACK FOUNDATIONS**								
4201030	**Membranes**								
420103020	**1000 gauge polythene; including side and end laps:**								
420103020A	horizontal; over 300 mm wide	m²	0.05	0.60	–	0.91	1.51	2.12	3.02
420103020B	vertical; not exceeding 150 mm high	m	0.08	0.96	–	0.91	1.87	2.62	3.73
420103020C	vertical; 150 – 300 mm high	m	0.09	1.08	–	0.91	1.99	2.79	3.97
420103020D	vertical; over 300 mm high	m²	0.10	1.20	–	0.91	2.11	2.95	4.21
420103030	**Polypropylene sheet geogrid; including side and end laps:**								
420103030A	horizontal; over 300 mm wide	m²	0.10	1.20	–	2.00	3.20	4.46	6.38
420103030B	vertical; not exceeding 150 mm high	m	0.05	0.60	–	2.00	2.60	3.63	5.19
420103030C	vertical; 150 – 300 mm high	m	0.08	0.96	–	2.00	2.96	4.13	5.90
420103030D	vertical; over 300 mm high	m²	0.09	1.08	–	2.00	3.08	4.30	6.14

Permanent Way

Rail 2003 Ref		Unit	Labour Hours	Labour Cost £	Plant Cost £	Materials Cost £	Unit Rate Base Cost £	Unit Rate Green Zone £	Unit Rate Red Zone £

430 PLAIN LINE

43010 TRACKWORK

4301010 Take up and clear away

430101010 Dismantle and set aside component parts; include setting aside and off-site storage and disposal of redundant materials:

Ref	Description	Unit	Hours	Labour	Plant	Mat	Base	Green	Red
430101010A	bull head rails on timber sleepers, jointed	m	0.20	5.52	2.10	–	7.62	10.68	15.25
430101010B	flat bottom rails on timber sleepers, jointed	m	0.20	5.52	2.10	–	7.62	10.68	15.25
430101010C	flat bottom rails on timber sleepers, welded	m	0.25	6.80	2.52	–	9.32	13.04	18.63
430101010D	flat bottom rails on concrete sleepers; jointed	m	0.22	6.07	2.32	–	8.39	11.74	16.77
430101010E	flat bottom rails on concrete sleepers; welded	m	0.28	7.48	2.78	–	10.26	14.36	20.50
430101010F	rail built buffer stop	Nr	10.35	281.14	176.37	–	457.51	640.51	915.02

4301030 Material supply

430103010 Rails; flat bottom; new; ref:

Ref	Description	Unit				Mat	Base	Green	Red
430103010A	98FB	m	–	–	–	25.97	25.97	36.36	51.94
430103010B	109	m	–	–	–	31.20	31.20	43.68	62.40
430103010C	110A	m	–	–	–	31.21	31.21	43.70	62.43
430103010D	113A	m	–	–	–	25.98	25.98	36.37	51.95
430103010E	113A, CWR	m	–	–	–	31.21	31.21	43.70	62.43
430103010F	113A, twist rail	m	–	–	–	37.21	37.21	52.09	74.41

430103015 Rails; flat bottom; serviceable; ref:

Ref	Description	Unit				Mat	Base	Green	Red
430103015A	110A	m	–	–	–	19.06	19.06	26.69	38.12
430103015B	113A	m	–	–	–	19.06	19.06	26.69	38.12

430103020 Rails; bull head; serviceable; ref:

Ref	Description	Unit				Mat	Base	Green	Red
430103020A	95BH	m	–	–	–	27.20	27.20	38.08	54.40

Permanent Way

Rail 2003 Ref		Unit	Labour Hours	Labour Cost £	Plant Cost £	Materials Cost £	Unit Rate Base Cost £	Unit Rate Green Zone £	Unit Rate Red Zone £
430	**PLAIN LINE**								
4301030	**Material supply**								
430103025	**Rails; new; ref:**								
430103025A	UIC60............................	m	–	–	–	27.72	27.72	38.80	55.43
430103030	**Sleepers; concrete; new; ref:**								
430103030A	EF28............................	Nr	–	–	–	37.20	37.20	52.08	74.40
430103030B	F10.............................	Nr	–	–	–	29.65	29.65	41.50	59.29
430103030C	F16.............................	Nr	–	–	–	30.02	30.02	42.03	60.05
430103030D	F19.............................	Nr	–	–	–	31.22	31.22	43.71	62.44
430103030E	F23.............................	Nr	–	–	–	31.60	31.60	44.24	63.20
430103030F	F24.............................	Nr	–	–	–	34.43	34.43	48.20	68.86
430103030G	F27.............................	Nr	–	–	–	38.73	38.73	54.22	77.46
430103030H	F40.............................	Nr	–	–	–	32.50	32.50	45.50	64.99
430103030I	F40; drilled for AWS..............	Nr	–	–	–	39.72	39.72	55.61	79.45
430103030J	F40; factory modified for lateral resistance end plates..............	Nr	–	–	–	54.29	54.29	76.01	108.59
430103030K	G44.............................	Nr	–	–	–	40.50	40.50	56.70	81.00
430103035	**Sleepers; concrete; serviceable; ref:**								
430103035A	F24.............................	Nr	–	–	–	8.79	8.79	12.31	17.59
430103035B	F27.............................	Nr	–	–	–	28.74	28.74	40.23	57.47
430103040	**Sleepers; hardwood; new; size:**								
430103040A	2600 × 250 × 125 mm............	Nr	–	–	–	44.88	44.88	62.84	89.77
430103040B	2600 × 250 × 125 mm; including PAN6 baseplates...................	Nr	–	–	–	87.31	87.31	122.23	174.62
430103040C	2600 × 250 × 125 mm; including VN baseplates........................	Nr	–	–	–	79.14	79.14	110.80	158.29
430103040D	2600 × 300 × 150 mm............	Nr	–	–	–	58.73	58.73	82.23	117.47
430103040E	2600 × 300 × 150 mm; including PAN6 baseplates...................	Nr	–	–	–	67.91	67.91	95.07	135.81
430103040F	2600 × 300 × 150 mm; including VN baseplates........................	Nr	–	–	–	68.85	68.85	96.39	137.70

Permanent Way

Rail 2003 Ref		Unit	Labour Hours £	Labour Cost £	Plant Cost £	Materials Cost £	Unit Rate Base Cost £	Unit Rate Green Zone £	Unit Rate Red Zone £
430	**PLAIN LINE**								
4301030	**Material supply**								
430103045	Sleepers; hardwood; serviceable; size:								
430103045A	2600 × 300 × 150 mm	Nr	–	–	–	8.79	8.79	12.31	17.59
430103048	Sleepers; steel; new; ref:								
430103048A	standard	Nr	–	–	–	39.89	39.89	55.85	79.79
430103050	Baseplates; type:								
430103050A	PAN11	Nr	–	–	–	17.26	17.26	24.16	34.52
430103050B	V	Nr	–	–	–	20.39	20.39	28.54	40.77
430103050C	VN	Nr	–	–	–	22.71	22.71	31.79	45.41
430103060	Clips; pandrol; ref:								
430103060A	E1809	Nr	–	–	–	1.79	1.79	2.51	3.58
430103060B	E1809; corrosion resistant	Nr	–	–	–	2.41	2.41	3.37	4.82
430103060C	E1810	Nr	–	–	–	2.64	2.64	3.69	5.27
430103060D	401A	Nr	–	–	–	1.88	1.88	2.63	3.76
430103060E	PR401A	Nr	–	–	–	4.00	4.00	5.60	8.00
430103060F	PR402A	Nr	–	–	–	5.31	5.31	7.44	10.62
430103060G	PR427A	Nr	–	–	–	5.38	5.38	7.53	10.75
430103060H	Pandrol Fastclip	Nr	–	–	–	5.52	5.52	7.72	11.03
430103070	Fishplates; type:								
430103070A	plain 4-hole	pair	–	–	–	46.43	46.43	65.00	92.85
430103070B	extra for broached hole	Nr	–	–	–	1.37	1.37	1.92	2.75
430103070C	junction fb/bh 4-hole	pair	–	–	–	52.68	52.68	73.75	105.36
430103070D	lift 4-hole 3 mm	pair	–	–	–	52.20	52.20	73.08	104.40
430103070E	junction lift fb/bh 4-hole 3 mm	pair	–	–	–	57.75	57.75	80.85	115.50

Permanent Way

Rail 2003 Ref		Unit	Labour Hours	Labour Cost	Plant Cost	Materials Cost	Unit Rate	Unit Rate	Unit Rate
				£	£	£	Base Cost £	Green Zone £	Red Zone £

430 **PLAIN LINE**

4301050 **Lay only**

430105010 Plain line; flat bottom; new; on concrete:

430105010A	jointed	m	0.97	23.33	10.71	–	34.04	47.66	68.09
430105010B	CWR	m	0.97	23.23	11.63	–	34.86	48.80	69.72
430105010C	F40; 26/18.29 m; welded joints	m	0.77	18.57	9.14	–	27.71	38.79	55.42
430105010D	F40; 28/18.29 m; welded joints	m	0.77	18.57	9.14	–	27.71	38.79	55.42

430105015 Plain line; flat bottom; re-used; on concrete:

430105015A	jointed	m	0.97	23.33	10.71	–	34.04	47.66	68.09
430105015B	CWR	m	0.97	23.23	11.63	–	34.86	48.80	69.72
430105015C	F40; 28/18.29 m; welded joints	m	0.77	18.57	9.14	–	27.71	38.79	55.42
430105015D	F40; 28/18.29 m; welded joints	m	0.77	18.57	9.14	–	27.71	38.79	55.42

430105020 Plain line; bullhead; re-used; on concrete:

| 430105020A | 95BH | m | 0.97 | 23.33 | 10.71 | – | 34.04 | 47.66 | 68.09 |

430105025 Plain line; new; on concrete:

| 430105025A | UIC60 | m | 0.97 | 23.33 | 10.71 | – | 34.04 | 47.66 | 68.09 |

430105030 Plain line; flat bottom; new; on hardwood:

430105030A	jointed	m	1.14	27.23	12.72	–	39.95	55.93	79.90
430105030B	CWR	m	1.05	25.32	12.46	–	37.78	52.90	75.56
430105030C	PAN6; 28/18.29 m; welded joints	m	0.97	23.23	11.63	–	34.86	48.80	69.72
430105030D	VN; 28/18.29 m; welded joints	m	0.97	23.23	11.63	–	34.86	48.80	69.72
430105030E	part welded in association with switches and crossings	m	0.70	16.88	8.31	–	25.19	35.26	50.38
430105030F	form curve; radius over 300 m	m	0.05	2.01	–	–	2.01	2.81	4.02
430105030G	form curve; radius not exceeding 300 m	m	0.08	3.22	–	–	3.22	4.50	6.43

Permanent Way

Rail 2003 Ref		Unit	Labour Hours	Labour Cost	Plant Cost	Materials Cost	Unit Rate	Unit Rate	Unit Rate
				£	£	£	Base Cost £	Green Zone £	Red Zone £
430	**PLAIN LINE**								
4301050	**Lay only**								
430105035	Plain line; flat bottom; re-used; on hardwood:								
430105035A	jointed	m	1.14	27.23	12.72	–	39.95	55.93	79.90
430105035B	CWR	m	1.05	25.32	12.46	–	37.78	52.90	75.56
430105035C	PAN6; 28/18.29 m; welded joints	m	0.97	23.23	11.63	–	34.86	48.80	69.72
430105035D	VN; 28/18.29 m; welded joints	m	0.97	23.23	11.63	–	34.86	48.80	69.72
430105035E	form curve; radius over 300 m	m	0.05	2.01	–	–	2.01	2.81	4.02
430105035F	form curve; radius not exceeding 300 m	m	0.08	3.22	–	–	3.22	4.50	6.43
430105040	Plain line; bullhead; re-used; on hardwood:								
430105040A	95BH jointed	m	0.97	23.33	10.71	–	34.04	47.66	68.09
430105045	Check rail on plain line; re-used; on hardwood:								
430105045A	form curve; radius not exceeding 300 m	m	1.57	37.66	17.41	–	55.07	77.10	110.14
430105047	Plain line; new; on concrete:								
430105047A	UIC60	m	0.97	23.33	10.71	–	34.04	47.66	68.09
430105048	Plain line; new; on steel:								
430105048A	UIC60	m	1.20	28.66	13.39	–	42.05	58.87	84.11
4301060	**Lift, pack and slue**								
430106050	Lifting, packing and slueing:								
430106050A	fb rail; slue max 50 mm	m	0.44	12.80	–	–	12.80	17.92	25.60
430106050B	fb rail; slue max 200 mm	m	0.44	12.80	–	–	12.80	17.92	25.60
430106050C	fb rail; slue max 250 mm	m	0.44	12.80	–	–	12.80	17.92	25.60
430106050D	fb rail; slue max 380 mm	m	0.44	12.80	–	–	12.80	17.92	25.60
430106050E	fb rail; slue max 80 mm; lift max 100 mm	m	0.55	15.89	–	–	15.89	22.24	31.77

Permanent Way

Rail 2003 Ref		Unit	Labour Hours	Labour Cost	Plant Cost	Materials Cost	Unit Rate	Unit Rate	Unit Rate
				£	£	£	Base Cost £	Green Zone £	Red Zone £

470   **SWITCHES AND CROSSINGS**

47010   **TRACKWORK**

4701010   Take up and clear away

470101010   Turnouts and crossovers; dismantle and set aside component parts; include off-site storage and disposal of redundant materials:

Ref	Description	Unit	Labour Hours	Labour Cost	Plant Cost	Materials Cost	Base Cost	Green Zone	Red Zone
470101010A	turnout; bullhead rail on timber sleepers; jointed	Nr	7.59	362.83	–	–	362.83	507.96	725.65
470101010B	tandem turnout; bullhead rail on timber sleeepers; jointed	Nr	727.50	727.50	–	–	727.50	1018.50	1455.00
470101010C	crossover; bullhead rail on timber sleepers; jointed	Nr	865.75	865.75	–	–	865.75	1212.05	1731.50
470101010D	diamond crossover; bull head on timber sleepers; jointed	Nr	385.75	385.75	–	–	385.75	540.05	771.50
470101010E	scissors crossover; bullhead on timber sleepers; jointed	Nr	2938.50	2938.50	–	–	2938.50	4113.90	5877.00
470101010F	catch point; bullhead on timber sleepers; jointed	Nr	335.50	335.50	–	–	335.50	469.70	671.00

470101015   Turnouts and crossings; flat bottom rail; dismantle and set aside component parts; include off-site storage and disposal of redundant materials:

Ref	Description	Unit	Labour Hours	Labour Cost	Plant Cost	Materials Cost	Base Cost	Green Zone	Red Zone
470101015A	turnout; on concrete sleepers; jointed	Nr	432.75	432.75	–	–	432.75	605.85	865.50
470101015B	tandem turnout; on concrete sleepers; jointed	Nr	454.75	454.75	–	–	454.75	636.65	909.50
470101015C	crossing; on concrete sleepers; jointed	Nr	477.50	477.50	–	–	477.50	668.50	955.00
470101015D	diamond crossover; on concrete sleeper; jointed	Nr	499.00	499.00	–	–	499.00	698.60	998.00
470101015E	scissors crossover; on concrete sleepers; jointed	Nr	525.00	525.00	–	–	525.00	735.00	1050.00
470101015F	catch point; on concrete sleepers; jointed	Nr	500.00	500.00	–	–	500.00	700.00	1000.00

Permanent Way

Rail 2003 Ref		Unit	Labour Hours	Labour Cost	Plant Cost	Materials Cost	Unit Rate	Unit Rate	Unit Rate
				£	£	£	Base Cost £	Green Zone £	Red Zone £
470	**SWITCHES AND CROSSINGS**								
4701030	**Material supply; catch points**								
470103010	Crossovers; flat bottom; on hardwood; ref:								
470103010A	CV91/4	Nr	–	–	–	38641.33	38641.33	54097.86	77282.65
470103010B	CV10	Nr	–	–	–	32136.92	32136.92	44991.69	64273.84
470103010C	CV10; modified	Nr	–	–	–	40386.28	40386.28	56540.79	80772.55
470103010D	CV13	Nr	–	–	–	34334.29	34334.29	48068.00	68668.57
470103010E	CV13; baseplates type V; modified to receive heaters	Nr	–	–	–	56170.03	56170.03	78638.04	112340.05
470103010F	DV103/4 baseplates type V; modified to receive heaters	m	–	–	–	55153.48	55153.48	77214.87	110306.96
470103020	Crossovers; flat bottom; on concrete; ref:								
470103020A	EV21	Nr	–	–	–	57687.33	57687.33	80762.26	115374.66
470103025	Diamond crossover								
470103025A	plain diamond	Nr	–	–	–	94610.62	94610.62	132454.87	189221.25
470103025B	switched diamond	Nr	–	–	–	151377.00	151377.00	211927.80	302753.99
470103030	Scissors; flat bottom; on hardwood; ref:								
470103030A	CV10 cast manganese crossing	Nr	–	–	–	114421.07	114421.07	160189.49	228842.13
470103030B	CV10 cast manganese crossing; baseplates type V; modified to receive heaters	Nr	–	–	–	187207.85	187207.85	262090.99	374415.70
470103035	Catch points; on hardwood; ref:								
470103035A	AV trap	Nr	–	–	–	5780.15	5780.15	8092.21	11560.30
470103035B	Bv inclined	Nr	–	–	–	7271.93	7271.93	10180.71	14543.87
470103035C	Bv switch	Nr	–	–	–	9173.79	9173.79	12843.30	18347.58
470103040	Catch points; flat bottom; on concrete; ref:								
470103040A	Av trap	Nr	–	–	–	1850.00	1850.00	2590.00	3700.00
470103040B	Bv inclined	Nr	–	–	–	1625.00	1625.00	2275.00	3250.00
470103040C	Bv switch	Nr	–	–	–	875.00	875.00	1225.00	1750.00

Permanent Way

Rail 2003 Ref		Unit	Labour Hours £	Labour Cost £	Plant Cost £	Materials Cost £	Unit Rate Base Cost £	Unit Rate Green Zone £	Unit Rate Red Zone £
470	**SWITCHES AND CROSSINGS**								
4701030	Material supply; catch points								
470103045	Catch points; UIC60; on concrete; ref:								
470103045A	Av trap	Nr	–	–	–	2150.00	2150.00	3010.00	4300.00
470103045B	Bv inclined	Nr	–	–	–	1892.50	1892.50	2649.50	3785.00
470103045C	Bv switch	Nr	–	–	–	1015.00	1015.00	1421.00	2030.00
470103065	Turnouts; flat bottom; on hardwood; ref:								
470103065A	BV8	Nr	–	–	–	18885.68	18885.68	26439.95	37771.36
470103065B	BV8; baseplates type V; modified to receive lever and lock	Nr	–	–	–	22222.72	22222.72	31111.81	44445.45
470103065C	BV8; baseplates type V; modified to receive heaters	Nr	–	–	–	22015.29	22015.29	30821.40	44030.58
470103065D	BV 9 1/4	Nr	–	–	–	18474.25	18474.25	25863.95	36948.50
470103065E	BV 9 1/4, baseplates type V; modified to receive lever and lock	Nr	–	–	–	21826.32	21826.32	30556.84	43652.63
470103065F	CV8	Nr	–	–	–	21147.30	21147.30	29606.22	42294.60
470103065G	CV8; baseplates type V; modified to receive heaters	Nr	–	–	–	25432.07	25432.07	35604.89	50864.13
470103065H	CV9 1/4	Nr	–	–	–	23700.83	23700.83	33181.16	47401.66
470103065I	CV9 1/4; baseplates type V; modified to receive heaters	Nr	–	–	–	22271.49	22271.49	31180.09	44542.98
470103065J	CV10	Nr	–	–	–	20329.40	20329.40	28461.16	40658.80
470103065K	CV13	Nr	–	–	–	27532.86	27532.86	38546.01	55065.73
470103065L	CV13; baseplates type V; modified to receive heaters	Nr	–	–	–	28662.95	28662.95	40128.12	57325.89
470103065M	DV10 3/4	Nr	–	–	–	34295.12	34295.12	48013.16	68590.23
470103065N	DV10 3/4; baseplates type V; modified to receive heaters	Nr	–	–	–	28750.15	28750.15	40250.21	57500.30
470103065O	DV10 3/4; baseplates type V; modified to receive strip heaters, one rolled manganese half set	Nr	–	–	–	35702.76	35702.76	49983.87	71405.52
470103071	Turnouts; flat bottom rail; 113A; on concrete; ref. Bv:								
470103071A	type Bv 8	Nr	–	–	–	29592.00	29592.00	41428.80	59184.00

Permanent Way

Rail 2003 Ref		Unit	Labour Hours	Labour Cost £	Plant Cost £	Materials Cost £	Unit Rate Base Cost £	Unit Rate Green Zone £	Unit Rate Red Zone £
470	**SWITCHES AND CROSSINGS**								
4701030	**Material supply; catch points**								
470103072	Turnouts; flat bottom rail; 113A; on concrete; ref. Cv:								
470103072A	type Cv 8	Nr	–	–	–	31834.00	31834.00	44567.60	63668.00
470103072B	type Cv 9.25	Nr	–	–	–	34076.00	34076.00	47706.40	68152.00
470103072C	type Cv 10	Nr	–	–	–	35421.00	35421.00	49589.40	70842.00
470103072D	type Cv 10.75	Nr	–	–	–	36766.00	36766.00	51472.40	73532.00
470103072E	type Cv 13	Nr	–	–	–	40802.00	40802.00	57122.80	81604.00
470103072F	type Cv 15	Nr	–	–	–	43940.00	43940.00	61516.00	87880.00
470103073	Turnouts; flat bottom rail; 113A; on concrete; ref. Dv:								
470103073A	type Dv 9.25	Nr	–	–	–	36766.00	36766.00	51472.40	73532.00
470103073B	type Dv 10	Nr	–	–	–	38080.00	38080.00	53312.00	76160.00
470103073C	type Dv 10.75	Nr	–	–	–	39905.00	39905.00	55867.00	79810.00
470103073D	type Dv 13	Nr	–	–	–	43940.00	43940.00	61516.00	87880.00
470103073E	type Dv 15	Nr	–	–	–	47527.00	47527.00	66537.80	95054.00
470103073F	type Dv 16	Nr	–	–	–	49321.00	49321.00	69049.40	98642.00
470103074	Turnouts; flat bottom rail; 113A; on concrete; ref. Ev:								
470103074A	type Ev 10.75	Nr	–	–	–	45285.00	45285.00	63399.00	90570.00
470103074B	type Ev 13	Nr	–	–	–	49769.00	49769.00	69676.60	99538.00
470103074C	type Ev 15	Nr	–	–	–	53804.00	53804.00	75325.60	107608.00
470103074D	type Ev 16	Nr	–	–	–	56046.00	56046.00	78464.40	112092.00
470103074E	type Ev 18.5	Nr	–	–	–	60530.00	60530.00	84742.00	121060.00
470103074F	type Ev 21	Nr	–	–	–	65014.00	65014.00	91019.60	130028.00
470103075	Turnouts; flat bottom rail; 113A; on concrete; ref. Fv:								
470103075A	type Fv 15	Nr	–	–	–	59185.00	59185.00	82859.00	118370.00
470103075B	type Fv 16	Nr	–	–	–	60978.00	60978.00	85369.20	121956.00
470103075C	type Fv 18.5	Nr	–	–	–	66359.00	66359.00	92902.60	132718.00
470103075D	type Fv 21	Nr	–	–	–	70842.00	70842.00	99178.80	141684.00
470103075E	type Fv 24	Nr	–	–	–	76671.00	76671.00	107339.40	153342.00
470103075F	type Fv 28	Nr	–	–	–	82716.00	82716.00	115802.40	165432.00

Permanent Way

Rail 2003 Ref		Unit	Labour Hours £	Labour Cost £	Plant Cost £	Materials Cost	Unit Rate Base Cost £	Unit Rate Green Zone £	Unit Rate Red Zone £
470	**SWITCHES AND CROSSINGS**								
4701030	**Material supply; catch points**								
470103076	Turnouts; flat bottom rail; 113A on concrete; ref. Gv:								
470103076A	type Gv 18.5	Nr	–	–	–	74878.00	74878.00	104829.20	149756.00
470103076B	type Gv 21	Nr	–	–	–	79810.00	79810.00	111734.00	159620.00
470103076C	type Gv 24	Nr	–	–	–	86087.00	86087.00	120521.80	172174.00
470103076D	type Gv 28	Nr	–	–	–	93709.00	93709.00	131192.60	187418.00
470103076E	type Gv 32.365	Nr	–	–	–	101780.00	101780.00	142492.00	203560.00
470103077	Turnouts; flat bottom rail; 113A: on concrete; ref. Hv:								
470103077A	type Hv 28	Nr	–	–	–	108505.00	108505.00	151907.00	217010.00
470103077B	type Hv 32.365	Nr	–	–	–	117473.00	117473.00	164462.20	234946.00
470103077C	type Hv 45.75	Nr	–	–	–	132717.00	132717.00	185803.80	265434.00
470103078	Turnouts; flat bottom rail; 113A; on concrete; ref. SGv:								
470103078A	type SGv 16	Nr	–	–	–	65014.00	65014.00	91019.60	130028.00
470103078B	type SGv 18.5	Nr	–	–	–	70394.00	70394.00	98551.60	140788.00
470103078C	type SGv 21	Nr	–	–	–	75326.00	75326.00	105456.40	150652.00
470103078D	type SGv 24	Nr	–	–	–	81155.00	81155.00	113617.00	162310.00
470103078E	type SGv 28	Nr	–	–	–	88777.00	88777.00	124287.80	177554.00
470103078F	type SGv 32.365	Nr	–	–	–	95951.00	95951.00	134331.40	191902.00
4701031	**Material supply; crossovers**								
470103111	Crossovers; flat bottom rail; 113A; on concrete; ref. Bv:								
470103111A	type Bv 8	Nr	–	–	–	53356.00	53356.00	74698.40	106712.00
470103121	Crossovers; flat bottom rail; 113A; on concrete; ref. Cv:								
470103121A	type Cv 8	Nr	–	–	–	57840.00	57840.00	80976.00	115680.00
470103121B	type Cv 9.25	Nr	–	–	–	61427.00	61427.00	85997.80	122854.00
470103121C	type Cv 10	Nr	–	–	–	63668.00	63668.00	89135.20	127336.00
470103121D	type Cv 10.75	Nr	–	–	–	65910.00	65910.00	92274.00	131820.00
470103121E	type Cv 13	Nr	–	–	–	72187.00	72187.00	101061.80	144374.00
470103121F	type Cv 15	Nr	–	–	–	77119.00	77119.00	107966.60	154238.00

Permanent Way

Rail 2003 Ref		Unit	Labour Hours	Labour Cost £	Plant Cost £	Materials Cost £	Unit Rate Base Cost £	Unit Rate Green Zone £	Unit Rate Red Zone £
470	**SWITCHES AND CROSSINGS**								
4701031	**Material supply; crossovers**								
470103131	Crossovers; flat bottom rail; 113A; on concrete; ref. Dv:								
470103131A	type Dv 9.25	Nr	–	–	–	66807.00	66807.00	93529.80	133614.00
470103131B	type Dv 10	Nr	–	–	–	69049.00	69049.00	96668.60	138098.00
470103131C	type Dv 10.75	Nr	–	–	–	71739.00	71739.00	100434.60	143478.00
470103131D	type Dv 13	Nr	–	–	–	78465.00	78465.00	109851.00	156930.00
470103131E	type Dv 15	Nr	–	–	–	84742.00	84742.00	118638.80	169484.00
470103131F	type Dv 16	Nr	–	–	–	87432.00	87432.00	122404.80	174864.00
470103141	Crossovers; flat bottom rail; 113A; on concrete; ref. Ev:								
470103141A	type Ev 10.75	Nr	–	–	–	82500.00	82500.00	115500.00	165000.00
470103141B	type Ev 13	Nr	–	–	–	90122.00	90122.00	126170.80	180244.00
470103141C	type Ev 15	Nr	–	–	–	96848.00	96848.00	135587.20	193696.00
470103141D	type Ev 16	Nr	–	–	–	99986.00	99986.00	139980.40	199972.00
470103141E	type Ev 18.5	Nr	–	–	–	107609.00	107609.00	150652.60	215218.00
470103141F	type Ev 21	Nr	–	–	–	114782.00	114782.00	160694.80	229564.00
470103151	Crossovers; flat bottom rail; 113A; on concrete; ref. Fv:								
470103151A	type Fv 15	Nr	–	–	–	107160.00	107160.00	150024.00	214320.00
470103151B	type Fv 16	Nr	–	–	–	110756.00	110756.00	155058.40	221512.00
470103151C	type Fv 18.5	Nr	–	–	–	118818.00	118818.00	166345.20	237636.00
470103151D	type Fv 21	Nr	–	–	–	126888.00	126888.00	177643.20	253776.00
470103151E	type Fv 24	Nr	–	–	–	135856.00	135856.00	190198.40	271712.00
470103151F	type Fv 28	Nr	–	–	–	146771.00	146771.00	205479.40	293542.00
470103161	Crossovers; flat bottom rail; 113A; on concrete; ref. Gv:								
470103161A	type Gv 18.5	Nr	–	–	–	135856.00	135856.00	190198.40	271712.00
470103161B	type Gv 21	Nr	–	–	–	144823.00	144823.00	202752.20	289646.00
470103161C	type Gv 24	Nr	–	–	–	154687.00	154687.00	216561.80	309374.00
470103161D	type Gv 28	Nr	–	–	–	167242.00	167242.00	234138.80	334484.00
470103161E	type Gv 32.365	Nr	–	–	–	173070.00	173070.00	242298.00	346140.00

Permanent Way

Rail 2003 Ref		Unit	Labour Hours £	Labour Cost £	Plant Cost £	Materials Cost £	Unit Rate Base Cost £	Unit Rate Green Zone £	Unit Rate Red Zone £
470	**SWITCHES AND CROSSINGS**								
4701031	Material supply; crossovers								
470103171	Crossovers; flat bottom rail; 113A; on concrete; ref. Hv:								
470103171A	type Hv 28	Nr	–	–	–	196834.00	196834.00	275567.60	393668.00
470103171B	type Hv 32.365	Nr	–	–	–	211630.00	211630.00	296282.00	423260.00
470103171C	type Hv 45.75	Nr	–	–	–	237636.00	237636.00	332690.40	475272.00
470103181	Crossovers; flat bottom rail; 113A; on concrete; ref. SGv:								
470103181A	type SGv 16	Nr	–	–	–	118369.00	118369.00	165716.60	236738.00
470103181B	type SGv 18.5	Nr	–	–	–	127337.00	127337.00	178271.80	254674.00
470103181C	type SGv 21	Nr	–	–	–	135856.00	135856.00	190198.40	271712.00
470103181D	type SGv 24	Nr	–	–	–	145272.00	145272.00	203380.80	290544.00
470103181E	type SGv 28	Nr	–	–	–	156929.00	156929.00	219700.60	313858.00
470103181F	type SGv 32.365	Nr	–	–	–	168587.00	168587.00	236021.80	337174.00
4701032	Material supply; double junctions								
470103201	Double junctions; flat bottom rail; 113A; on concrete:								
470103201A	switch type Av 7	Nr	–	–	–	73760.00	73760.00	103264.00	147520.00
470103201B	switch type Bv 8	Nr	–	–	–	84162.00	84162.00	117826.80	168324.00
470103201C	switch type Cv 9.25	Nr	–	–	–	105874.00	105874.00	148223.60	211748.00
470103201D	switch type Dv 10.75	Nr	–	–	–	121454.00	121454.00	170035.60	242908.00
470103201E	switch type Ev 15	Nr	–	–	–	154395.00	154395.00	216153.00	308790.00
470103201F	switch type Fv 18.5	Nr	–	–	–	188114.00	188114.00	263359.60	376228.00
470103201G	switch type Gv 24	Nr	–	–	–	247495.00	247495.00	346493.00	494990.00
470103201H	switch type Hv 32.365	Nr	–	–	–	331563.00	331563.00	464188.20	663126.00
470103201I	switch type SGv 21	Nr	–	–	–	216893.00	216893.00	303650.20	433786.00
4701033	Material supply; RT double junctions								
470103321	RT single junctions; ref: RT60C; type:								
470103321A	RT60C 8.25	Nr	–	–	–	46415.00	46415.00	64981.00	92830.00
470103321B	RT60C 9.5	Nr	–	–	–	51822.00	51822.00	72550.80	103644.00
470103321C	RT60C 11	Nr	–	–	–	55850.00	55850.00	78190.00	111700.00

Permanent Way

Rail 2003 Ref		Unit	Labour Hours £	Labour Cost £	Plant Cost £	Materials Cost £	Unit Rate Base Cost £	Unit Rate Green Zone £	Unit Rate Red Zone £
470	**SWITCHES AND CROSSINGS**								
4701033	**Material supply; RT double junctions**								
470103331	RT single junctions; ref: RT60D; type:								
470103331A	RT60D 12.5	Nr	–	–	–	65674.00	65674.00	91943.60	131348.00
470103341	RT single junctions; ref: RT60E; type:								
470103341A	RT60E 13.5	Nr	–	–	–	75263.00	75263.00	105368.20	150526.00
470103351	RT single junctions; ref: RT60F; type:								
470103351A	RT60F 17.25	Nr	–	–	–	96299.00	96299.00	134818.60	192598.00
470103361	RT single junctions; ref: RT60G; type:								
470103361A	RT60G 23.5	Nr	–	–	–	133883.00	133883.00	187436.20	267766.00
470103361B	RT60G 33.3	Nr	–	–	–	173126.00	173126.00	242376.40	346252.00
470103371	RT single junctions; ref: RT60H; type:								
470103371A	RT60H 43	Nr	–	–	–	224185.00	224185.00	313859.00	448370.00
4701034	**Material supply; RT circular turnouts**								
470103421	RT circular turnouts; ref: RT60C; type:								
470103421A	RT60C 7	Nr	–	–	–	32731.00	32731.00	45823.40	65462.00
470103421B	RT60C 8.25	Nr	–	–	–	35421.00	35421.00	49589.40	70842.00
470103421C	RT60C 9.5	Nr	–	–	–	37663.00	37663.00	52728.20	75326.00
470103431	RT circular turnouts; ref: RT60D; type:								
470103431A	RT60D 8.25	Nr	–	–	–	37663.00	37663.00	52728.20	75326.00
470103431B	RT60D 9.5	Nr	–	–	–	40353.00	40353.00	56494.20	80706.00
470103431C	RT60D 11	Nr	–	–	–	43940.00	43940.00	61516.00	87880.00
470103431D	RT60D 12.5	Nr	–	–	–	47079.00	47079.00	65910.60	94158.00
470103441	RT circular turnouts; ref: RT60E; type:								
470103441A	RT60E 9.5	Nr	–	–	–	45285.00	45285.00	63399.00	90570.00
470103441B	RT60E 11	Nr	–	–	–	49321.00	49321.00	69049.40	98642.00
470103441C	RT60E 12.5	Nr	–	–	–	52459.00	52459.00	73442.60	104918.00
470103441D	RT60E 13.5	Nr	–	–	–	54701.00	54701.00	76581.40	109402.00
470103441E	RT60E 15.75	Nr	–	–	–	59633.00	59633.00	83486.20	119266.00

Permanent Way

Rail 2003 Ref		Unit	Labour Hours £	Labour Cost £	Plant Cost £	Materials Cost £	Unit Rate Base Cost £	Unit Rate Green Zone £	Unit Rate Red Zone £
470	**SWITCHES AND CROSSINGS**								
4701034	**Material supply; RT circular turnouts**								
470103451	RT circular turnouts; ref: RT60F; type:								
470103451A	RT60F 12.5	Nr	–	–	–	58288.00	58288.00	81603.20	116576.00
470103451B	RT60F 13.5	Nr	–	–	–	60530.00	60530.00	84742.00	121060.00
470103451C	RT60F 15.75	Nr	–	–	–	65910.00	65910.00	92274.00	131820.00
470103451D	RT60F 17.25	Nr	–	–	–	69049.00	69049.00	96668.60	138098.00
470103461	RT circular turnouts; ref: RT60G; type:								
470103461A	RT60G 21.5	Nr	–	–	–	92812.00	92812.00	129936.80	185624.00
470103461B	RT60G 23.5	Nr	–	–	–	96461.00	96461.00	135045.40	192922.00
470103461C	RT60G 26.75	Nr	–	–	–	104918.00	104918.00	146885.20	209836.00
470103471	RT circular turnouts; ref: RT60H; type:								
470103471A	RT60H 31.25	Nr	–	–	–	129130.00	129130.00	180782.00	258260.00
470103491	RT circular turnouts; ref: RT60J; type:								
470103491A	RT60J 39.25	Nr	–	–	–	161413.00	161413.00	225978.20	322826.00
4701035	**Material supply; RT transition turnouts**								
470103521	RT transition turnouts; ref: RT60C; type:								
470103521A	RT60C 11	Nr	–	–	–	39456.00	39456.00	55238.40	78912.00
470103531	RT transition turnouts; ref: RT60D; type:								
470103531A	RT60D 13.5	Nr	–	–	–	47079.00	47079.00	65910.60	94158.00
470103541	RT transition turnouts; ref: RT60E; type:								
470103541A	RT60E 17.25	Nr	–	–	–	60530.00	60530.00	84742.00	121060.00

Permanent Way

Rail 2003 Ref		Unit	Labour Hours	Labour Cost £	Plant Cost £	Materials Cost £	Unit Rate Base Cost £	Unit Rate Green Zone £	Unit Rate Red Zone £
470	**SWITCHES AND CROSSINGS**								
4701035	**Material supply; RT transition turnouts**								
470103551	RT transition turnouts; ref: RT60F; type:								
470103551A	RT60F 21.5	Nr	–	–	–	74878.00	74878.00	104829.20	149756.00
470103561	RT transition turnouts; ref: RT60G; type:								
470103561A	RT60G 33.5	Nr	–	–	–	113886.00	113886.00	159440.40	227772.00
470103571	RT transition turnouts; ref: Rt60H; type:								
470103571A	RT60H 43	Nr	–	–	–	147962.00	147962.00	207146.80	295924.00
470103591	RT transition turnouts; ref: RT60J; type:								
470103591A	RT60J 54	Nr	–	–	–	185176.00	185176.00	259246.40	370352.00
4701036	**Material supply; RT transition crossovers**								
470103621	RT transition crossovers; ref: RT60C; type:								
470103621A	RT60C 11	Nr	–	–	–	70394.00	70394.00	98551.60	140788.00
470103631	RT transition crossovers; ref: RT60C; type:								
470103631A	RT60D 13.5	Nr	–	–	–	83397.00	83397.00	116755.80	166794.00

Permanent Way

Rail 2003 Ref		Unit	Labour Hours £	Labour Cost £	Plant Cost £	Materials Cost £	Unit Rate Base Cost £	Unit Rate Green Zone £	Unit Rate Red Zone £
470	**SWITCHES AND CROSSINGS**								
4701036	Material supply; RT transition crossovers								
470103651	RT transition crossovers; ref: RT60F; type:								
470103651A	RT60F 21.5	Nr	–	–	–	133614.00	133614.00	187059.60	267228.00
470103661	RT transition crossovers; ref: RT60G; type:								
470103661A	RT60G 33.5	Nr	–	–	–	201766.00	201766.00	282472.40	403532.00
470103671	RT transition crossovers; ref: RT60H; type:								
470103671A	RT60H 43	Nr	–	–	–	262744.00	262744.00	367841.60	525488.00
470103691	RT transition crossovers; ref: RT60J; type:								
470103691A	RT60J 54	Nr	–	–	–	329103.00	329103.00	460744.20	658206.00
4701037	Material supply; RT double junctions								
470103721	RT double junctions; ref: RT60C; type:								
470103721A	RT60C 8.25	Nr	–	–	–	87730.00	87730.00	122822.00	175460.00
470103731	RT double junctions; ref: RT60D; type:								
470103731A	RT60D 9.5	Nr	–	–	–	100710.00	100710.00	140994.00	201420.00
470103741	RT double junctions; ref: RT60E; type:								
470103741A	RT60E 12.5	Nr	–	–	–	131381.00	131381.00	183933.40	262762.00
470103751	RT double junctions; ref: RT60F; type:								
470103751A	RT60F 15.75	Nr	–	–	–	166629.00	166629.00	233280.60	333258.00

Permanent Way

Rail 2003 Ref		Unit	Labour Hours	Labour Cost £	Plant Cost £	Materials Cost £	Unit Rate Base Cost £	Unit Rate Green Zone £	Unit Rate Red Zone £
470	**SWITCHES AND CROSSINGS**								
4701037	**Material supply; RT double junctions**								
470103761	RT double junctions; ref: RT60G; type:								
470103761A	RT60G 23.5	Nr	–	–	–	245023.00	245023.00	343032.20	490046.00
470103771	RT double junctions; ref: RT60H; type:								
470103771A	RT60H 31.25	Nr	–	–	–	322254.00	322254.00	451155.60	644508.00
470103791	RT double junctions; ref: RT60J; type:								
470103791A	RT60J 39.25	Nr	–	–	–	403125.00	403125.00	564375.00	806250.00
4701040	**Lay only; crossovers**								
470104010	Crossovers; flat bottom; on hardwood sleepers; ref:								
470104010A	CV 9.25	Nr	53.00	4150.96	–	–	4150.96	5811.34	8301.92
470104010B	CV10	Nr	64.00	5007.92	–	–	5007.92	7011.08	10015.83
470104010C	CV13	Nr	74.00	5791.23	–	–	5791.23	8107.72	11582.46
470104010D	CV13; part welded	Nr	24.50	1913.17	–	–	1913.17	2678.43	3826.33
470104010E	DV 10.75; part welded	Nr	26.30	2060.18	–	–	2060.18	2884.25	4120.36
470104012	Crossovers; flat bottom; on concrete; ref:								
470104012A	EV21	Nr	59.20	4636.29	–	–	4636.29	6490.80	9272.57
470104013	Diamond crossover								
470104013A	plain diamond	Nr	193.25	15137.70	–	–	15137.70	21192.78	30275.40
470104013B	switched diamond	Nr	290.00	22706.55	–	–	22706.55	31789.17	45413.10
470104014	Scissors; flat bottom; on hardwood; ref:								
470104014A	CV10; jointed	Nr	143.25	11214.93	–	–	11214.93	15700.90	22429.86
470104014B	CV10; part welded	Nr	51.75	4051.29	–	–	4051.29	5671.81	8102.58

Permanent Way

Rail 2003 Ref		Unit	Labour Hours	Labour Cost £	Plant Cost £	Materials Cost £	Unit Rate Base Cost £	Unit Rate Green Zone £	Unit Rate Red Zone £
470	**SWITCHES AND CROSSINGS**								
4701040	**Lay only; crossovers**								
470104020	Turnouts; flat bottom; on hardwood; ref:								
470104020A	BV8	Nr	17.60	707.64	1258.04	–	1965.68	2751.95	3931.36
470104020B	BV8; part welded	Nr	7.30	292.95	520.79	–	813.74	1139.24	1627.48
470104020C	BV 9.25	Nr	17.00	683.40	1206.24	–	1889.64	2645.50	3779.28
470104020D	BV 9.25; part welded	Nr	7.60	305.52	545.46	–	850.98	1191.37	1701.96
470104020E	CV8; part welded	Nr	40.00	316.61	563.75	–	880.36	1232.50	1760.72
470104020F	CV 9.25	Nr	17.50	703.50	1244.82	–	1948.32	2727.65	3896.65
470104020G	CV 9.25; part welded	Nr	7.50	301.50	539.61	–	841.11	1177.56	1682.23
470104020H	CV10	Nr	18.00	723.60	1270.05	–	1993.65	2791.11	3987.30
470104020I	CV13; part welded	Nr	8.80	353.76	627.84	–	981.60	1374.25	1963.21
470104030	Catch points; on hardwood; ref:								
470104030A	AV trap	Nr	11.25	452.25	807.51	–	1259.76	1763.66	2519.52
470104030B	inclined 'B' f/b	Nr	4.50	180.90	321.99	–	502.89	704.04	1005.77
470104030C	switch BV	Nr	4.25	170.85	303.68	–	474.53	664.34	949.06
470104040	Catch points; flat bottom; on concrete; ref:								
470104040A	Av trap	Nr	27.00	1085.35	–	–	1085.35	1519.48	2170.69
470104040B	Bv inclined	Nr	31.00	1246.14	–	–	1246.14	1744.59	2492.28
470104040C	Bv switch	Nr	30.00	1205.94	–	–	1205.94	1688.32	2411.88
470104045	Catch points; UIC60; on concrete; ref:								
470104045A	Av trap	Nr	30.00	1205.94	–	–	1205.94	1688.32	2411.88
470104045B	Bv inclined	Nr	35.00	1406.93	–	–	1406.93	1969.70	2813.86
470104045C	Bv switch	Nr	33.00	1326.53	–	–	1326.53	1857.15	2653.07
470104071	Turnouts; flat bottom rail; 113A; on concrete; ref. Bv:								
470104071A	type Bv 8	Nr	330.00	13266.00	23470.00	–	36736.00	51430.40	73472.00
470104072	Turnouts; flat bottom rail; 113A; on concrete; ref. Cv:								
470104072A	type Cv 8	Nr	355.00	14270.00	25248.00	–	39518.00	55325.20	79036.00
470104072B	type Cv 9.25	Nr	380.00	15275.00	27026.00	–	42301.00	59221.40	84602.00
470104072C	type Cv 10	Nr	395.00	15878.00	28092.00	–	43972.00	61560.80	87944.00
470104072D	type Cv 10.75	Nr	410.00	16481.00	29159.00	–	45640.00	63896.00	91280.00
470104072E	type Cv 13	Nr	455.00	18290.00	32360.00	–	50650.00	70910.00	101300.00
470104072F	type Cv 15	Nr	490.00	19697.00	34849.00	–	54546.00	76364.40	109092.00

Permanent Way

Rail 2003 Ref		Unit	Labour Hours	Labour Cost £	Plant Cost £	Materials Cost £	Unit Rate Base Cost £	Unit Rate Green Zone £	Unit Rate Red Zone £

470 SWITCHES AND CROSSINGS

4701040 Lay only; crossovers

470104073 Turnouts; flat bottom rail; 113A; on concrete; ref. Dv:

470104073A	type Dv 9.25	Nr	410.00	16481.00	29159.00	–	45640.00	63896.00	91280.00
470104073B	type Dv 10	Nr	425.00	17071.00	30202.00	–	47273.00	66182.20	94546.00
470104073C	type Dv 10.75	Nr	445.00	17888.00	31649.00	–	49537.00	69351.80	99074.00
470104073D	type Dv 13	Nr	490.00	19697.00	34849.00	–	54546.00	76364.40	109092.00
470104073E	type Dv 15	Nr	530.00	21305.00	37694.00	–	58999.00	82598.60	117998.00
470104073F	type Dv 16	Nr	550.00	22109.00	39116.00	–	61225.00	85715.00	122450.00

470104074 Turnouts; flat bottom rail; 113A; on concrete; ref. Ev:

470104074A	type Ev 10.75	Nr	505.00	20300.00	35916.00	–	56216.00	78702.40	112432.00
470104074B	type Ev 13	Nr	555.00	22310.00	39472.00	–	61782.00	86494.80	123564.00
470104074C	type Ev 15	Nr	600.00	24119.00	42672.00	–	66791.00	93507.40	133582.00
470104074D	type Ev 16	Nr	625.00	25124.00	44450.00	–	69574.00	97403.60	139148.00
470104074E	type Ev 18.5	Nr	675.00	27134.00	48006.00	–	75140.00	105196.00	150280.00
470104074F	type Ev 21	Nr	725.00	29144.00	51562.00	–	80706.00	112988.40	161412.00

470104075 Turnouts; flat bottom rail; 113A; on concrete; ref. Fv:

470104075A	type Fv 15	Nr	660.00	26531.00	46940.00	–	73471.00	102859.40	146942.00
470104075B	type Fv 16	Nr	680.00	27335.00	48362.00	–	75697.00	105975.80	151394.00
470104075C	type Fv 18.5	Nr	740.00	29747.00	52629.00	–	82376.00	115326.40	164752.00
470104075D	type Fv 21	Nr	790.00	31757.00	56185.00	–	87942.00	123118.80	175884.00
470104075E	type Fv 24	Nr	855.00	34370.00	60808.00	–	95178.00	133249.20	190356.00
470104075F	type Fv 28	Nr	922.00	37080.00	65603.00	–	102683.00	143756.20	205366.00

470104076 Turnouts; flat bottom rail; 113A; on concrete; ref. Gv:

470104076A	type Gv 18.5	Nr	835.00	33566.00	59386.00	–	92952.00	130132.80	185904.00
470104076B	type Gv 21	Nr	890.00	35777.00	63297.00	–	99074.00	138703.60	198148.00
470104076C	type Gv 24	Nr	960.00	38591.00	68276.00	–	106867.00	149613.80	213734.00
470104076D	type Gv 28	Nr	1045.00	42008.00	74321.00	–	116329.00	162860.60	232658.00
470104076E	type Gv 32.365	Nr	1135.00	45625.00	80722.00	–	126347.00	176885.80	252694.00

Permanent Way

Rail 2003 Ref		Unit	Labour Hours	Labour Cost £	Plant Cost £	Materials Cost £	Unit Rate Base Cost £	Unit Rate Green Zone £	Unit Rate Red Zone £
470	**SWITCHES AND CROSSINGS**								
4701040	**Lay only; crossovers**								
470104077	Turnouts; flat bottom rail; 113A; on concrete; ref. Hv:								
470104077A	type Hv 28	Nr	1210.00	48640.00	86056.00	–	134696.00	188574.40	269392.00
470104077B	type Hv 32.365	Nr	1310.00	52660.00	93168.00	–	145828.00	204159.20	291656.00
470104077C	type Hv 45.75	Nr	1480.00	59494.00	105259.00	–	164753.00	230654.20	329506.00
470104078	Turnouts; flat bottom rail; 113A; on concrete; ref. SGv								
470104078A	type SGv 16	Nr	725.00	29144.00	51562.00	–	80706.00	112988.40	161412.00
470104078B	type SGv 18.5	Nr	785.00	31556.00	55830.00	–	87386.00	122340.40	174772.00
470104078C	type SGv 21	Nr	840.00	33767.00	59741.00	–	93508.00	130911.20	187016.00
470104078D	type SGv 24	Nr	905.00	36380.00	64364.00	–	100744.00	141041.60	201488.00
470104078E	type SGv 28	Nr	990.00	39797.00	70409.00	–	110206.00	154288.40	220412.00
470104078F	type SGv 32.365	Nr	1070.00	43012.00	76099.00	–	119111.00	166755.40	238222.00
4701041	**Lay only; crossovers**								
470104111	Crossovers; flat bottom rail; 113A; on concrete; ref. Bv:								
470104111A	type Bv 8	Nr	595.00	23918.00	42317.00	–	66235.00	92729.00	132470.00
470104121	Crossovers; flat bottom rail; 113A; on concrete; ref. Cv:								
470104121A	type Cv 8	Nr	645.00	25928.00	45873.00	–	71801.00	100521.40	143602.00
470104121B	type Cv 9.25	Nr	685.00	27536.00	48718.00	–	76254.00	106755.60	152508.00
470104121C	type Cv 10	Nr	710.00	28541.00	50496.00	–	79037.00	110651.80	158074.00
470104121D	type Cv 10.75	Nr	735.00	29546.00	52274.00	–	81820.00	114548.00	163640.00
470104121E	type Cv 13	Nr	805.00	32360.00	57252.00	–	89612.00	125456.80	179224.00
470104121F	type Cv 15	Nr	860.00	34571.00	61164.00	–	95735.00	134029.00	191470.00
470104131	Crossovers; flat bottom rail; 113A; on concrete; ref. Dv:								
470104131A	type Dv 9.25	Nr	745.00	29948.00	52985.00	–	82933.00	116106.20	165866.00
470104131B	type Dv 10	Nr	770.00	30953.00	54763.00	–	85716.00	120002.40	171432.00
470104131C	type Dv 10.75	Nr	800.00	32159.00	56896.00	–	89055.00	124677.00	178110.00
470104131D	type Dv 13	Nr	875.00	35174.00	62231.00	–	97405.00	136367.00	194810.00
470104131E	type Dv 15	Nr	945.00	37988.00	67209.00	–	105197.00	147275.80	210394.00
470104131F	type Dv 16	Nr	975.00	39194.00	69343.00	–	108537.00	151951.80	217074.00

Permanent Way

Rail 2003 Ref		Unit	Labour Hours	Labour Cost £	Plant Cost £	Materials Cost £	Unit Rate Base Cost £	Unit Rate Green Zone £	Unit Rate Red Zone £
470	**SWITCHES AND CROSSINGS**								
4701041	**Lay only; crossovers**								
470104141	Crossovers; flat bottom rail; 113A; on concrete; ref. Ev:								
470104141A	type Ev 10.75	Nr	920.00	36983.00	65431.00	–	102414.00	143379.60	204828.00
470104141B	type Ev 13	Nr	1005.00	40400.00	71476.00	–	111876.00	156626.40	223752.00
470104141C	type Ev 15	Nr	1080.00	43414.00	76810.00	–	120224.00	168313.60	240448.00
470104141D	type Ev 16	Nr	1115.00	44821.00	79299.00	–	124120.00	173768.00	248240.00
470104141E	type Ev 18.5	Nr	1200.00	48238.00	85345.00	–	133583.00	187016.20	267166.00
470104141F	type Ev 21	Nr	1200.00	51454.00	91034.00	–	142488.00	199483.20	284976.00
470104151	Crossovers; flat bottom rail; 113A; on concrete; ref. Fv:								
470104151A	type Fv 15	Nr	1195.00	48037.00	84989.00	–	133026.00	186236.40	266052.00
470104151B	type Fv 16	Nr	1235.00	49649.00	87841.00	–	137490.00	192486.00	274980.00
470104151C	type Fv 18.5	Nr	1325.00	53263.00	94235.00	–	147498.00	206497.20	294996.00
470104151D	type Fv 21	Nr	1415.00	56881.00	100636.00	–	157517.00	220523.80	315034.00
470104151E	type Fv 24	Nr	1515.00	60901.00	107748.00	–	168649.00	236108.60	337298.00
470104151F	type Fv 28	Nr	1635.00	65794.00	116405.00	–	182199.00	255078.60	364398.00
470104161	Crossovers; flat bottom rails; 113A; on concrete; ref. GV:								
470104161A	type Gv 18.5	Nr	1515.00	60901.00	107748.00	–	168649.00	236108.60	337298.00
470104161B	type Gv 21	Nr	1615.00	64921.00	114860.00	–	179781.00	251693.40	359562.00
470104161C	type Gv 24	Nr	1725.00	69343.00	122683.00	–	192026.00	268836.40	384052.00
470104161D	type Gv 28	Nr	1865.00	74970.00	132640.00	–	207610.00	290654.00	415220.00
470104161E	type Gv 32.365	Nr	1930.00	77583.00	137263.00	–	214846.00	300784.40	429692.00
470104171	Crossovers; flat bottom rails; 113A; on concrete; ref. Hv:								
470104171A	type Hv 28	Nr	2195.00	88236.00	156110.00	–	244346.00	342084.40	488692.00
470104171B	type Hv 32.365	Nr	2360.00	94869.00	167845.00	–	262714.00	367799.60	525428.00
470104171C	type Hv 45.75	Nr	2650.00	106526.00	188470.00	–	294996.00	412994.40	589992.00
470104181	Crossovers; flat bottom rail; 113A; on concrete; ref. SGv:								
470104181A	type SGv 16	Nr	1320.00	53062.00	93879.00	–	146941.00	205717.40	293882.00
470104181B	type SGv 18.5	Nr	1420.00	57082.00	100991.00	–	158073.00	221302.20	316146.00
470104181C	type SGv 21	Nr	1515.00	60901.00	107748.00	–	168649.00	236108.60	337298.00
470104181D	type SGv 24	Nr	1620.00	65122.00	115215.00	–	180337.00	252471.80	360674.00
470104181E	type SGv 28	Nr	1750.00	70348.00	124461.00	–	194809.00	272732.60	389618.00
470104181F	type SGv 32.365	Nr	1880.00	75573.00	133707.00	–	209280.00	292992.00	418560.00

Permanent Way

Rail 2003 Ref		Unit	Labour Hours	Labour Cost £	Plant Cost £	Materials Cost £	Unit Rate Base Cost £	Unit Rate Green Zone £	Unit Rate Red Zone £

470 SWITCHES AND CROSSINGS

4701042 Lay only; double junctions

470104201 Double junctions; flat bottom rail; 113A; on concrete:

470104201A	switch type Av 7	Nr	822.50	33065.00	58499.00	–	91564.00	128189.60	183128.00
470104201B	switch type Bv 8	Nr	938.50	37728.00	66749.00	–	104477.00	146267.80	208954.00
470104201C	switch type Cv 9.25	Nr	1100.50	47461.00	83969.00	–	131430.00	184002.00	262860.00
470104201D	switch type Dv 10.75	Nr	1355.00	54445.00	96326.00	–	150771.00	211079.40	301542.00
470104201E	switch type Ev 15	Nr	1721.50	69212.00	122451.00	–	191663.00	268328.20	383326.00
470104201F	switch type Fv 18.5	Nr	2097.50	84327.00	149194.00	–	233521.00	326929.40	467042.00
470104201G	switch type Gv 24	Nr	2760.00	110946.00	196289.00	–	307235.00	430129.00	614470.00
470104201H	switch type Hv 32.365	Nr	3697.50	148632.00	262964.00	–	411596.00	576234.40	823192.00
470104201I	switch type SGv 21	Nr	2418.50	97227.00	172017.00	–	269244.00	376941.60	538488.00

4701043 Lay only; RT single junctions

470104321 RT single junctions; ref: RT60C; type:

470104321A	RT60C 8.25	Nr	517.50	20807.00	36812.00	–	57619.00	80666.60	115238.00
470104321B	RT60C 9.5	Nr	578.00	23231.00	41100.00	–	64331.00	90063.40	128662.00
470104321C	RT60C 11	Nr	623.00	25036.00	44295.00	–	69331.00	97063.40	138662.00

470104331 RT single junctions; ref: RT60D; type:

470104331A	RT60D 12.5	Nr	732.50	29440.00	52086.00	–	81526.00	114136.40	163052.00

470104341 RT single junctions; ref: RT60E; type:

470104341A	RT60E 13.5	Nr	839.50	33738.00	59691.00	–	93429.00	130800.60	186858.00

470104351 RT single junctions; ref: RT60F; type:

470104351A	RT60F 17.25	Nr	1076.00	43168.00	76375.00	–	119543.00	167360.20	239086.00

Permanent Way

Rail 2003 Ref		Unit	Labour Hours	Labour Cost £	Plant Cost £	Materials Cost £	Unit Rate Base Cost £	Unit Rate Green Zone £	Unit Rate Red Zone £
470	**SWITCHES AND CROSSINGS**								
4701043	**Lay only; RT single junctions**								
470104361	RT single junctions; ref: RT60G; type:								
470104361A	RT60G 23.5	Nr	1493.00	60016.00	106183.00	–	166199.00	232678.60	332398.00
470104361B	RT60G 33.3	Nr	1930.50	77608.00	137307.00	–	214915.00	300881.00	429830.00
470104371	RT single junctions; ref: RT60H; type:								
470104371A	RT60H 43	Nr	2500.00	100496.00	177802.00	–	278298.00	389617.20	556596.00
4701044	**Lay only; RT circular turnouts**								
470104421	RT circular turnouts; ref: RT60C; type:								
470104421A	RT60C 7	Nr	365.00	14672.00	25959.00	–	40631.00	56883.40	81262.00
470104421B	RT60C 8.25	Nr	395.00	15878.00	28093.00	–	43971.00	61559.40	87942.00
470104421C	RT60C 9.5	Nr	420.00	16883.00	29871.00	–	46754.00	65455.60	93508.00
470104431	RT circular turnouts; ref: RT60D; type:								
470104431A	RT60D 8.25	Nr	420.00	16883.00	29871.00	–	46754.00	65455.60	93508.00
470104431B	RT60D 9.5	Nr	450.00	18089.00	32004.00	–	50093.00	70130.20	100186.00
470104431C	RT60D 11	Nr	490.00	19697.00	34849.00	–	54546.00	76364.40	109092.00
470104431D	RT60D 12.5	Nr	525.00	21104.00	37338.00	–	58442.00	81818.80	116884.00
470104441	RT circular turnouts; ref: RT60E; type:								
470104441A	RT60E 9.5	Nr	505.00	20300.00	35916.00	–	56216.00	78702.40	112432.00
470104441B	RT60E 11	Nr	550.00	22109.00	39116.00	–	61225.00	85715.00	122450.00
470104441C	RT60E 12.5	Nr	585.00	23516.00	41606.00	–	65122.00	91170.80	130244.00
470104441D	RT60E 13.5	Nr	610.00	24521.00	43384.00	–	67905.00	95067.00	135810.00
470104441E	RT60E 15.75	Nr	665.00	26732.00	47295.00	–	74027.00	103637.80	148054.00
470104451	RT circular turnouts; ref: RT60F; type:								
470104451A	RT60F 12.5	Nr	650.00	26129.00	46228.00	–	72357.00	101299.80	144714.00
470104451B	RT60F 13.5	Nr	675.00	27134.00	48006.00	–	75140.00	105196.00	150280.00
470104451C	RT60F 15.75	Nr	735.00	29546.00	52274.00	–	81820.00	114548.00	163640.00
470104451D	RT60F 17.25	Nr	770.00	30953.00	54763.00	–	85716.00	120002.40	171432.00

Permanent Way

Rail 2003 Ref		Unit	Labour Hours	Labour Cost £	Plant Cost £	Materials Cost £	Unit Rate Base Cost £	Unit Rate Green Zone £	Unit Rate Red Zone £
470	**SWITCHES AND CROSSINGS**								
4701044	**Lay only; RT circular turnouts**								
470104461	RT circular turnouts; ref: RT60G; type:								
470104461A	RT60G 21.5	Nr	1035.00	41606.00	73610.00	–	115216.00	161302.40	230432.00
470104461B	RT60G 23.5	Nr	1075.50	43241.00	76504.00	–	119745.00	167643.00	239490.00
470104461C	RT60G 26.75	Nr	1170.00	47032.00	83211.00	–	130243.00	182340.20	260486.00
470104471	RT circular turnouts; ref: RT60H; type:								
470104471A	RT60H 31.25	Nr	1440.00	57886.00	102414.00	–	160300.00	224420.00	320600.00
470104491	RT circular turnouts; ref: RT60J; type:								
470104491A	RT60J 39.25	Nr	1800.00	72357.00	128017.00	–	200374.00	280523.60	400748.00
4701045	**Lay only; RT transition turnouts**								
470104521	RT transition turnouts; ref: RT60C; type:								
470104521A	RT60C 11	Nr	440.00	17687.00	31293.00	–	48980.00	68572.00	97960.00
470104531	RT transition turnouts; ref: RT60D; type:								
470104531A	RT60D 13.5	Nr	525.00	21104.00	37338.00	–	58442.00	81818.80	116884.00
470104541	RT transition turnouts; ref: RT60E; type:								
470104541A	RT60E 17.25	Nr	675.00	27134.00	48006.00	–	75140.00	105196.00	150280.00
470104551	RT transition turnouts; ref: RT60F; type:								
470104551A	RT60F 21.5	Nr	835.00	33566.00	59386.00	–	92952.00	130132.80	185904.00
470104561	RT transition turnouts; ref: RT60G; type:								
470104561A	RT60G 33.5	Nr	1280.00	51052.00	90323.00	–	141375.00	197925.00	282750.00

Permanent Way

Rail 2003 Ref		Unit	Labour Hours	Labour Cost £	Plant Cost £	Materials Cost £	Unit Rate Base Cost £	Unit Rate Green Zone £	Unit Rate Red Zone £
470	**SWITCHES AND CROSSINGS**								
4701045	**Lay only; RT transition turnouts**								
470104571	RT transition turnouts; ref: RT60H; type:								
470104571A	RT60H 43	Nr	1650.00	66328.00	117349.00	–	183677.00	257147.80	367354.00
470104591	RT transition turnouts; ref: RT60J; type:								
470104591A	RT60J 54	Nr	2065.00	83010.00	146864.00	–	229874.00	321823.60	459748.00
4701046	**Lay only; RT transition crossovers**								
470104621	RT transition crossovers; ref: RT60C; type:								
470104621A	RT60C 11	Nr	785.00	31556.00	55830.00	–	87386.00	122340.40	174772.00
470104631	RT transition crossovers; ref: RT60D; type:								
470104631A	RT60D 13.5	Nr	930.00	37385.00	66142.00	–	103527.00	144937.80	207054.00
470104641	RT transition crossovers; ref: RT60E; type:								
470104641A	RT60E 17.25	Nr	1195.00	48037.00	84989.00	–	133026.00	186236.40	266052.00
470104651	RT transition crossovers; ref: RT60F; type:								
470104651A	RT60F 21.5	Nr	1490.00	59896.00	105970.00	–	165866.00	232212.40	331732.00
470104661	RT transition crossovers; ref: RT60G; type:								
470104661A	RT60G 33.5	Nr	2250.00	90447.00	160021.00	–	250468.00	350655.20	500936.00

Permanent Way

Rail 2003 Ref		Unit	Labour Hours	Labour Cost	Plant Cost	Materials Cost	Unit Rate	Unit Rate	Unit Rate
				£	£	£	Base Cost £	Green Zone £	Red Zone £
470	**SWITCHES AND CROSSINGS**								
4701046	Lay only; RT transition crossovers								
470104671	RT transition crossovers; ref: RT60H; type:								
470104671A	RT60H 43	Nr	2930.00	117782.00	208383.00	–	326165.00	456631.00	652330.00
470104691	RT transition crossovers; ref: RT60J; type:								
470104691A	RT60J 54	Nr	3670.00	147529.00	261013.00	–	408542.00	571958.80	817084.00
4701047	Lay only; RT double junctions								
470104721	RT double junctions; ref: RT60C; type:								
470104721A	RT60C 8.25	Nr	975.00	39327.00	69579.00	–	108906.00	152468.40	217812.00
470104731	RT double junctions; ref: RT60D; type:								
470104731A	RT60D 9.5	Nr	1125.00	45146.00	79873.00	–	125019.00	175026.60	250038.00
470104741	RT double junctions; ref: RT60E; type:								
470104741A	RT60E 12.5	Nr	1465.00	58895.00	104199.00	–	163094.00	228331.60	326188.00
470104751	RT double junctions; ref: RT60F; type:								
470104751A	RT60F 15.75	Nr	1860.00	74696.00	132154.00	–	206850.00	289590.00	413700.00
470104761	RT double junctions; ref: RT60G; type:								
470104761A	RT60G 23.5	Nr	2730.00	109838.00	194328.00	–	304166.00	425832.40	608332.00
470104771	RT double junctions; ref: RT60H; type:								
470104771A	RT60H 31.25	Nr	3595.00	144458.00	255581.00	–	400039.00	560054.60	800078.00
470104791	RT double junctions; ref: RT60J; type:								
470104791A	RT60J 39.25	Nr	4495.00	180711.00	319720.00	–	500431.00	700603.40	1000862.00

Permanent Way

Rail 2003 Ref		Unit	Labour Hours	Labour Cost £	Plant Cost £	Materials Cost £	Unit Rate Base Cost £	Unit Rate Green Zone £	Unit Rate Red Zone £

480 ANCILLARIES

48010 TRACKWORK

4801010 Take up and clear away

480101010 Set aside:

480101010A	rail built buffer stop	Nr	2.00	95.60	–	–	95.60	133.84	191.20
480101010B	switch heater	Nr	0.50	23.90	–	–	23.90	33.46	47.80
480101010C	switch hand lever	Nr	0.25	11.95	–	–	11.95	16.73	23.90

4801020 Material supply

480102010 Sundries; generally:

480102010A	rail built buffer stop	Nr	–	–	–	1885.23	1885.23	2639.32	3770.46
480102010B	switch hand lever	Nr	–	–	–	248.84	248.84	348.37	497.67
480102010C	switch lock	Nr	–	–	–	665.11	665.11	931.16	1330.22
480102010D	adjustment switch	Nr	–	–	–	2344.53	2344.53	3282.34	4689.05
480102010E	lateral resistance end plate	Nr	–	–	–	19.94	19.94	27.92	39.88
480102010F	UIC60 to 113A transition rail section 9.146 m	Nr	–	–	–	2750.00	2750.00	3850.00	5500.00

480102050 Joints; generally:

480102050A	IRJ; type 66	Nr	–	–	–	72.78	72.78	101.89	145.56
480102050B	IRJ; site glued	Nr	–	–	–	180.47	180.47	252.66	360.94
480102050C	IRJ; Mk III 6-hole; shop fitted	Nr	–	–	–	390.78	390.78	547.09	781.56
480102050D	IRJ; Mk III 4-hole; shop fitted	Nr	–	–	–	271.85	271.85	380.59	543.71

4801030 Fix only

480103010 Sundries; generally:

480103010A	rail built buffer stop	Nr	24.00	313.28	–	–	313.28	438.59	626.56
480103010B	switch hand lever	Nr	2.60	34.34	–	–	34.34	48.08	68.68
480103010C	switch lock	Nr	8.00	104.42	–	–	104.42	146.19	208.84
480103010D	adjustment switch	Nr	24.00	313.28	–	–	313.28	438.59	626.56
480103010E	lateral resistance end plate	Nr	1.20	15.66	–	–	15.66	21.93	31.33
480103010F	UIC60 to 113A transition rail section 9.146m	Nr	20.40	266.29	–	–	266.29	372.80	532.58

Permanent Way

Rail 2003 Ref		Unit	Labour Hours	Labour Cost £	Plant Cost £	Materials Cost £	Unit Rate Base Cost £	Unit Rate Green Zone £	Unit Rate Red Zone £
480	**ANCILLARIES**								
4801030	**Fix only**								
480103030	**Joints; generally:**								
480103030A	IRJ; type 66	Nr	1.20	15.66	–	–	15.66	21.93	31.33
480103030B	IRJ; site glued	Nr	8.00	104.42	–	–	104.42	146.19	208.84
480103030C	IRJ; Mk III 6-hole; shop fitted	Nr	4.00	52.22	–	–	52.22	73.11	104.44
480103030D	IRJ; Mk III 4-hole; shop fitted	Nr	4.00	52.22	–	–	52.22	73.11	104.44
480103030E	weld; Thermic process	Nr	7.20	93.98	63.50	–	157.48	220.48	314.97
480103030F	weld; Thermic process; joint to existing	Nr	6.60	86.15	65.85	–	152.00	212.80	304.00
480103040	**Stress relieving; generally:**								
480103040A	pre-weld heat treatment	Nr	–	–	–	107.23	107.23	150.12	214.45
480103040B	post weld heat treatment	Nr	–	–	–	107.23	107.23	150.12	214.45

Permanent Way

Rail 2003 Ref		Unit	Labour Hours	Labour Cost £	Plant Cost £	Materials Cost £	Unit Rate Base Cost £	Unit Rate Green Zone £	Unit Rate Red Zone £

490 TRACK DRAINAGE

49010 PIPEWORK

4901010 HepSeal; virtrified clay pipes; BS EN295: 1991; spigot and socket flexible joints

490101010 Nominal bore 100 mm; in trenches; depth:

490101010A	not exceeding 1.5 m	m	1.68	20.08	3.98	7.52	31.58	44.22	63.17
490101010B	1.5 – 2.0 m	m	1.91	22.83	3.98	7.68	34.49	48.29	68.99
490101010C	2.0 – 2.5 m	m	2.17	25.94	5.26	7.84	39.04	54.64	78.08
490101010D	2.5 – 3.0 m	m	2.44	29.17	8.88	8.00	46.05	64.46	92.08

490101020 Nominal bore 150 mm; in trenches; depth:

490101020A	not exceeding 1.5 m	m	1.73	20.68	3.98	9.44	34.10	47.74	68.20
490101020B	1.5 – 2.0 m	m	1.96	23.43	5.26	9.59	38.28	53.59	76.57
490101020C	2.0 – 2.5 m	m	2.20	26.30	7.23	9.75	43.28	60.61	86.57
490101020D	2.5 – 3.0 m	m	2.49	29.76	8.88	9.91	48.55	67.98	97.11

490101030 Nominal bore 225 mm; in trenches; depth:

490101030A	not exceeding 1.5 m	m	1.93	23.07	5.23	17.38	45.68	63.96	91.36
490101030B	1.5 – 2.0 m	m	2.18	26.06	6.86	17.54	50.46	70.63	100.90
490101030C	2.0 – 2.5 m	m	2.48	29.64	9.53	17.70	56.87	79.62	113.75
490101030D	2.5 – 3.0 m	m	2.75	32.87	11.18	17.86	61.91	86.67	123.80

490101040 Nominal bore 300 mm; in trenches; depth:

490101040A	not exceeding 1.5 m	m	2.17	25.94	6.48	26.57	58.99	82.58	117.96
490101040B	1.5 – 2.0 m	m	2.44	29.17	8.45	26.73	64.35	90.08	128.69
490101040C	2.0 – 2.5 m	m	2.78	33.23	11.66	26.89	71.78	100.49	143.56
490101040D	2.5 – 3.0 m	m	3.08	36.81	13.48	27.05	77.34	108.28	154.68

490101050 Nominal bore 400 mm; in trenches; depth:

490101050A	not exceeding 1.5 m	m	2.54	30.36	9.13	52.92	92.41	129.37	184.82
490101050B	1.5 – 2.0 m	m	2.86	34.19	11.98	53.08	99.25	138.95	198.49
490101050C	2.0 – 2.5 m	m	3.30	39.44	16.43	53.24	109.11	152.76	218.23
490101050D	2.5 – 3.0 m	m	3.64	43.51	19.65	53.40	116.56	163.17	233.11

Permanent Way

Rail 2003 Ref		Unit	Labour Hours	Labour Cost £	Plant Cost £	Materials Cost £	Unit Rate Base Cost £	Unit Rate Green Zone £	Unit Rate Red Zone £

490 TRACK DRAINAGE

4901010 HepSeal; virtrified clay pipes; BS EN295: 1991; spigot and socket flexible joints

490101060 Nominal bore 450 mm; in trenches; depth:

490101060A	not exceeding 1.5 m	m	2.62	31.32	16.13	68.25	115.70	161.98	231.40
490101060B	1.5 – 2.0 m	m	2.94	35.14	18.98	68.41	122.53	171.55	245.06
490101060C	2.0 – 2.5 m	m	3.38	40.40	23.43	68.57	132.40	185.36	264.80
490101060D	2.5 – 3.0 m	m	3.72	44.46	26.65	68.73	139.84	195.78	279.68

490101070 Nominal bore 500 mm; in trenches; depth:

490101070A	not exceeding 1.5 m	m	2.35	28.09	18.25	75.70	122.04	170.85	244.08
490101070B	1.5 – 2.0 m	m	2.70	32.27	23.03	75.86	131.16	183.63	262.33
490101070C	2.0 – 2.5 m	m	3.17	37.89	26.43	76.02	140.34	196.48	280.68
490101070D	2.5 – 3.0 m	m	3.55	42.43	30.01	76.18	148.62	208.07	297.24

490101080 Nominal bore 600 mm; in trenches; depth:

490101080A	not exceeding 1.5 m	m	2.93	35.02	20.19	119.55	174.76	244.68	349.53
490101080B	1.5 – 2.0 m	m	3.33	39.80	23.76	119.71	183.27	256.58	366.55
490101080C	2.0 – 2.5 m	m	3.82	45.66	29.43	119.87	194.96	272.95	389.94
490101080D	2.5 – 3.0 m	m	4.24	50.68	33.18	120.03	203.89	285.46	407.80

4901020 HepLine; vitrified clay subsoil drainage pipes; perforated; BS 65: 1991

490102010 Nominal bore 225 mm; in trenches; depth:

490102010A	not exceeding 1.5 m	m	1.93	23.07	5.23	14.73	43.03	60.25	86.07
490102010B	1.5 – 2.0 m	m	2.18	26.06	6.86	14.89	47.81	66.92	95.61
490102010C	2.0 – 2.5 m	m	2.48	29.64	9.53	15.05	54.22	75.92	108.46
490102010D	2.5 – 3.0 m	m	2.75	32.87	11.18	15.21	59.26	82.96	118.51

490102020 Nominal bore 300 mm; in trenches; depth:

490102020A	not exceeding 1.5 m	m	2.17	25.94	6.48	26.98	59.40	83.15	118.78
490102020B	1.5 – 2.0 m	m	2.44	29.17	8.45	27.14	64.76	90.66	129.51
490102020C	2.0 – 2.5 m	m	2.78	33.23	11.66	27.30	72.19	101.06	144.38
490102020D	2.5 – 3.0 m	m	3.08	36.81	13.48	27.46	77.75	108.85	155.50

Permanent Way

Rail 2003 Ref		Unit	Labour Hours	Labour Cost	Plant Cost	Materials Cost	Unit Rate	Unit Rate	Unit Rate
				£	£	£	Base Cost £	Green Zone £	Red Zone £
490	**TRACK DRAINAGE**								
4901020	HepLine; vitrified clay subsoil drainage pipes; perforated; BS 65: 1991								
490102030	Nominal bore 400 mm; in trenches; depth:								
490102030A	not exceeding 1.5 m	m	2.54	30.36	9.13	84.56	124.05	173.67	248.10
490102030B	1.5 – 2.0 m	m	2.86	34.19	11.98	84.72	130.89	183.24	261.77
490102030C	2.0 – 2.5 m	m	3.30	39.44	16.43	84.88	140.75	197.06	281.51
490102030D	2.5 – 3.0 m	m	3.64	43.51	19.65	85.04	148.20	207.47	296.39
490102040	Nominal bore 450 mm; in trenches; depth:								
490102040A	not exceeding 1.5 m	m	2.62	31.32	16.13	107.47	154.92	216.89	309.84
490102040B	1.5 – 2.0 m	m	2.94	35.14	18.98	107.63	161.75	226.46	323.51
490102040C	2.0 – 2.5 m	m	3.38	40.40	23.43	107.79	171.62	240.27	343.25
490102040D	2.5 – 3.0 m	m	3.72	44.46	26.65	107.95	179.06	250.69	358.13
4901030	Concrete pipes; BS 5911; rebated flexible joints with mastic seal								
490103010	Nominal bore 300 mm; in trenches; depth:								
490103010A	not exceeding 1.5 m	m	2.17	25.94	6.48	8.78	41.20	57.68	82.39
490103010B	1.5 – 2.0 m	m	2.44	29.17	8.45	8.94	46.56	65.18	93.12
490103010C	2.0 – 2.5 m	m	2.78	33.23	11.66	9.10	53.99	75.58	107.99
490103010D	2.5 – 3.0 m	m	3.08	36.81	13.48	9.26	59.55	83.38	119.11
490103020	Nominal bore 375 mm; in trenches; depth:								
490103020A	not exceeding 1.5 m	m	2.54	30.36	9.13	11.37	50.86	71.21	101.73
490103020B	1.5 – 2.0 m	m	2.86	34.19	11.98	11.53	57.70	80.78	115.40
490103020C	2.0 – 2.5 m	m	3.30	39.44	16.43	11.69	67.56	94.60	135.14
490103020D	2.5 – 3.0 m	m	3.64	43.51	19.65	11.85	75.01	105.01	150.02
490103030	Nominal bore 450 mm; in trenches; depth:								
490103030A	not exceeding 1.5 m	m	2.62	31.32	16.13	13.68	61.13	85.59	122.27
490103030B	1.5 – 2.0 m	m	2.94	35.14	18.98	13.84	67.96	95.16	135.93
490103030C	2.0 – 2.5 m	m	3.38	40.40	23.43	14.00	77.83	108.97	155.67
490103030D	2.5 – 3.0 m	m	3.72	44.46	26.65	14.16	85.27	119.39	170.55

Permanent Way

Rail 2003 Ref		Unit	Labour Hours	Labour Cost £	Plant Cost £	Materials Cost £	Unit Rate Base Cost £	Unit Rate Green Zone £	Unit Rate Red Zone £

490   **TRACK DRAINAGE**

4901030   Concrete pipes; BS 5911; rebated flexible joints with mastic seal

490103040   Nominal bore 600 mm; in trenches; depth:

490103040A	1.5 – 2.0 m	m	2.93	35.02	20.19	25.20	80.41	112.58	160.82
490103040B	2.0 – 2.5 m	m	3.33	39.80	23.76	25.36	88.92	124.48	177.84
490103040C	2.5 – 3.0 m	m	3.82	45.66	29.43	25.52	100.61	140.86	201.23
490103040D	3.0 – 3.5 m	m	4.24	50.68	33.18	25.68	109.54	153.36	219.09

4901040   Concrete pipes; land drainage system; ogee joints

490104010   Nominal bore 225 mm; in trenches; depth:

490104010A	not exceeding 1.5 m	m	1.93	23.07	5.23	5.23	33.53	46.96	67.08
490104010B	1.5 – 2.0 m	m	2.18	26.06	6.86	5.39	38.31	53.63	76.62
490104010C	2.0 – 2.5 m	m	2.48	29.64	9.53	5.55	44.72	62.62	89.47
490104010D	2.5 – 3.0 m	m	2.75	32.87	11.18	5.71	49.76	69.67	99.52

490104020   Nominal bore 300 mm; in trenches; depth:

490104020A	not exceeding 1.5 m	m	2.17	25.94	6.48	7.97	40.39	56.53	80.75
490104020B	1.5 – 2.0 m	m	2.44	29.17	8.45	8.13	45.75	64.04	91.48
490104020C	2.0 – 2.5 m	m	2.78	33.23	11.66	8.29	53.18	74.44	106.35
490104020D	2.5 – 3.0 m	m	3.08	36.81	13.48	8.45	58.74	82.23	117.47

490104030   Nominal bore 375 mm; in trenches; depth:

490104030A	not exceeding 1.5 m	m	2.54	30.36	9.13	10.28	49.77	69.68	99.54
490104030B	1.5 – 2.0 m	m	2.86	34.19	11.98	10.44	56.61	79.25	113.21
490104030C	2.0 – 2.5 m	m	3.30	39.44	16.43	10.60	66.47	93.06	132.95
490104030D	2.5 – 3.0 m	m	3.64	43.51	19.65	10.75	73.91	103.48	147.83

490104040   Nominal bore 450 mm; in trenches; depth:

490104040A	1.5 – 2.0 m	m	2.62	31.32	16.13	12.32	59.77	83.68	119.55
490104040B	2.0 – 2.5 m	m	2.94	35.14	18.98	12.48	66.60	93.25	133.21
490104040C	2.5 – 3.0 m	m	3.38	40.40	23.43	12.64	76.47	107.07	152.95
490104040D	3.0 – 3.5 m	m	3.72	44.46	26.65	12.80	83.91	117.48	167.83

Permanent Way

Rail 2003 Ref		Unit	Labour Hours	Labour Cost £	Plant Cost £	Materials Cost £	Unit Rate Base Cost £	Unit Rate Green Zone £	Unit Rate Red Zone £

490 TRACK DRAINAGE

4901050 UPVC pipes; BS 5481; flexible joints

490105010 Nominal bore 200 mm; in trenches; depth:

490105010A	not exceeding 1.5 m	m	1.73	20.68	3.98	31.39	56.05	78.47	112.10
490105010B	1.5 – 2.0 m	m	1.96	23.43	5.26	31.55	60.24	84.32	120.47
490105010C	2.0 – 2.5 m	m	2.20	26.30	7.23	31.71	65.24	91.34	130.47
490105010D	2.5 – 3.0 m	m	2.49	29.76	8.88	31.87	70.51	98.71	141.01

490105020 Nominal bore 250 mm; in trenches; depth:

490105020A	not exceeding 1.5 m	m	1.93	23.07	5.23	47.33	75.63	105.89	151.27
490105020B	1.5 – 2.0 m	m	2.18	26.06	6.86	47.49	80.41	112.57	160.81
490105020C	2.0 – 2.5 m	m	2.48	29.64	9.53	47.65	86.82	121.56	173.66
490105020D	2.5 – 3.0 m	m	2.75	32.87	11.18	47.81	91.86	128.61	183.71

490105030 Nominal bore 315 mm; in trenches; depth:

490105030A	not exceeding 1.5 m	m	2.17	25.94	6.48	70.85	103.27	144.57	206.52
490105030B	1.5 – 2.0 m	m	2.44	29.17	8.45	71.01	108.63	152.07	217.24
490105030C	2.0 – 2.5 m	m	2.78	33.23	11.66	71.17	116.06	162.47	232.11
490105030D	2.5 – 3.0 m	m	3.08	36.81	13.48	71.33	121.62	170.27	243.23

490105040 Nominal bore 400 mm; in trenches; depth:

490105040A	not exceeding 1.5 m	m	2.54	30.36	9.13	111.94	151.43	212.00	302.86
490105040B	1.5 – 2.0 m	m	2.86	34.19	11.98	112.10	158.27	221.57	316.53
490105040C	2.0 – 2.5 m	m	3.30	39.44	16.43	112.25	168.12	235.39	336.27
490105040D	2.5 – 3.0 m	m	3.64	43.51	19.65	112.41	175.57	245.80	351.15

4901060 UPVC pipes; slotted; BS 5481; flexible joints

490106010 Nominal bore 200 mm; in trenches; depth:

490106010A	not exceeding 1.5 m	m	1.73	20.68	3.98	6.98	31.64	44.30	63.29
490106010B	1.5 – 2.0 m	m	1.96	23.43	5.26	7.14	35.83	50.15	71.66
490106010C	2.0 – 2.5 m	m	2.20	26.30	7.23	7.30	40.83	57.17	81.66
490106010D	2.5 – 3.0 m	m	2.49	29.76	8.88	7.46	46.10	64.54	92.20

Permanent Way

Rail 2003 Ref		Unit	Labour Hours	Labour Cost £	Plant Cost £	Materials Cost £	Unit Rate Base Cost £	Unit Rate Green Zone £	Unit Rate Red Zone £
490	**TRACK DRAINAGE**								
4901060	UPVC pipes; slotted; BS 5481; flexible joints								
490106020	Nominal bore 250 mm; in trenches; depth:								
490106020A	not exceeding 1.5 m	m	1.73	20.68	3.98	11.28	35.94	50.32	71.88
490106020B	1.5 – 2.0 m	m	1.96	23.43	5.26	11.44	40.13	56.17	80.25
490106020C	2.0 – 2.5 m	m	2.20	26.30	7.23	11.60	45.13	63.18	90.25
490106020D	2.5 – 3.0 m	m	2.49	29.76	8.88	11.76	50.40	70.56	100.79
490106030	Nominal bore 300 mm; in trenches; depth:								
490106030A	not exceeding 1.5 m	m	1.94	23.15	4.90	21.93	49.98	69.97	99.95
490106030B	1.5 – 2.0 m	m	2.53	30.19	6.53	22.29	59.01	82.63	118.04
490106030C	2.0 – 2.5 m	m	3.17	37.84	8.17	22.66	68.67	96.14	137.34
490106030D	2.5 – 3.0 m	m	3.80	45.45	9.80	23.03	78.28	109.58	156.54
49030	**PIPE FITTINGS**								
4903010	HepSeal vitrified clay fittings, BS EN295: 1991								
490301010	Bends; size:								
490301010A	100 mm	Nr	0.20	2.39	–	9.21	11.60	16.24	23.19
490301010B	150 mm	Nr	0.20	2.39	–	15.19	17.58	24.61	35.16
490301010C	225 mm	Nr	0.22	2.63	–	31.15	33.78	47.28	67.55
490301010D	300 mm	Nr	0.35	4.18	–	61.43	65.61	91.86	131.22
490301010E	400 mm	Nr	0.45	5.38	–	138.45	143.83	201.36	287.66
490301020	Junctions; size:								
490301020A	100 × 100 mm	Nr	0.25	2.99	–	12.79	15.78	22.09	31.57
490301020B	150 × 100 mm	Nr	0.25	2.99	–	18.92	21.91	30.67	43.82
490301020C	150 × 150 mm	Nr	0.25	2.99	–	19.83	22.82	31.95	45.65
490301020D	225 × 150 mm	Nr	0.27	3.23	–	47.47	50.70	70.97	101.38

Permanent Way

Rail 2003 Ref		Unit	Labour Hours	Labour Cost	Plant Cost	Materials Cost	Unit Rate	Unit Rate	Unit Rate
				£	£	£	Base Cost £	Green Zone £	Red Zone £
490	**TRACK DRAINAGE**								
4903020	**Concrete fittings, BS 5911**								
490302010	Bends; size:								
490302010A	300 mm	Nr	0.60	7.17	3.50	38.39	49.06	68.69	98.12
490302010B	375 mm	Nr	0.75	8.96	3.50	48.00	60.46	84.64	120.92
490302010C	450 mm	Nr	0.85	10.16	3.50	58.25	71.91	100.67	143.82
490302010D	600 mm	Nr	1.00	11.95	4.38	67.85	84.18	117.85	168.35
490302010E	750 mm	Nr	1.20	14.34	4.90	77.45	96.69	135.36	193.37
490302020	Junctions; size:								
490302020A	300 × 300 mm	Nr	0.70	8.37	3.50	84.73	96.60	135.23	193.18
490302020B	300 × 450 mm	Nr	0.80	9.56	3.50	95.32	108.38	151.72	216.75
490302020C	450 × 450 mm	Nr	0.90	10.76	3.50	63.54	77.80	108.92	155.60
490302020D	525 × 450 mm	Nr	1.10	13.15	4.38	73.15	90.68	126.94	181.34
490302020E	525 × 525 mm	Nr	1.30	15.54	4.90	82.74	103.18	144.45	206.35
4903030	**UPVC fittings; BS 5481; flexible joints**								
490303010	Bends; size:								
490303010A	200 mm	Nr	0.17	2.03	–	6.51	8.54	11.96	17.08
490303010B	250 mm	Nr	0.20	2.39	–	9.30	11.69	16.37	23.39
490303010C	315 mm	Nr	0.20	2.39	–	14.51	16.90	23.67	33.81
490303010D	400 mm	Nr	0.27	3.23	–	36.06	39.29	55.00	78.56
490303010E	450 mm	Nr	0.31	3.71	–	58.54	62.25	87.14	124.49
490303020	Junctions; size:								
490303020A	200 × 200 mm	Nr	0.20	2.39	–	9.16	11.55	16.17	23.10
490303020B	250 × 200 mm	Nr	0.25	2.99	–	19.27	22.26	31.16	44.52
490303020C	250 × 250 mm	Nr	0.30	3.59	–	22.64	26.23	36.71	52.45
490303020D	315 × 250 mm	Nr	0.25	2.99	–	19.68	22.67	31.73	45.33
490303020E	315 × 315 mm	Nr	0.35	4.18	–	25.12	29.30	41.02	58.60
490303020F	400 × 315 mm	Nr	0.35	4.18	–	50.31	54.49	76.29	108.98
490303020G	400 × 400 mm	Nr	0.35	4.18	–	51.86	56.04	78.46	112.09
490303020H	450 × 400 mm	Nr	0.37	4.42	–	51.86	56.28	78.79	112.56
490303020I	450 × 450 mm	Nr	0.37	4.42	–	64.60	69.02	96.63	138.05

Permanent Way

Rail 2003 Ref		Unit	Labour Hours	Labour Cost £	Plant Cost £	Materials Cost £	Unit Rate Base Cost £	Unit Rate Green Zone £	Unit Rate Red Zone £
490	**TRACK DRAINAGE**								
49035	**PIPE SURROUNDS**								
4903510	**Pipework bedding and surround**								
490351010	Imported granular material; 150mm bed and surround; nominal pipe bore:								
490351010A	not exceeding 200 mm	m	0.45	5.38	0.51	2.12	8.01	11.21	16.02
490351010B	200 – 300 mm	m	0.57	6.81	0.87	3.53	11.21	15.70	22.42
490351010C	300 – 600 mm	m	0.92	11.00	1.53	6.35	18.88	26.43	37.75
49040	**MANHOLES, CATCHPITS ETC**								
4904010	**Precast concrete manhole units; complete**								
490401010	1050 mm diameter; complete with Grade A cover and frame; depth to invert:								
490401010A	not exceeding 1.5 m	Nr	18.32	226.96	65.93	538.81	831.70	1164.37	1663.38
490401010B	1.5 – 2.0 m	Nr	26.94	329.99	88.68	644.17	1062.84	1487.98	2125.68
490401010C	2.0 – 2.5 m	Nr	31.20	382.15	99.46	743.97	1225.58	1715.82	2451.18
490401020	1200 mm diameter; complete with Grade A cover and frame; depth to invert:								
490401020A	not exceeding 1.5 m	Nr	18.50	229.11	76.60	592.90	898.61	1258.06	1797.23
490401020B	1.5 – 2.0 m	Nr	29.87	365.64	102.13	701.99	1169.76	1637.66	2339.52
490401020C	2.0 – 2.5 m	Nr	34.60	422.80	115.76	804.00	1342.56	1879.57	2685.10
4904020	**UPVC inspection chamber units; complete**								
490402010	475 mm diameter; standard steel push fit access cover and frame; depth to invert								
490402010A	not exceeding 1.5 m	Nr	1.80	21.51	–	100.25	121.76	170.45	243.51
4904030	**UPVC sump units; complete**								
490403010	375 mm diameter; untrapped; standard steel push fit access cover and frame; depth:								
490403010A	not exceeding 1.5 m	Nr	1.80	21.51	–	30.86	52.37	73.32	104.75

Permanent Way

Rail 2003 Ref		Unit	Labour Hours	Labour Cost £	Plant Cost £	Materials Cost £	Unit Rate Base Cost £	Unit Rate Green Zone £	Unit Rate Red Zone £
490	**TRACK DRAINAGE**								
4904030	UPVC sump units; complete								
490403020	510 mm diameter; untrapped; standard steel push fit access cover and frame; depth:								
490403020A	not exceeding 1.5 m	Nr	2.25	26.89	–	38.36	65.25	91.35	130.51
49060	**TRACK CROSSINGS**								
4906010	HepSeal; vitrified clay pipes; BS EN295: 1991; spigot and socket flexible joints;								
490601010	Nominal bore 225 mm; crossing width:								
490601010A	not exceeding 3.0 m	Nr	5.79	69.21	15.70	52.13	137.04	191.85	274.07
490601010B	3.0 – 5.0 m	Nr	4.65	55.58	26.17	86.88	168.63	236.08	337.25
490601010C	5.0 – 10.0 m	Nr	9.30	111.15	52.33	173.77	337.25	472.15	674.50
490601020	Nominal bore 300 mm; crossing width:								
490601020A	not exceeding 3.0 m	Nr	6.51	77.81	19.43	79.71	176.95	247.74	353.90
490601020B	3.0 – 5.0 m	Nr	10.85	129.69	32.38	132.86	294.93	412.89	589.83
490601020C	5.0 – 10.0 m	Nr	21.70	259.37	64.75	265.72	589.84	825.77	1179.68
49070	**SOAKAWAYS**								
4907010	Precast concrete manhole units; complete								
490701010	600 mm diameter; complete with Grade A cover and frame; depth to invert:								
490701010A	not exceeding 1.5 m	Nr	15.35	188.98	82.29	817.08	1088.35	1523.69	2176.71
490701010B	1.5 – 2.0 m	Nr	19.47	238.23	100.52	999.11	1337.86	1873.01	2675.73
490701010C	2.0 – 2.5 m	Nr	29.93	363.27	133.63	1213.07	1709.97	2393.95	3419.93
490701010D	2.5 – 3.0 m	Nr	35.22	426.50	152.84	1368.29	1947.63	2726.69	3895.28

Permanent Way

Rail 2003 Ref		Unit	Labour Hours	Labour Cost £	Plant Cost £	Materials Cost £	Unit Rate Base Cost £	Unit Rate Green Zone £	Unit Rate Red Zone £
490	**TRACK DRAINAGE**								
49080	**PIPE BEDDING**								
4908030	**Pipework surrounds**								
490803010	Sand; 100 mm cover; nominal pipe bore:								
490803010A	not exceeding 200 mm	m	0.32	3.82	0.29	1.31	5.42	7.60	10.85
490803010B	200 – 300 mm	m	0.44	5.26	0.73	2.13	8.12	11.36	16.23
490803010C	300 – 600 mm	m	0.71	8.49	1.31	3.85	13.65	19.10	27.28
490803015	Sand; 150 mm cover; nominal pipe bore:								
490803015A	not exceeding 250 mm	m	0.32	3.82	0.29	1.31	5.42	7.60	10.85
490803015B	200 – 300 mm	m	0.57	6.81	0.87	2.87	10.55	14.78	21.10
490803015C	300 – 600 mm	m	0.92	11.00	1.53	5.16	17.69	24.76	35.37
490803020	Imported granular material; 100 mm cover; nominal pipe bore:								
490803020A	not exceeding 200 mm	m	0.34	4.06	0.44	1.61	6.11	8.56	12.23
490803020B	200 – 300 mm	m	0.44	5.26	0.73	2.62	8.61	12.05	17.21
490803020C	300 – 600 mm	m	0.71	8.49	1.31	4.74	14.54	20.35	29.06
490803025	Imported granular material; 150 mm cover; nominal pipe bore:								
490803025A	not exceeding 200 mm	m	0.45	5.38	0.51	2.12	8.01	11.21	16.02
490803025B	200 – 300 mm	m	0.57	6.81	0.87	3.53	11.21	15.70	22.42
490803025C	300 – 600 mm	m	0.92	11.00	1.53	6.35	18.88	26.43	37.75
490803030	Mass concrete; grade C15P; 100 mm cover; nominal pipe bore:								
490803030A	not exceeding 200 mm	m	0.34	4.06	0.94	10.71	15.71	22.01	31.45
490803030B	200 – 300 mm	m	0.52	6.22	1.82	13.84	21.88	30.61	43.74
490803030C	300 – 600 mm								
490803035	Mass concrete; grade C15P; 150 mm cover; nominal pipe bore:								
490803035A	not exceeding 200 mm	m	0.38	4.54	1.02	12.50	18.06	25.28	36.11
490803035B	200 – 300 mm	m	0.58	6.93	2.03	15.62	24.58	34.42	49.18
490803035C	300 – 600 mm	m	1.08	12.91	4.50	25.44	42.85	59.99	85.71

Permanent Way

Rail 2003 Ref		Unit	Labour Hours	Labour Cost £	Plant Cost £	Materials Cost £	Unit Rate Base Cost £	Unit Rate Green Zone £	Unit Rate Red Zone £

490 TRACK DRAINAGE

4908030 Pipework surrounds

490803040 Reinforced concrete; grade C25P; 100 mm cover; nominal pipe bore:

490803040A	not exceeding 200 mm	m	0.56	6.81	0.94	19.40	27.15	38.01	54.31
490803040B	200 – 300 mm	m	0.74	8.96	1.82	23.02	33.80	47.32	67.59
490803040C	300 – 600 mm	m	1.30	15.65	4.50	36.49	56.64	79.30	113.28

490803045 Reinforced concrete; grade C25P; 150 mm cover; nominal pipe bore:

490803045A	not exceeding 200 mm	m	0.60	7.29	1.02	21.47	29.78	41.68	59.55
490803045B	200 – 300 mm	m	0.74	8.96	1.82	25.09	35.87	50.22	71.74
490803045C	300 – 600 mm	m	1.41	16.97	4.94	39.59	61.50	86.10	122.99

Telecommunications

Rail 2003 Ref		Unit	Labour Hours	Labour Cost	Plant Cost	Materials Cost	Unit Rate	Unit Rate	Unit Rate
				£	£	£	Base Cost £	Green Zone £	Red Zone £
5	**TELECOMMUNICATIONS**								
521	**ALL SYSTEMS, CABLING AND CABLE ROUTES**								
52101	**CABLING**								
5210101	**Copper**								
521010101	Supply; type:								
521010101A	2 pair	m	–	–	–	0.85	0.85	1.19	1.70
521010101B	10 pair	m	–	–	–	2.57	2.57	3.60	5.14
521010101C	30 pair	m	–	–	–	4.33	4.33	6.06	8.66
521010101D	50 pair	m	–	–	–	5.57	5.57	7.80	11.14
521010101E	75 pair	m	–	–	–	9.59	9.59	13.43	19.18
5210131	**Fibre**								
521013101	Supply; type:								
521013101C	single mode	m	–	–	–	9.51	9.51	13.31	19.02

Telecommunications

Rail 2003 Ref		Unit	Composite Cost £	Unit Rate Base Cost £	Unit Rate Green Zone £	Unit Rate Red Zone £
541	**OPERATIONAL SYSTEMS**					
54101	**EQUIPMENT**					
5410101	**Transmission**					
541010101	Telephones:					
541010101A	trackside CB telephone	Nr	– – –	347.87	487.02	695.74
541010101B	trackside auto telephone	Nr	– – –	386.53	541.14	773.06
541010101C	telephone pedestal	Nr	– – –	77.31	108.23	154.62
5410121	**Concentrators**					
541012101	Supply; to suit:					
541012101A	signal post telephone backrow system	Nr	– – – 27829.80	27829.80	38961.72	55659.60
541012101B	system line cards	Nr	– – – 386.53	386.53	541.14	773.06
541012101C	power system; 50v	Nr	– – – 15461.00	15461.00	21645.40	30922.00
5410141	**Information systems**					
541014101	Customer information system:					
541014101A	supply	Nr	– – – 5411.35	5411.35	7575.89	10822.70
541014101B	installation	Nr	– – – 2319.15	2319.15	3246.81	4638.30

Telecommunications

Rail 2003 Ref		Unit	Composite Cost £	Unit Rate Base Cost £	Unit Rate Green Zone £	Unit Rate Red Zone £

561 INFRASTRUCTURE SYSTEMS

56101 EXISTING LINE

5610101 Transmission

561010101 Equipment rack:

561010101A	size 0.6 × 0.3 × 2.2 m	Nr	–	–	–	1546.10	1546.10	2164.54	3092.20
561010101B	size 0.6 × 0.6 × 2.2 m	Nr	–	–	–	1855.32	1855.32	2597.45	3710.64
561010101C	location case	Nr	–	–	–	3478.75	3478.75	4870.25	6957.50
561010101D	FTE	Nr	–	–	–	927.66	927.66	1298.72	1855.32
561010101E	plug point	Nr	–	–	–	386.53	386.53	541.14	773.06
561010101F	signal post telephone	Nr	–	–	–	773.05	773.05	1082.27	1546.10
561010101G	termination	Nr	–	–	–	579.79	579.79	811.71	1159.58
561010101H	case label	Nr	–	–	–	61.84	61.84	86.58	123.68

561010106 Primary multiplex:

561010106B	drop and insert 30 cct	Nr	–	–	–	3865.25	3865.25	5411.35	7730.50
561010106D	ringer	Nr	–	–	–	695.75	695.75	974.05	1391.50
561010106E	termination 30 cct	Nr	–	–	–	1932.63	1932.63	2705.68	3865.26

561010111 Termination circuits:

561010111A	sub-stations	Nr	–	–	–	1091.49	1091.49	1528.09	2182.98
561010111B	exchange	Nr	–	–	–	1091.49	1091.49	1528.09	2182.98
561010111C	VF/E&M	Nr	–	–	–	1546.10	1546.10	2164.54	3092.20
561010111D	X.24/X.27	Nr	–	–	–	927.66	927.66	1298.72	1855.32
561010111E	9.6 kbit V24	Nr	–	–	–	927.66	927.66	1298.72	1855.32

561010116 Data patch panel:

561010116A	G 703 rack mounted	Nr	–	–	–	61.84	61.84	86.58	123.68
561010116C	connector	Nr	–	–	–	11.60	11.60	16.24	23.20
561010116D	distribution shelf	Nr	–	–	–	231.92	231.92	324.69	463.84
561010116E	IDF 1 vert	Nr	–	–	–	927.66	927.66	1298.72	1855.32
561010116F	IDF 2 vert	Nr	–	–	–	1236.88	1236.88	1731.63	2473.76

561010121 Multiplex:

| 561010121A | 2/8 | Nr | – | – | – | 1004.97 | 1004.97 | 1406.96 | 2009.94 |

Telecommunications

Rail 2003 Ref	Unit	Composite Cost £	Unit Rate Base Cost £	Unit Rate Green Zone £	Unit Rate Red Zone £
561	**INFRASTRUCTURE SYSTEMS**				
5610101	Transmission				
561010131	Optical line termination equipment:				
561010131B	8 Mbits .... Nr	– – –	2705.68	2705.68	3787.95	5411.36
561010151	Co-axial cable to DDF:				
561010151A	per location .... Nr	– – –	30.92	30.92	43.29	61.84
5610111	Information systems; cabling; type				
561011101	RG59 BU SWA; coaxial:				
561011101A	surface mounted.................. m	– – –	11.60	11.60	16.24	23.20
561011101B	run in ladders and tray............. m	– – –	6.57	6.57	9.20	13.14
561011101C	run in underground duct .... m	– – –	6.96	6.96	9.74	13.92
561011101D	termination including glands .... Nr	– – –	77.31	77.31	108.23	154.62
561011111	PVC insulated; 2.5 mm² SWA; two core:				
561011111A	surface mounted.................. m	– – –	10.05	10.05	14.07	20.10
561011111B	run in ladders and tray............. m	– – –	5.80	5.80	8.12	11.60
561011111C	run in underground duct .... m	– – –	6.18	6.18	8.65	12.36
561011111D	termination including glands .... Nr	– – –	77.31	77.31	108.23	154.62
561011121	Ref. 6491X, earth, 2.5 mm²; one core:				
561011121A	surface mounted.................. m	– – –	5.41	5.41	7.57	10.82
561011121B	run in ladders and tray............. m	– – –	4.64	4.64	6.50	9.28
561011121C	run in underground duct .... m	– – –	5.02	5.02	7.03	10.04
561011121D	termination including ferrule........ Nr	– – –	25.51	25.51	35.71	51.02
561011131	Ref. 6491X, earth, 16 mm²; one core:				
561011131A	surface mounted.................. m	– – –	6.96	6.96	9.74	13.92
561011131B	run in ladders and tray............. m	– – –	5.91	5.91	8.27	11.82
561011131C	run in underground duct .... m	– – –	6.26	6.26	8.76	12.52
561011131D	termination including ferrule........ Nr	– – –	25.12	25.12	35.17	50.24

Telecommunications

Rail 2003 Ref		Unit	Composite Cost £	Unit Rate Base Cost £	Unit Rate Green Zone £	Unit Rate Red Zone £

561 INFRASTRUCTURE SYSTEMS

5610121 Closed circuit television (CCTV)

561012101 Monitors:

Ref	Description	Unit				Base	Green	Red
561012101A	monitor complete	Nr	–	–	–	4638.30	4638.30	6493.62	9276.60
561012101B	PECA model 5143	Nr	–	–	–	1855.32	1855.32	2597.45	3710.64
561012101C	monitor housing	Nr	–	–	–	2319.15	2319.15	3246.81	4638.30
561012101D	double monitor housing	Nr	–	–	–	1391.49	1391.49	1948.09	2782.98
561012101E	two tier double monitor housing	Nr	–	–	–	1623.41	1623.41	2272.77	3246.82
561012101F	hydraulic pole, 2 section	Nr	–	–	–	2782.98	2782.98	3896.17	5565.96
561012101G	monitor pole; 1.5 m long	Nr	–	–	–	2164.54	2164.54	3030.36	4329.08
561012101H	core drill precast concrete platform; 50 mm dia	Nr	–	–	–	371.06	371.06	519.48	742.12
561012101I	core drill precast concrete platfrom; 75 mm dia	Nr	–	–	–	618.44	618.44	865.82	1236.88
561012101J	barrier fence to monitor pole	Nr	–	–	–	1546.10	1546.10	2164.54	3092.20
561012101K	video enhancer; Peca type 6040	Nr	–	–	–	927.66	927.66	1298.72	1855.32
561012101L	train mass detector	Nr	–	–	–	1159.58	1159.58	1623.41	2319.16

561012121 Cameras:

561012121A	camera complete	Nr	–	–	–	2937.59	2937.59	4112.63	5875.18
561012121B	Burle ref. TC652 BTX	Nr	–	–	–	541.14	541.14	757.60	1082.28
561012121C	auto iris lens; 16 mm	Nr	–	–	–	123.69	123.69	173.17	247.38
561012121D	auto iris lens; 25 mm	Nr	–	–	–	340.14	340.14	476.20	680.28
561012121E	camera housing	Nr	–	–	–	347.87	347.87	487.02	695.74
561012121F	hinged camera pole	Nr	–	–	–	1855.32	1855.32	2597.45	3710.64
561012121G	core drill precast concrete platform; 75 mm dia	Nr	–	–	–	541.14	541.14	757.60	1082.28
561012121H	camera pole extension arm; 500 mm long	Nr	–	–	–	309.22	309.22	432.91	618.44
561012121I	camera mounting bracket	Nr	–	–	–	1004.97	1004.97	1406.96	2009.94
561012121J	camera wall mounting bracket	Nr	–	–	–	1043.62	1043.62	1461.07	2087.24

561012131 Control room:

561012131A	monitor	Nr	–	–	–	500.00	500.00	700.00	1000.00
561012131B	VCR	Nr	–	–	–	830.00	830.00	1162.00	1660.00
561012131C	multiplexer	Nr	–	–	–	1995.00	1995.00	2793.00	3990.00
561012131D	installation	Nr	–	–	–	332.55	332.55	465.57	665.10

Telecommunications

Rail 2003 Ref		Unit	Composite Cost	Unit Rate	Unit Rate	Unit Rate
			£	Base Cost £	Green Zone £	Red Zone £
561	**INFRASTRUCTURE SYSTEMS**					
5610121	**Closed circuit television (CCTV)**					
561012141	Power supply:					
561012141A	distribution box...............	Nr	– – –	657.09	657.09	919.93	1314.18
561012141B	low voltage step-down transformer; 0.5 kVA 240/110v; complete	Nr	– – –	347.87	347.87	487.02	695.74
561012141C	cabling	Nr	– – –	4.00	4.00	5.60	8.00
561012141D	connection......................	Nr	– – –	20.00	20.00	28.00	40.00
561012161	Heated mirrors:					
561012161A	heated mirror complete	Nr	– – –	5411.35	5411.35	7575.89	10822.70
561012161B	landscape hooded membrane; 5 × 4 .	Nr	– – –	4870.22	4870.22	6818.31	9740.44
561012161C	portrait hooded membrane; 4 × 5....	Nr	– – –	5643.27	5643.27	7900.58	11286.54
561012161D	landscape hooded membrane; 4 × 3 .	Nr	– – –	4019.86	4019.86	5627.80	8039.72
561012161E	portrait hooded membrane; 3 × 4....	Nr	– – –	4329.08	4329.08	6060.71	8658.16
561012161F	mirror pole; 3.5 m high	Nr	– – –	695.75	695.75	974.05	1391.50
561012161G	mirror pole; 4.5 m high	Nr	– – –	773.05	773.05	1082.27	1546.10
561012161H	flange plate and fixing bolts.........	Nr	– – –	618.44	618.44	865.82	1236.88
561012161I	mirror yoke and safety collar.........	Nr	– – –	231.92	231.92	324.69	463.84
561012171	Video cabling:					
561012171A	cabling	Nr	– – –	2.75	2.75	3.85	5.50
561012171B	connection......................	Nr	– – –	29.50	29.50	41.30	59.00
561012176	Cable containment:					
561012176A	conduit.........................	Nr	– – –	15.50	15.50	21.70	31.00
561012176B	trunking	Nr	– – –	15.50	15.50	21.70	31.00
561012191	CCTV all-in installation costs including survey, drawings, testing, management and on-site overheads; station type:					
561012191A	category B – 30 fixed cameras	Nr	– – –	70783.25	70783.25	99096.55	141566.50
561012191B	category C – 20 fixed cameras	Nr	– – –	47798.50	47798.50	66917.90	95597.00
561012191C	category D – 12 fixed cameras	Nr	– – –	30912.50	30912.50	43277.50	61825.00
561012191D	category E – 9 fixed cameras	Nr	– – –	22695.00	22695.00	31773.00	45390.00

Telecommunications

Rail 2003 Ref		Unit	Composite Cost £	Unit Rate Base Cost £	Unit Rate Green Zone £	Unit Rate Red Zone £
561	**INFRASTRUCTURE SYSTEMS**					
5610131	**Customer Information Systems (CIS)**					
561013121	Display system; 15" monitors:					
561013121A	monitor	Nr	– – –	367.50	367.50	514.50	735.00
561013121B	monitor housing	Nr	– – –	568.65	568.65	796.11	1137.30
561013121C	monitor housing bracket	Nr	– – –	221.75	221.75	310.45	443.50
561013121D	installation	Nr	– – –	88.75	88.75	124.25	177.50
561013123	Display systems; 25" monitors:					
561013123A	monitor	Nr	– – –	487.75	487.75	682.85	975.50
561013123B	monitor housing	Nr	– – –	704.00	704.00	985.60	1408.00
561013123C	monitor housing bracket	Nr	– – –	221.75	221.75	310.45	443.50
561013123D	installation	Nr	– – –	116.50	116.50	163.10	233.00
561013125	Display systems; 28" monitors:					
561013125A	monitor	Nr	– – –	521.00	521.00	729.40	1042.00
561013125B	monitor housing	Nr	– – –	705.00	705.00	987.00	1410.00
561013125C	monitor housing bracket	Nr	– – –	221.75	221.75	310.45	443.50
561013125D	installation	Nr	– – –	116.50	116.50	163.10	233.00
561013130	LED display systems:					
561013130A	single sided 3-line 32 char LED display.	Nr	– – –	1855.00	1855.00	2597.00	3710.00
561013130B	double sided 3-line 32 char LED display.	Nr	– – –	3635.00	3635.00	5089.00	7270.00
561013130C	LED display bracket	Nr	– – –	232.50	232.50	325.50	465.00
561013130E	LED display installation – single sided.	Nr	– – –	166.25	166.25	232.75	332.50
561013130F	LED display installation – double sided.	Nr	– – –	194.00	194.00	271.60	388.00
561013135	CIS control equipment:					
561013135A	cabinet	Nr	– – –	800.00	800.00	1120.00	1600.00
561013135B	desking	Nr	– – –	1110.00	1110.00	1554.00	2220.00
561013135D	control p.c. hardware	Nr	– – –	1662.75	1662.75	2327.85	3325.50
561013135E	control p.c. software	Nr	– – –	3880.00	3880.00	5432.00	7760.00
561013135F	display driver (4 monitor)	Nr	– – –	1995.00	1995.00	2793.00	3990.00
561013135G	installation	Nr	– – –	277.00	277.00	387.80	554.00

Telecommunications

Rail 2003 Ref		Unit	Composite Cost £	Unit Rate Base Cost £	Unit Rate Green Zone £	Unit Rate Red Zone £
561	**INFRASTRUCTURE SYSTEMS**					
5610131	**Customer Information Systems (CIS)**					
561013140	Power supply cabling:					
561013140A	power cable	m	– – –	7.75	7.75	10.85	15.50
561013140B	power cable connection	Nr	– – –	19.50	19.50	27.30	39.00
561013145	CIS data cabling:					
561013145A	data cable	m	– – –	2.75	2.75	3.85	5.50
561013145B	data cable connection	Nr	– – –	29.50	29.50	41.30	59.00
561013191	CIS all-in installation costs including survey, drawings, management, on-site overheads, testing and commissioning; station type:					
561013191A	category B – 24 monitors	Nr	– – –	123161.00	123161.00	172425.40	246322.00
561013191B	category C – 15 monitors	Nr	– – –	74837.00	74837.00	104771.80	149674.00
561013191C	category D – 10 monitors	Nr	– – –	42856.00	42856.00	59998.40	85712.00
561013191D	category E – 5 monitors	Nr	– – –	29350.00	29350.00	41090.00	58700.00
5610141	**Public Address System (PAS)**					
561014111	Core public address system:					
561014111A	microphone	Nr	– – –	500.00	500.00	700.00	1000.00
561014111B	zone amplifier	Nr	– – –	1665.00	1665.00	2331.00	3330.00
561014111C	digital voice announcer	Nr	– – –	1220.00	1220.00	1708.00	2440.00
561014111E	ambient noise sensing (per zone)	Nr	– – –	166.25	166.25	232.75	332.50
561014111F	platform announce switch/microphone	Nr	– – –	232.75	232.75	325.85	465.50
561014111G	installation	Nr	– – –	277.50	277.50	388.50	555.00
561014121	Loudspeakers; supply and installation:					
561014121A	projection speaker	Nr	– – –	40.00	40.00	56.00	80.00
561014121B	ceiling mounted speaker	Nr	– – –	22.50	22.50	31.50	45.00
561014121C	column mounted speaker	Nr	– – –	105.00	105.00	147.00	210.00

Telecommunications

Rail 2003 Ref		Unit	Composite Cost £	Unit Rate Base Cost £	Unit Rate Green Zone £	Unit Rate Red Zone £
561	**INFRASTRUCTURE SYSTEMS**					
5610141	**Public Address System (PAS)**					
561014131	**PAS power supply cabling:**					
561014131A	power supply cable	m	– – –	4.00	5.60	8.00
561014131B	power supply cable connection	Nr	– – –	20.00	28.00	40.00
561014141	**PAS communication cabling:**					
561014141A	communication cable	m	– – –	3.90	5.46	7.80
561014141B	communication cable connection	Nr	– – –	6.65	9.31	13.30
561014191	**PA System all-in installation costs including survey, drawings, management on-site costs, and testing and commissioning; station type:**					
561014191A	category B – 10 zone	Nr	– – – 60225.00	60225.00	84315.00	120450.00
561014191B	category C – 6 zone	Nr	– – – 31206.00	31206.00	43688.40	62412.00
561014191C	category D – 2 zone	Nr	– – – 19717.00	19717.00	27603.80	39434.00
561014191D	category E – 1 zone	Nr	– – – 10941.00	10941.00	15317.40	21882.00
5610181	**Ancillaries**					
561018181	**Miscellaneous items:**					
561018181A	LLPA interface	Nr	– – – 25124.13	25124.13	35173.78	50248.26
561018181B	spares; allowance	item	– – – 1623.41	1623.41	2272.77	3246.82
561018181C	system integration	item	– – – 7730.50	7730.50	10822.70	15461.00
561018181D	optical patch cords	Nr	– – – 154.61	154.61	216.45	309.22
561018181E	GDT	Nr	– – – 38.65	38.65	54.11	77.30
561018181F	TMS adapters	Nr	– – – 2473.76	2473.76	3463.26	4947.52
561018181G	socket strip	Nr	– – – 309.22	309.22	432.91	618.44
561018181H	service terminals	Nr	– – – 695.75	695.75	974.05	1391.50

Property – New Build

Rail 2003 Ref		Unit	Specialist Cost £	Unit Rate Base Cost £	Unit Rate Green Zone £	Unit Rate Red Zone £
6	**PROPERTY – NEW BUILD**					
630	**PLATFORMS**					
63010	**SOLID FILL CONSTRUCTION**					
6301010	Single Track					
630101010	Macadam surfacing on 250 mm sub-base of consolidated hardcore fill; paving thickness:					
630101010A	100 mm thick	m^2	– – –	509.00	509.00	712.60	1018.00
630101020	Block paved surfacing on sand bed and 200 mm sub-base of consolidated hardcore fill; paving thickness:					
630101020A	65 mm thick	m^2	– – –	547.00	547.00	765.80	1094.00
630101020B	80 mm thick	m^2	– – –	577.00	577.00	807.80	1154.00
6301020	Island Position					
630102010	Macadam surfacing on 250 mm sub-base of consolidated hardcore fill; paving thickness:					
630102010A	100 mm thick	m^2	– – –	545.00	545.00	763.00	1090.00
630102020	Block paved surfacing on sand bed and 200 mm sub-base of consolidated hardcore fill; paving thickness:					
630102020A	65 mm thick	m^2	– – –	583.00	583.00	816.20	1166.00
630102020B	80 mm thick	m^2	– – –	613.00	613.00	858.20	1226.00

Property – New Build

Rail 2003 Ref		Unit	Specialist Cost £	Unit Rate Base Cost £	Unit Rate Green Zone £	Unit Rate Red Zone £
630	**PLATFORMS**					
63020	**CROSS WALL CONSTRUCTION**					
6302010	Single Track					
630201010	Macadam surfacing on concrete topping to beam and pot structural floor element; paving thickness:					
630201010A	100 mm thick	m^2	–	391.00	547.40	782.00
630201020	Block paved surfacing on sand bed and concrete topping to beam and pot structural floor element; paving thickness:					
630201020A	65 mm thick	m^2	–	429.00	600.60	858.00
630201020B	80 mm thick	m^2	–	459.00	642.60	918.00
6302020	Island Position					
630202010	Macadam surfacing on concrete topping to beam and pot structural floor element; paving thickness:					
630202010A	100 mm thick	m^2	–	419.00	586.60	838.00
630202020	Block paved surfacing on concrete topping to beam and pot structural floor element; paving thickness:					
630202020A	65 mm thick	m^2	–	457.00	639.80	914.00
630202020B	80 mm thick	m^2	–	489.00	684.60	978.00

Property – New Build

Rail 2003 Ref		Unit	Labour Hours	Labour Cost £	Plant Cost £	Materials Cost £	Unit Rate Base Cost £	Unit Rate Green Zone £	Unit Rate Red Zone £
650	**FACILITIES**								
65010	**STORAGE AND REPAIR FACILITIES** Typical elemental cost analyses								
6501010	**Repair and Amenities Building**								
650101005	**Substructure**								
650101005A	average cost	m²	–	123.78	29.34	126.30	279.42	391.19	558.84
650101010	**Frame**								
650101010A	average cost	m²	–	39.69	13.07	193.79	246.55	345.18	493.10
650101015	**Roof**								
650101015A	average cost	m²	–	76.14	1.97	119.14	197.25	276.16	394.50
650101020	**External Walls**								
650101020A	average cost	m²	–	47.64	–	50.99	98.63	138.09	197.26
650101025	**Windows and Doors**								
650101025A	average cost	m²	–	11.24	–	54.51	65.75	92.05	131.50
650101030	**Internal Walls**								
650101030A	average cost	m²	–	17.45	–	15.42	32.87	46.02	65.74
650101035	**Finishes**								
650101035A	average cost	m²	–	27.29	–	38.47	65.76	92.07	131.52
650101040	**Sanitary Finishes**								
650101040A	average cost	m²	–	23.60	–	9.27	32.87	46.02	65.74
650101045	**Mechanical**								
650101045A	average cost	m²	–	87.21	0.23	142.67	230.11	322.15	460.22

Property – New Build

Rail 2003 Ref		Unit	Labour Hours	Labour Cost £	Plant Cost £	Materials Cost £	Unit Rate Base Cost £	Unit Rate Green Zone £	Unit Rate Red Zone £
650	**FACILITIES**								
6501010	**Repair and Amenities Building**								
650101050	**Electrical**								
650101050A	average cost	m²	–	59.60	–	55.46	115.06	161.08	230.12
650101055	**Drainage and External Work**								
650101055A	average cost	m²	–	86.34	21.24	171.84	279.42	391.20	558.84
6501020	**Service and Amenities Building**								
650102005	**Substructure**								
650102005A	average cost	m²	–	135.42	32.10	138.17	305.69	427.97	611.38
650102010	**Frame**								
650102010A	average cost	m²	–	44.04	14.50	214.98	273.52	382.93	547.04
650102015	**Roof**								
650102015A	average cost	m²	–	80.73	2.09	126.33	209.15	292.81	418.30
650102020	**External Walls**								
650102020A	average cost	m²	–	38.85	–	41.59	80.44	112.62	160.88
650102025	**Windows and Doors**								
650102025A	average cost	m²	–	8.25	–	40.01	48.26	67.56	96.52
650102030	**Internal Walls**								
650102030A	average cost	m²	–	8.55	–	7.55	16.10	22.54	32.20
650102035	**Finishes**								
650102035A	average cost	m²	–	33.38	–	47.06	80.44	112.61	160.88

Property – New Build

Rail 2003 Ref		Unit	Labour Hours	Labour Cost £	Plant Cost £	Materials Cost £	Unit Rate Base Cost £	Unit Rate Green Zone £	Unit Rate Red Zone £
650	**FACILITIES**								
6501020	**Service and Amenities Building**								
650102040	**Sanitary Finishes**								
650102040A	average cost	m²	–	23.10	–	9.07	32.17	45.04	64.34
650102045	**Mechanical**								
650102045A	average cost	m²	–	79.27	0.21	129.67	209.15	292.81	418.30
650102050	**Electrical**								
650102050A	average cost	m²	–	58.34	–	54.28	112.62	157.67	225.24
650102055	**Drainage and External Work**								
650102055A	average cost	m²	–	74.57	18.34	148.42	241.33	337.87	482.66
6501030	**Wheel Lathe Building**								
650103005	**Substructure**								
650103005A	average cost	m²	–	269.68	63.92	275.16	608.76	852.26	1217.52
650103010	**Frame**								
650103010A	average cost	m²	–	84.01	27.66	410.13	521.80	730.51	1043.60
650103015	**Roof**								
650103015A	average cost	m²	–	201.41	5.22	315.17	521.80	730.52	1043.60
650103020	**External Walls**								
650103020A	average cost	m²	–	147.02	–	157.37	304.39	426.15	608.78
650103025	**Windows and Doors**								
650103025A	average cost	m²	–	29.74	–	144.19	173.93	243.51	347.86

Property – New Build

Rail 2003 Ref		Unit	Labour Hours	Labour Cost £	Plant Cost £	Materials Cost £	Unit Rate Base Cost £	Unit Rate Green Zone £	Unit Rate Red Zone £
650	**FACILITIES**								
6501030	**Wheel Lathe Building**								
650103030	**Internal Walls**								
650103030A	average cost	m^2	–	46.18	–	40.79	86.97	121.76	173.94
650103035	**Finishes**								
650103035A	average cost	m^2	–	72.18	–	101.75	173.93	243.50	347.86
650103040	**Sanitary Fittings**								
650103040A	average cost	m^2	–	62.44	–	24.52	86.96	121.75	173.92
650103045	**Mechanical**								
650103045A	average cost	m^2	–	296.64	0.78	485.27	782.69	1095.77	1565.38
650103050	**Electrical**								
650103050A	average cost	m^2	–	202.72	–	188.63	391.35	547.89	782.70
650103055	**Drainage and External Work**								
650103055A	average cost	m^2	–	214.98	52.88	427.88	695.74	974.03	1391.48
6501040	**Bogie Drop Building**								
650104005	**Substructure**								
650104005A	average cost	m^2	–	235.70	55.87	240.49	532.06	744.89	1064.12
650104010	**Frame**								
650104010A	average cost	m^2	–	97.90	32.23	477.94	608.07	851.30	1216.14
650104015	**Roof**								
650104015A	average cost	m^2	–	146.70	3.80	229.54	380.04	532.06	760.08

Property – New Build

Rail 2003 Ref		Unit	Labour Hours	Labour Cost £	Plant Cost £	Materials Cost £	Unit Rate Base Cost £	Unit Rate Green Zone £	Unit Rate Red Zone £
650	**FACILITIES**								
6501040	**Bogie Drop Building**								
650104020	**External Walls**								
650104020A	average cost	m²	–	73.43	–	78.60	152.03	212.84	304.06
650104025	**Windows and Doors**								
650104025A	average cost	m²	–	19.50	–	94.51	114.01	159.61	228.02
650104030	**Internal Walls**								
650104030A	average cost	m²	–	20.18	–	17.82	38.00	53.20	76.00
650104035	**Finishes**								
650104035A	average cost	m²	–	47.31	–	66.69	114.00	159.60	228.00
650104040	**Sanitary Fittings**								
650104040A	average cost	m²	–	54.57	–	21.43	76.00	106.40	152.00
650104045	**Mechanical**								
650104045A	average cost	m²	–	273.67	0.72	447.69	722.08	1010.92	1444.16
650104050	**Electrical**								
650104050A	average cost	m²	–	196.86	–	183.18	380.04	532.05	760.08
650104055	**Drainage and External Work**								
650104055A	average cost	m²	–	211.38	51.99	420.71	684.08	957.71	1368.16

Property – New Build

Rail 2003 Ref		Unit	Labour Hours	Labour Cost £	Plant Cost £	Materials Cost £	Unit Rate Base Cost £	Unit Rate Green Zone £	Unit Rate Red Zone £
650	**FACILITIES**								
65020	**SECURITY FACILITIES** Typical elemental cost analyses								
6502010	**Security Centre**								
650201005	**Substructure**								
650201005A	average cost	m²	–	192.56	45.64	196.47	434.67	608.54	869.34
650201010	**Frame**								
650201010A	average cost	m²	–	59.98	19.75	292.84	372.57	521.60	745.14
650201015	**Roof**								
650201015A	average cost	m²	–	143.81	3.73	225.03	372.57	521.59	745.14
650201020	**External Walls**								
650201020A	average cost	m²	–	89.98	–	96.31	186.29	260.80	372.58
650201025	**Windows and Doors**								
650201025A	average cost	m²	–	21.24	–	102.96	124.20	173.88	248.40
650201030	**Internal Walls**								
650201030A	average cost	m²	–	65.95	–	58.25	124.20	173.88	248.40
650201035	**Finishes**								
650201035A	average cost	m²	–	77.31	–	108.98	186.29	260.80	372.58
650201040	**Sanitary Fittings**								
650201040A	average cost	m²	–	89.17	–	35.02	124.19	173.87	248.38
650201045	**Mechanical**								
650201045A	average cost	m²	–	141.20	0.37	230.99	372.56	521.59	745.12

Property – New Build

Rail 2003 Ref		Unit	Labour Hours	Labour Cost £	Plant Cost £	Materials Cost £	Unit Rate Base Cost £	Unit Rate Green Zone £	Unit Rate Red Zone £
650	**FACILITIES**								
6502010	**Security Centre**								
650201050	**Electrical**								
650201050A	average cost	m^2	–	144.74	–	134.68	279.42	391.19	558.84
650201055	**Drainage and External Work**								
650201055A	average cost	m^2	–	163.09	40.11	324.60	527.80	738.92	1055.60

Property – Refurbishment

Rail 2003 Ref		Unit	Labour Hours	Labour Cost £	Plant Cost £	Materials Cost £	Unit Rate Base Cost £	Unit Rate Green Zone £	Unit Rate Red Zone £
65	**PROPERTY – REFURBISHMENT**								
6501	**DEMOLITIONS AND REMOVALS**								
650110	**GENERAL DEMOLITIONS AND REMOVALS**								
65011010	Concrete structural elements								
6501101010	Reinforced concrete walls and attached columns:								
6501101010A	not exceeding 150 mm thick	m³	12.00	143.45	71.14	–	214.59	300.42	429.17
6501101010B	150 – 300 mm thick	m³	14.00	167.36	82.99	–	250.35	350.49	500.69
6501101010C	exceeding 300 mm thick	m³	18.00	215.17	106.70	–	321.87	450.63	643.75
6501101020	Reinforced concrete ground slabs:								
6501101020A	not exceeding 150 mm thick	m³	6.00	71.72	35.57	–	107.29	150.21	214.59
6501101020B	150 – 300 mm thick	m³	8.00	95.63	47.42	–	143.05	200.27	286.11
6501101020C	exceeding 300 mm thick	m³	10.00	119.54	59.28	–	178.82	250.35	357.64
6501101030	Reinforced suspended slab and attached beams:								
6501101030A	not exceeding 150 mm thick	m³	10.00	119.54	59.28	–	178.82	250.35	357.64
6501101030B	150 – 300 mm thick	m³	16.00	191.26	94.85	–	286.11	400.56	572.23
6501101030C	exceeding 300 mm thick	m³	19.00	227.13	112.63	–	339.76	475.66	679.51
6501101040	Reinforced concrete isolated beams:								
6501101040A	not exceeding 0.1 m² sectional area	m³	12.00	143.45	71.14	–	214.59	300.42	429.17
6501101040B	0.1 – 0.25 m²	m³	14.00	167.36	82.99	–	250.35	350.49	500.69
6501101040C	exceeding 0.25 m²	m³	19.00	227.13	112.63	–	339.76	475.66	679.51
6501101050	Reinforced concrete isolated columns:								
6501101050A	not exceeding 0.1 m² sectional area	m³	10.00	119.54	59.28	–	178.82	250.35	357.64
6501101050B	0.1 – 0.25 m²	m³	13.00	155.40	77.06	–	232.46	325.45	464.93
6501101050C	exceeding 0.25 m²	m³	18.00	215.17	106.70	–	321.87	450.63	643.75

Property – Refurbishment

Rail 2003 Ref		Unit	Labour Hours	Labour Cost	Plant Cost	Materials Cost	Unit Rate	Unit Rate	Unit Rate
				£	£	£	Base Cost £	Green Zone £	Red Zone £
6501	**DEMOLITIONS AND REMOVALS**								
65011020	Demolition of masonry structural elements								
6501102010	Brick walls:								
6501102010A	half brick thick	m²	0.80	9.56	0.62	–	10.18	14.25	20.36
6501102010B	one brick thick	m²	1.40	16.74	1.08	–	17.82	24.94	35.63
6501102010C	one and a half brick thick	m²	2.10	25.10	1.62	–	26.72	37.41	53.45
6501102010D	two brick thick	m²	2.70	32.28	2.08	–	34.36	48.10	68.71
6501102020	Block walls:								
6501102020A	100 mm thick	m²	0.60	7.17	0.46	–	7.63	10.69	15.26
6501102020B	150 mm thick	m²	0.70	8.37	0.54	–	8.91	12.47	17.82
6501102020C	200 mm thick	m²	0.80	9.56	0.62	–	10.18	14.25	20.36
6501102030	Attached chimney breasts and the like:								
6501102030A	half brick thick	m²	0.90	10.76	0.69	–	11.45	16.03	22.91
6501102030B	one brick thick	m²	1.60	19.13	1.23	–	20.36	28.51	40.72
6501102030C	one and a half brick thick	m²	2.30	27.49	1.77	–	29.26	40.97	58.53
6501102030D	two brick thick	m²	2.80	33.47	2.16	–	35.63	49.88	71.25
6501102040	Isolated chimney stacks:								
6501102040A	common brickwork within building	m³	9.60	114.76	3.70	–	118.46	165.84	236.92
6501102040B	common brickwork above roof slope	m³	8.40	100.41	3.24	–	103.65	145.11	207.30
6501102040C	facing brickwork within building	m³	10.50	125.52	3.85	–	129.37	181.11	258.74
6501102040D	facing brickwork above roof slope	m³	8.70	104.00	3.39	–	107.39	150.35	214.78
65011030	Removal of roofs, ceilings and timber floors								
6501103010	Pitched roofs; placing debris in rubbish skips:								
6501103010A	tiles or slates on battens and felt; rafters, hangers and collars	m²	0.37	4.42	–	–	4.42	6.19	8.85
6501103010B	plasterboard ceiling	m²	0.15	1.79	–	–	1.79	2.51	3.59

Property – Refurbishment

Rail 2003 Ref		Unit	Labour Hours	Labour Cost	Plant Cost	Materials Cost	Unit Rate	Unit Rate	Unit Rate
				£	£	£	Base Cost £	Green Zone £	Red Zone £
6501	**DEMOLITIONS AND REMOVALS**								
65011030	Removal of roofs, ceilings and timber floors								
6501103020	Flat roofs; placing debris in rubbish skips:								
6501103020A	built up felt roofing on ply or chipboard decking; joists and insulation; plasterboard ceiling	m^2	0.45	5.38	–	–	5.38	7.53	10.76
6501103030	Timber suspended floors; placing debris in rubbish skips:								
6501103030A	softwood boarding on joists; plasterboard ceiling under	m^2	0.22	2.63	–	–	2.63	3.68	5.26
6501103030B	chipboard flooring on joists; plasterboard ceiling under	m^2	0.25	2.99	–	–	2.99	4.18	5.98
65011040	Removal of internal partitions								
6501104040	Stud partitions, faced both sides; placing debris in rubbish skips:								
6501104040A	lath and plaster finish	m^2	0.65	7.77	0.94	–	8.71	12.19	17.42
6501104040B	plasterboard finish	m^2	0.60	7.17	0.94	–	8.11	11.35	16.22
6501104040C	softwood boarded finish	m^2	0.55	6.58	0.75	–	7.33	10.26	14.65
65011050	Taking down, cleaning and setting aside materials for re-use								
6501105010	Facing brickwork; placing debris in rubbish skips:								
6501105010A	half brick wall	m^2	1.20	14.35	–	–	14.35	20.08	28.69
6501105010B	one brick wall	m^2	2.10	25.10	–	–	25.10	35.14	50.21
6501105010C	one and a half brick wall	m^2	3.15	37.66	–	–	37.66	52.72	75.31
6501105020	Roof coverings; placing debris in rubbish skips:								
6501105020A	roof slating	m^2	0.25	2.99	–	–	2.99	4.18	5.98
6501105020B	roof tiling	m^2	0.20	2.39	–	–	2.39	3.35	4.78

Property – Refurbishment

Rail 2003 Ref		Unit	Labour Hours	Labour Cost	Plant Cost	Materials Cost	Unit Rate	Unit Rate	Unit Rate
				£	£	£	Base Cost £	Green Zone £	Red Zone £
6501	**DEMOLITIONS AND REMOVALS**								
650130	**REMOVAL OF BUILDING FABRIC FIXTURES AND FITTINGS**								
65013010	**Staircase units**								
6501301010	Straight flight timber staircase; placing debris in rubbish skip:								
6501301010A	900 mm wide, 2600 mm rise	Nr	2.20	26.30	4.70	–	31.00	43.39	61.99
6501301010B	1200 mm wide, 2600 mm rise	Nr	2.60	31.08	6.57	–	37.65	52.71	75.31
6501301020	Two flight timber staircase with landing; placing debris in rubbish skip:								
6501301020A	900 mm wide, 2600 mm rise	Nr	3.20	38.25	4.70	–	42.95	60.12	85.90
6501301020B	1200 mm wide, 2600 mm rise	Nr	3.75	44.83	8.45	–	53.28	74.59	106.56
6501301030	Timber balustrading; placing debris in rubbish skips:								
6501301030A	horizontal	Nr	0.25	2.99	0.84	–	3.83	5.36	7.67
6501301030B	raking	Nr	0.35	4.18	0.84	–	5.02	7.04	10.06
65013020	**Kitchen units**								
6501302010	Base units; placing debris in rubbish skips:								
6501302010A	500 × 600 × 900 mm	Nr	0.18	2.15	1.50	–	3.65	5.11	7.30
6501302010B	1000 × 600 × 900 mm	Nr	0.25	2.99	2.54	–	5.53	7.73	11.05
6501302010C	1500 × 600 × 900 mm	Nr	0.25	2.99	3.01	–	6.00	8.39	11.99
6501302020	Wall units; placing debris in rubbish skips:								
6501302020A	500 × 300 × 600 mm	Nr	0.25	2.99	0.47	–	3.46	4.84	6.92
6501302020B	1000 × 300 × 600 mm	Nr	0.30	3.59	1.41	–	5.00	6.99	9.99
6501302020C	500 × 300 × 900 mm	Nr	0.30	3.59	0.75	–	4.34	6.07	8.67
6501302020D	1000 × 300 × 900 mm	Nr	0.35	4.18	2.07	–	6.25	8.75	12.50
6501302030	Storage housing units; placing debris in rubbish skips:								
6501302030A	1200 × 600 × 900 mm	Nr	0.25	2.99	3.01	–	6.00	8.39	11.99
6501302030B	600 × 600 × 1950 mm	Nr	0.30	3.59	3.29	–	6.88	9.62	13.74

Property – Refurbishment

Rail 2003 Ref		Unit	Labour Hours	Labour Cost £	Plant Cost £	Materials Cost £	Unit Rate Base Cost £	Unit Rate Green Zone £	Unit Rate Red Zone £
6501	**DEMOLITIONS AND REMOVALS**								
65013020	Kitchen units								
6501302040	Worktops; placing debris in rubbish skips:								
6501302040A	500 mm wide	Nr	0.25	2.99	0.47	–	3.46	4.84	6.92
6501302040B	600 mm wide	Nr	0.25	2.99	0.56	–	3.55	4.97	7.11
650140	REMOVAL OF WINDOW, DOORS AND FRAMES								
65014010	Timber doors, frames and linings								
6501401010	Doors; placing debris in rubbish skips:								
6501401010A	external	Nr	0.80	9.56	–	–	9.56	13.39	19.13
6501401010B	internal	Nr	0.55	6.58	–	–	6.58	9.21	13.15
6501401020	Door frames; placing debris in rubbish skips:								
6501401020A	900 × 2100 mm	Nr	0.50	5.98	–	–	5.98	8.37	11.95
6501401020B	1200 × 2100 mm	Nr	0.70	8.37	–	–	8.37	11.72	16.74
6501401020C	1500 × 2100 mm	Nr	0.85	10.16	–	–	10.16	14.23	20.32
6501401030	Door linings; placing debris in rubbish skips:								
6501401030A	32 × 138 mm	m	0.13	1.55	–	–	1.55	2.18	3.11
6501401030B	50 × 175 mm	m	0.15	1.79	–	–	1.79	2.51	3.59
65014020	Window frames and linings								
6501402010	Timber window frames; placing debris in rubbish skips:								
6501402010A	not exceeding 0.5 m^2	Nr	0.75	8.97	–	–	8.97	12.55	17.93
6501402010B	0.5 – 1.0 m^2	Nr	1.00	11.95	–	–	11.95	16.74	23.91
6501402010C	1.0 – 2.0 m^2	Nr	1.25	14.94	–	–	14.94	20.92	29.89
6501402010D	2.0 – 3.0 m^2	Nr	1.50	17.93	–	–	17.93	25.10	35.86

Property – Refurbishment

Rail 2003 Ref		Unit	Labour Hours	Labour Cost £	Plant Cost £	Materials Cost £	Unit Rate Base Cost £	Unit Rate Green Zone £	Unit Rate Red Zone £
6501	**DEMOLITIONS AND REMOVALS**								
650145	**REMOVAL OF SUNDRY JOINERY ITEMS**								
65014510	Softwood timber second fixings								
6501451010	Architraves; placing debris in rubbish skips:								
6501451010A	19 × 25 mm	m	0.08	0.96	–	–	0.96	1.34	1.91
6501451010B	25 × 38 mm	m	0.09	1.08	–	–	1.08	1.51	2.15
6501451010C	25 × 50 mm	m	0.09	1.08	–	–	1.08	1.51	2.15
6501451020	Skirtings; placing debris in rubbish skips:								
6501451020A	19 × 100 mm	m	0.11	1.32	–	–	1.32	1.84	2.63
6501451020B	25 × 150 mm	m	0.12	1.43	–	–	1.43	2.01	2.87
6501451020C	25 × 175 mm	m	0.13	1.55	–	–	1.55	2.18	3.11
6501451030	Picture and dado rails; placing debris in rubbish skips:								
6501451030A	19 × 25 mm	m	0.09	1.08	–	–	1.08	1.51	2.15
6501451030B	25 × 38 mm	m	0.10	1.20	–	–	1.20	1.67	2.39
6501451030C	25 × 50 mm	m	0.11	1.32	–	–	1.32	1.84	2.63
650150	**REMOVAL OF PLUMBING AND HEATING INSTALLATIONS**								
65015010	Fixtures and fittings								
6501501010	General heating items; placing debris in rubbish skips:								
6501501010A	back boiler	Nr	0.50	5.98	1.13	–	7.11	9.95	14.20
6501501010B	bolier and flue pipe	Nr	0.68	8.13	1.41	–	9.54	13.35	19.08
6501501010C	radiator	Nr	0.50	5.98	0.47	–	6.45	9.03	12.89
6501501010D	circulating pump	Nr	0.50	5.98	0.47	–	6.45	9.03	12.89
6501501010E	oil tank and supply pipe	Nr	2.50	29.89	20.66	–	50.55	70.76	101.09
6501501010F	drain down heating systems; not exceeding 10 radiators	Nr	1.00	11.95	–	–	11.95	16.74	23.91

Property – Refurbishment

Rail 2003 Ref		Unit	Labour Hours	Labour Cost £	Plant Cost £	Materials Cost £	Unit Rate Base Cost £	Unit Rate Green Zone £	Unit Rate Red Zone £
6501	**DEMOLITIONS AND REMOVALS**								
65015010	**Fixtures and fittings**								
6501501020	General plumbing fittings and pipework placing debris in rubbish skips:								
6501501020A	feed and expansion tank	Nr	0.50	5.98	0.94	–	6.92	9.68	13.83
6501501020B	cold water storage tank	Nr	0.80	9.56	3.38	–	12.94	18.12	25.89
6501501020C	hot water cylinder..................	Nr	1.00	11.95	1.69	–	13.64	19.11	27.29
6501501020D	pipework to fittings; not exceeding 5 m run	Nr	0.50	5.98	1.13	–	7.11	9.95	14.20
6501501020E	pipework to fittings; exceeding 5 m run.	Nr	0.80	9.56	1.03	–	10.59	14.84	21.20
6501501020F	drain down water supply system	Nr	1.00	11.95	–	–	11.95	16.74	23.91
6501501020G	disconnect and make safe mains water supply	Nr	0.70	8.37	–	2.00	10.37	14.52	20.74
65015030	**Sanitary fixtures and fittings**								
6501503030	Sanitaryware, complete with taps and traps; placing debris in rubbish skips:								
6501503030A	W.C. suite	Nr	0.50	5.98	0.94	–	6.92	9.68	13.83
6501503030B	bidet.............................	Nr	0.50	5.98	0.94	–	6.92	9.68	13.83
6501503030C	bath	Nr	0.75	8.97	4.70	–	13.67	19.12	27.32
6501503030D	shower tray	Nr	0.30	3.59	2.35	–	5.94	8.31	11.87
6501503030E	hand basin.......................	Nr	0.50	5.98	0.94	–	6.92	9.68	13.83
6501503030F	sink..............................	Nr	0.50	5.98	1.41	–	7.39	10.34	14.77
650160	**REMOVAL OF ELECTRICAL INSTALLATIONS**								
65016010	**Fixtures and fittings**								
6501601010	General electrical fixtures and fittings placing debris in rubbish skips:								
6501601010A	distribution and switch boards	Nr	0.50	5.98	2.82	–	8.80	12.31	17.58
6501601010B	meters	Nr	0.30	3.59	0.09	–	3.68	5.15	7.36
6501601010C	light fittings	Nr	0.15	1.79	–	–	1.79	2.51	3.59
6501601010D	flush power sockets	Nr	0.20	2.39	–	–	2.39	3.35	4.78
6501601010E	surface mounted power sockets	Nr	0.20	2.39	–	–	2.39	3.35	4.78
6501601010F	immersion heaters	Nr	0.25	2.99	–	–	2.99	4.18	5.98
6501601010G	night storage heaters	Nr	0.45	5.38	1.88	–	7.26	10.16	14.52

Property – Refurbishment

Rail 2003 Ref		Unit	Labour Hours	Labour Cost	Plant Cost	Materials Cost	Unit Rate	Unit Rate	Unit Rate
				£	£	£	Base Cost £	Green Zone £	Red Zone £
6501	**DEMOLITIONS AND REMOVALS**								
65016010	**Fixtures and fittings**								
6501601020	**General disconnection:**								
6501601020A	cap off and make safe buried cables; not exceeding 30 Amp	Nr	1.00	11.95	–	5.00	16.95	23.74	33.91
6501601020B	disconnect and make safe mains supply	Nr	0.75	8.97	–	2.00	10.97	15.35	21.93
650170	**REMOVAL OF BUILDING FABRIC FINISHES**								
65017010	**Removal of floor, wall and ceiling finishes**								
6501701010	**Floor finishes; placing debris in rubbish skips:**								
6501701010A	sand and cement screed	Nr	0.80	9.56	0.75	–	10.31	14.44	20.63
6501701010B	quarry tiles	Nr	0.80	9.56	0.47	–	10.03	14.05	20.07
6501701010C	ceramic tiles	Nr	0.55	6.58	0.28	–	6.86	9.60	13.71
6501701010D	plastic floor tiles	Nr	0.65	7.77	0.28	–	8.05	11.27	16.10
6501701010E	carpeting	Nr	0.10	1.20	0.94	–	2.14	2.98	4.27
6501701010F	asphalt	Nr	0.75	8.97	0.47	–	9.44	13.21	18.87
6501701010G	linoleum	Nr	0.08	0.96	0.19	–	1.15	1.60	2.29
6501701020	**Wall finishes; placing debris in rubbish skips:**								
6501701020A	plaster	Nr	0.60	7.17	0.28	–	7.45	10.43	14.90
6501701020B	render	Nr	0.85	10.16	0.28	–	10.44	14.62	20.88
6501701020C	plasterboard	Nr	0.35	4.18	0.38	–	4.56	6.39	9.12
6501701020D	lath and plaster	Nr	0.45	5.38	0.47	–	5.85	8.19	11.70
6501701020E	asphalt	Nr	0.65	7.77	0.28	–	8.05	11.27	16.10
6501701020F	ceramic tiles	Nr	0.45	5.38	0.28	–	5.66	7.92	11.32
6501701020G	softwood matching	Nr	0.50	5.98	0.28	–	6.26	8.76	12.51
6501701030	**Ceiling finishes; placing debris in rubbish skips:**								
6501701030A	lath and plaster	Nr	0.45	5.38	0.47	–	5.85	8.19	11.70
6501701030B	plasterboard	Nr	0.40	4.78	0.38	–	5.16	7.22	10.31
6501701030C	softwood matching	Nr	0.50	5.98	0.28	–	6.26	8.76	12.51
6501701030D	rigid sheeting	Nr	0.25	2.99	0.28	–	3.27	4.57	6.54

Property – Refurbishment

Rail 2003 Ref		Unit	Composite Cost £	Unit Rate Base Cost £	Unit Rate Green Zone £	Unit Rate Red Zone £
6501	**DEMOLITIONS AND REMOVALS**					
650180	**GENERAL DEMOLITIONS**					
65018010	**Footbridges**					
6501801010	Timber structures including foundations; removing debris from site:					
6501801010A	single track span	m	–	3478.73	4870.22	6957.46
6501801010B	double track span	m	–	3565.96	4992.34	7131.92
6501801010C	four track span	m	–	9044.69	12662.57	18089.38
6501801020	Concrete structures including foundations; removing debris from site:					
6501801020A	single track span	m	–	5218.09	7305.33	10436.18
6501801020B	double track span	m	–	8348.94	11688.52	16697.88
6501801020C	four track span	m	–	13219.16	18506.82	26438.32
6501801030	Steel structures including foundations; removing debris from site:					
6501801030A	single track span	m	–	3130.85	4383.19	6261.70
6501801030B	double track span	m	–	4870.22	6818.31	9740.44
6501801030C	four track span	m	–	8348.94	11688.52	16697.88
65018020	**Staircases**					
6501802010	Straight flight complete with balustrading, including foundations; removing debris form site:					
6501802010A	1200 mm wide × 3000 mm rise	Nr	–	618.44	865.82	1236.88
6501802010B	1500 mm wide × 3000 mm rise	Nr	–	773.05	1082.27	1546.10
65018030	**Ramps**					
6501803010	Concrete construction including foundations; removing debris from site:					
6501803010A	generally	m^2	–	46.38	64.93	92.76

Property – Refurbishment

Rail 2003 Ref		Unit	Composite Cost £	Unit Rate Base Cost £	Unit Rate Green Zone £	Unit Rate Red Zone £
6501	**DEMOLITIONS AND REMOVALS**					
65018040	**Platforms**					
6501804010	Asphalt paving on consolidated sub-base; including foundations; removing debris from site:					
6501804010A	generally	m^2	– – –	51.02	51.02	71.43	102.04
6501804020	Asphalt paving on precast concrete base; including foundations; removing debris from site:					
6501804020A	generally	m^2	– – –	80.40	80.40	112.56	160.80
6501804030	Block paving on consolidated sub-base; including foundations; removing debris from site:					
6501804030A	generally	m^2	– – –	54.11	54.11	75.75	108.22
6501804040	Block paving on precast concrete base; including foundations; removing debris from site:					
6501804040A	generally	m^2	– – –	85.03	85.03	119.04	170.06

Property – Refurbishment

Rail 2003 Ref		Unit	Labour Hours	Labour Cost £	Plant Cost £	Materials Cost £	Unit Rate Base Cost £	Unit Rate Green Zone £	Unit Rate Red Zone £
6502	**EXCAVATION AND EARTHWORKS**								
650210	**GENERAL EXCAVATION**								
65021010	**Excavation by machine**								
6502101010	Oversite excavation to remove topsoil; average depth:								
6502101010A	150 mm	m³	–	–	0.35	–	0.35	0.49	0.70
6502101010B	300 mm	m³	–	–	0.53	–	0.53	0.74	1.05
6502101020	Excavation to reduce levels; depth not exceeding:								
6502101020A	0.25 m	m³	–	–	1.40	–	1.40	1.96	2.80
6502101020B	1.00 m	m³	–	–	1.05	–	1.05	1.47	2.10
6502101020C	2.00 m	m³	–	–	1.58	–	1.58	2.21	3.15
6502101030	Basement excavations; depth not exceeding:								
6502101030A	0.25 m	m³	0.24	2.87	4.20	–	7.07	9.90	14.14
6502101030B	1.00 m	m³	0.20	2.39	3.50	–	5.89	8.25	11.78
6502101030C	2.00 m	m³	0.24	2.87	4.20	–	7.07	9.90	14.14
6502101030D	4.00 m	m³	0.28	3.35	4.90	–	8.25	11.54	16.49
6502101040	Trench excavation to receive foundations, pile caps and ground beams; depth not exceeding:								
6502101040A	0.25 m	m³	0.30	3.59	5.25	–	8.84	12.37	17.67
6502101040B	1.00 m	m³	0.26	3.11	4.55	–	7.66	10.72	15.31
6502101040C	2.00 m	m³	0.30	3.59	5.25	–	8.84	12.37	17.67
6502101050	Pit excavation to receive foundation bases; depth not exceeding:								
6502101050A	0.25 m	m³	0.35	4.18	6.13	–	10.31	14.44	20.62
6502101050B	1.00 m	m³	0.31	3.71	5.43	–	9.14	12.79	18.26
6502101050C	2.00 m	m³	0.35	4.18	6.13	–	10.31	14.44	20.62

Property – Refurbishment

Rail 2003 Ref		Unit	Labour Hours	Labour Cost £	Plant Cost £	Materials Cost £	Unit Rate Base Cost £	Unit Rate Green Zone £	Unit Rate Red Zone £
6502	**EXCAVATION AND EARTHWORKS**								
65021020	**Excavation by hand**								
6502102010	Oversite excavation to remove topsoil; average depth:								
6502102010A	150 mm	m^2	0.18	2.15	–	–	2.15	3.01	4.30
6502102010B	300 mm	m^2	0.32	3.82	–	–	3.82	5.35	7.65
6502102020	Excavation to reduce levels; depth not exceeding:								
6502102020A	0.25 m	m^3	1.50	17.93	–	–	17.93	25.10	35.85
6502102020B	1.00 m	m^3	2.00	23.90	–	–	23.90	33.46	47.80
6502102020C	2.00 m	m^3	2.30	27.49	–	–	27.49	38.48	54.97
6502102030	Basement excavation; depth not exceeding:								
6502102030A	0.25 m	m^3	2.00	23.90	–	–	23.90	33.46	47.80
6502102030B	1.00 m	m^3	2.40	28.68	–	–	28.68	40.15	57.36
6502102030C	2.00 m	m^3	2.65	31.67	–	–	31.67	44.34	63.34
6502102030D	4.00 m	m^3	3.70	44.22	–	–	44.22	61.90	88.43
6502102040	Trench excavation to receive foundations; pile caps and ground beams; depth not exceeding:								
6502102040A	0.25 m	m^3	2.30	27.49	–	–	27.49	38.48	54.97
6502102040B	1.00 m	m^3	2.85	34.06	–	–	34.06	47.68	68.12
6502102040C	2.00 m	m^3	3.30	39.44	–	–	39.44	55.21	78.87
6502102050	Pit excavation to receive foundation bases; depth not exceeding:								
6502102050A	0.25 m	m^3	2.80	33.46	–	–	33.46	46.84	66.92
6502102050B	1.00 m	m^3	3.45	41.23	–	–	41.23	57.72	82.46
6502102050C	2.00 m	m^3	3.80	45.41	–	–	45.41	63.57	90.82

Property – Refurbishment

Rail 2003 Ref		Unit	Labour Hours	Labour Cost £	Plant Cost £	Materials Cost £	Unit Rate Base Cost £	Unit Rate Green Zone £	Unit Rate Red Zone £
6502	**EXCAVATION AND EARTHWORKS**								
65021030	Breaking out obstructions with hand held mechanical tools								
6502103010	Extra over excavation for breaking out:								
6502103010A	soft rock or brickwork	m^3	3.60	43.02	25.14	–	68.16	95.43	136.32
6502103010B	hard rock	m^3	6.40	76.48	44.69	–	121.17	169.64	242.35
6502103010C	plain concrete	m^3	5.00	59.75	34.92	–	94.67	132.54	189.34
6502103010D	reinforced concrete	m^3	7.20	86.04	50.28	–	136.32	190.85	272.64
65021040	Breaking out pavings with hand held compressor tools								
6502104010	Excavation in tarmacadam paving:								
6502104010A	not exceeding 150 mm thick	m^2	0.40	4.78	1.40	–	6.18	8.65	12.35
6502104010B	150 – 300 mm thick	m^2	0.70	8.37	2.79	–	11.16	15.62	22.32
6502104010C	300 – 450 mm thick	m^2	1.00	11.95	4.19	–	16.14	22.60	32.28
6502104020	Excavation in plain concrete paving:								
6502104020A	not exceeding 150 mm thick	m^2	0.60	7.17	2.79	–	9.96	13.95	19.93
6502104020B	150 – 300 mm thick	m^2	0.90	10.76	4.19	–	14.95	20.93	29.89
6502104020C	300 – 450 mm thick	m^2	1.20	14.34	5.59	–	19.93	27.90	39.85
6502104030	Excavation in reinforced concrete paving:								
6502104030A	not exceeding 150 mm thick	m^2	0.80	9.56	4.48	–	14.04	19.66	28.09
6502104030B	150 – 300 mm thick	m^2	1.30	15.54	7.28	–	22.82	31.94	45.63
6502104030C	300 – 450 mm thick	m^2	2.00	23.90	11.47	–	35.37	49.52	70.74

Property – Refurbishment

Rail 2003 Ref		Unit	Labour Hours	Labour Cost £	Plant Cost £	Materials Cost £	Unit Rate Base Cost £	Unit Rate Green Zone £	Unit Rate Red Zone £
6502	**EXCAVATION AND EARTHWORKS**								
650220	**SERVICE TRENCHES**								
65022010	Excavation of trenches by machine for service pipes, cables and the like, including disposal and filling								
6502201010	Trenches to suit pipes not exceeding 200 mm diameter, average depth:								
6502201010A	0.50 m	m	0.12	1.43	1.45	–	2.88	4.04	5.77
6502201010B	0.75 m	m	0.15	1.79	1.98	–	3.77	5.28	7.54
6502201010C	1.00 m	m	0.22	2.63	2.53	–	5.16	7.23	10.33
6502201020	Trenches to suit pipes exceeding 200 mm diameter, average depth:								
6502201020A	0.50 m	m	0.18	2.15	2.17	–	4.32	6.04	8.63
6502201020B	0.75 m	m	0.23	2.75	3.04	–	5.79	8.11	11.58
6502201020C	1.00 m	m	0.36	4.30	4.48	–	8.78	12.30	17.58
65022020	Excavation of trenches by hand for service pipes, cables and the like, including disposal and filling								
6502202010	Trenches to suit pipes not exceeding 200 mm diameter, average depth:								
6502202010A	0.50 m	m	1.32	15.77	0.05	–	15.82	22.15	31.65
6502202010B	0.75 m	m	1.78	21.27	0.05	–	21.32	29.85	42.64
6502202010C	1.00 m	m	2.23	26.65	0.08	–	26.73	37.43	53.47
6502202020	Trenches to suit pipes exceeding 200 mm diameter, average depth:								
6502202020A	0.50 m	m	2.20	26.29	0.07	–	26.36	36.90	52.71
6502202020B	0.75 m	m	2.96	35.37	0.07	–	35.44	49.61	70.87
6502202020C	1.00 m	m	3.72	44.45	0.11	–	44.56	62.39	89.13

Property – Refurbishment

Rail 2003 Ref		Unit	Labour Hours	Labour Cost £	Plant Cost £	Materials Cost £	Unit Rate Base Cost £	Unit Rate Green Zone £	Unit Rate Red Zone £
6502	**EXCAVATION AND EARTHWORKS**								
65022030	Beds for service pipes, ducts, cables and the like								
6502203010	50 mm sand bed to pipes:								
6502203010A	not exceeding 200 mm diameter	m	0.06	0.72	0.07	0.25	1.04	1.45	2.08
6502203010B	exceeding 200 mm diameter	m	0.08	0.96	0.15	0.41	1.52	2.11	3.02
65022040	Beds and coverings for service pipes, ducts, cables and the like								
6502204010	150 mm bed and 150 mm covering of pea shingle to pipes:								
6502204010A	not exceeding 200 mm diameter	m	0.21	2.51	1.02	2.62	6.15	8.61	12.29
6502204020	150 mm bed and 150 mm covering of C10P concrete to pipes:								
6502204020A	not exceeding 200 mm diameter	m	0.39	4.66	2.03	13.84	20.53	28.76	41.08
650230	**EARTHWORK SUPPORT**								
65023010	Timber earthwork support to firm ground								
6502301010	To sides of excavation not exceeding 2.0 m apart, depth not exceeding:								
6502301010A	1.00 m	m^2	1.00	11.95	–	0.95	12.90	18.07	25.81
6502301010B	2.00 m	m^2	1.20	14.35	–	1.27	15.62	21.85	31.22
6502301010C	4.00 m	m^2	1.60	19.13	–	1.90	21.03	29.44	42.05
6502301020	To sides of excavation 2.0 – 4.00 m apart, depth not exceeding:								
6502301020A	1.00 m	m^2	1.40	16.74	–	1.58	18.32	25.64	36.63
6502301020B	2.00 m	m^2	1.60	19.13	–	1.90	21.03	29.44	42.05
6502301020C	4.00 m	m^2	2.00	23.91	–	2.53	26.44	37.01	52.88

Property – Refurbishment

Rail 2003 Ref		Unit	Labour Hours	Labour Cost £	Plant Cost £	Materials Cost £	Unit Rate Base Cost £	Unit Rate Green Zone £	Unit Rate Red Zone £
6502	**EXCAVATION AND EARTHWORKS**								
65023020	Timber earthwork support to loose ground								
6502302010	To sides of excavation not exceeding 2.0 m apart, depth not exceeding:								
6502302010A	1.00 m	m²	1.32	15.78	–	3.82	19.60	27.43	39.19
6502302010B	2.00 m	m²	1.56	18.65	–	4.13	22.78	31.89	45.56
6502302010C	4.00 m	m²	1.92	22.95	–	4.76	27.71	38.80	55.43
6502302020	To sides of excavation 2.0 – 4.00 m apart, depth not exceeding:								
6502302020A	1.00 m	m²	1.84	22.00	–	4.44	26.44	37.01	52.88
6502302020B	2.00 m	m²	2.04	24.39	–	4.76	29.15	40.81	58.30
6502302020C	4.00 m	m²	2.32	27.73	–	5.40	33.13	46.38	66.26
65023030	Steel trench sheeting to firm ground								
6502303010	To sides of excavation not exceeding 2.0 m apart, depth not exceeding:								
6502303010A	1.00 m	m²	0.60	7.17	4.04	0.63	11.84	16.57	23.67
6502303010B	2.00 m	m²	0.60	7.17	4.54	0.63	12.34	17.28	24.67
6502303010C	4.00 m	m²	0.60	7.17	4.98	0.63	12.78	17.89	25.54
6502303020	To sides of excavation 2.0 – 4.00 apart, depth not exceeding:								
6502303020A	1.00 m	m²	1.00	11.95	4.04	1.25	17.24	24.15	34.50
6502303020B	2.00 m	m²	1.00	11.95	4.82	1.25	18.02	25.25	36.06
6502303020C	4.00 m	m²	1.20	14.35	5.03	1.25	20.63	28.88	41.26
65023040	Steel trench sheeting to loose ground								
6502304010	To sides of excavation not exceeding 2.0 m apart, depth not exceeding:								
6502304010A	1.00 m	m²	0.60	7.17	10.67	0.63	18.47	25.86	36.93
6502304010B	2.00 m	m²	0.60	7.17	15.04	0.63	22.84	31.98	45.68
6502304010C	4.00 m	m²	0.60	7.17	24.31	0.63	32.11	44.95	64.21

Property – Refurbishment

Rail 2003 Ref		Unit	Labour Hours	Labour Cost £	Plant Cost £	Materials Cost £	Unit Rate Base Cost £	Unit Rate Green Zone £	Unit Rate Red Zone £
6502	**EXCAVATION AND EARTHWORKS**								
65023040	Steel trench sheeting to loose ground								
6502304020	To sides of excavation 2.0 – 4.00 m apart, depth not exceeding:								
6502304020A	1.00 m	m²	1.00	11.95	10.67	1.25	23.87	33.44	47.76
6502304020B	2.00 m	m²	1.00	11.95	15.04	1.25	28.24	39.56	56.51
6502304020C	4.00 m	m²	1.20	14.35	24.31	1.25	39.91	55.87	79.82
650250	**DISPOSAL AND FILLING**								
65025010	Off site disposal of material arising from earthworks								
6502501010	Removed, including providing a suitable tip:								
6502501010A	hand loading	m³	1.37	16.37	7.05	–	23.42	32.79	46.84
6502501010B	machine loading	m³	–	–	8.28	–	8.28	11.59	16.55
65025020	On site disposal of material arising from earthworks								
6502502010	Backfilled into excavation, compacting in 250 mm layers:								
6502502010A	by hand	m³	1.30	15.54	0.25	–	15.79	22.10	31.57
6502502010B	by machine	m³	0.30	3.59	4.63	–	8.22	11.50	16.42
6502502020	Backfilled in making up levels, compacting in 250 mm layers; wheeling average:								
6502502020A	25 m	m³	1.66	19.84	0.45	–	20.29	28.41	40.58
6502502020B	50 m	m³	2.00	23.90	0.45	–	24.35	34.10	48.71
6502502020C	75 m	m³	2.33	27.84	0.45	–	28.29	39.62	56.60
6502502020D	100 m	m³	2.66	31.79	0.45	–	32.24	45.14	64.48
6502502030	Backfilled in making up levels, compacting in 250 mm layers; transporting average:								
6502502030A	25 m	m³	0.27	3.23	2.87	–	6.10	8.54	12.19
6502502030B	50 m	m³	0.33	3.94	3.31	–	7.25	10.15	14.50
6502502030C	75 m	m³	0.40	4.78	3.81	–	8.59	12.03	17.19
6502502030D	100 m	m³	0.47	5.62	4.32	–	9.94	13.91	19.87

Property – Refurbishment

Rail 2003 Ref		Unit	Labour Hours	Labour Cost £	Plant Cost £	Materials Cost £	Unit Rate Base Cost £	Unit Rate Green Zone £	Unit Rate Red Zone £
6502	**EXCAVATION AND EARTHWORKS**								
65025020	On site disposal of material arising from earthworks								
6502502040	Backfilled oversite, compacting in 250 mm layers; wheeling average:								
6502502040A	25 m	m³	1.86	22.23	0.59	–	22.82	31.95	45.64
6502502040B	50 m	m³	2.20	26.29	0.59	–	26.88	37.64	53.77
6502502040C	75 m	m³	2.53	30.23	0.59	–	30.82	43.16	61.66
6502502040D	100 m	m³	2.86	34.18	0.59	–	34.77	48.68	69.54
6502502050	Backfilled oversite, compacting in 250 mm layers; transporting average:								
6502502050A	25 m	m³	0.35	4.18	4.18	–	8.36	11.71	16.72
6502502050B	50 m	m³	0.41	4.90	4.61	–	9.51	13.32	19.02
6502502050C	75 m	m³	0.49	5.86	5.19	–	11.05	15.47	22.09
6502502050D	100 m	m³	0.55	6.57	5.63	–	12.20	17.08	24.41
65025050	Filling to excavations with imported materials								
6502505010	Filled into excavation; by machine; compacting in layers:								
6502505010A	sand	m³	0.30	3.59	3.75	10.25	17.59	24.61	35.16
6502505010B	hardcore	m³	0.35	4.18	6.42	15.13	25.73	36.02	51.46
6502505010C	hoggin	m³	0.30	3.59	5.50	11.47	20.56	28.78	41.12
6502505010D	MOT type 1	m³	0.33	3.94	6.05	13.62	23.61	33.05	47.22
6502505010E	MOT type 2	m³	0.33	3.94	6.05	13.20	23.19	32.47	46.39
6502505020	Filled into excavation; by hand; compacting in layers:								
6502505020A	sand	m³	0.80	9.56	0.25	10.25	20.06	28.07	40.11
6502505020B	hardcore	m³	1.65	19.72	0.29	15.13	35.14	49.20	70.28
6502505020C	hoggin	m³	0.90	10.76	0.25	11.47	22.48	31.47	44.96
6502505020D	MOT type 1	m³	1.63	19.48	0.28	13.62	33.38	46.72	66.74
6502505020E	MOT type 2	m³	1.63	19.48	0.28	13.20	32.96	46.14	65.91

Property – Refurbishment

Rail 2003 Ref		Unit	Labour Hours	Labour Cost £	Plant Cost £	Materials Cost £	Unit Rate Base Cost £	Unit Rate Green Zone £	Unit Rate Red Zone £
6502	**EXCAVATION AND EARTHWORKS**								
65025050	**Filling to excavations with imported materials**								
6502505030	Filled in making up levels; by machine; compacting in layers:								
6502505030A	sand	m^3	0.30	3.59	3.75	10.25	17.59	24.61	35.16
6502505030B	hardcore	m^3	0.35	4.18	6.42	15.13	25.73	36.02	51.46
6502505030C	hoggin	m^3	0.30	3.59	5.50	11.47	20.56	28.78	41.12
6502505030D	MOT type 1	m^3	0.33	3.94	6.40	13.62	23.96	33.54	47.92
6502505030E	MOT type 2	m^3	0.33	3.94	6.40	13.20	23.54	32.96	47.09
6502505040	Filled in making up levels; by hand; compacting in layers:								
6502505040A	sand	m^3	0.83	9.92	0.73	10.25	20.90	29.25	41.78
6502505040B	hardcore	m^3	1.46	17.45	0.73	15.13	33.31	46.63	66.60
6502505040C	hoggin	m^3	0.93	11.11	0.73	11.47	23.31	32.64	46.63
6502505040D	MOT type 1	m^3	1.40	16.73	0.73	13.62	31.08	43.50	62.14
6502505040E	MOT type 2	m^3	1.40	16.73	0.73	13.20	30.66	42.92	61.31
6502505050	Filled into oversite; by machine; compacting in layers:								
6502505050A	sand	m^3	0.17	2.03	2.70	10.25	14.98	20.96	29.95
6502505050B	hardcore	m^3	0.17	2.03	3.58	15.13	20.74	29.03	41.47
6502505050C	hoggin	m^3	0.17	2.03	3.23	11.47	16.73	23.42	33.46
6502505050D	MOT type 1	m^3	0.17	2.03	3.23	13.62	18.88	26.42	37.74
6502505050E	MOT type 2	m^3	0.17	2.03	11.78	13.20	27.01	37.81	54.02
6502505060	Filled into oversite; by hand; compacting in layers:								
6502505060A	sand	m^3	1.03	12.31	0.95	10.25	23.51	32.90	47.01
6502505060B	hardcore	m^3	1.66	19.84	0.95	15.13	35.92	50.28	71.83
6502505060C	hoggin	m^3	1.13	13.50	0.95	11.47	25.92	36.30	51.86
6502505060D	MOT type 1	m^3	1.60	19.12	0.95	13.62	33.69	47.16	67.37
6502505060E	MOT type 2	m^3	1.60	19.12	0.95	13.20	33.27	46.58	66.54
6502505070	Blinding to hardcore:								
6502505070A	25 mm sand	m^2	0.04	0.48	–	0.25	0.73	1.01	1.45
6502505070B	50 mm sand	m^2	0.06	0.72	–	0.49	1.21	1.69	2.41

Property – Refurbishment

Rail 2003 Ref		Unit	Labour Hours	Labour Cost £	Plant Cost £	Materials Cost £	Unit Rate Base Cost £	Unit Rate Green Zone £	Unit Rate Red Zone £

6503   **CONCRETE WORK**

650310   **DEMOLITIONS AND ALTERATIONS**

65031010   Demolition of concrete structural elements including removal from site

6503101001   Reinforced concrete walls and attached columns:

Ref	Description	Unit	Hrs	Lab	Plant	Mat	Base	Green	Red
6503101001A	not exceeding 150 mm thick	m³	12.00	143.45	83.80	–	227.25	318.15	454.50
6503101001B	150 – 450 mm thick	m³	14.00	167.36	97.77	–	265.13	371.18	530.25
6503101001C	exceeding 450 mm	m³	18.00	215.17	125.70	–	340.87	477.22	681.75

6503101011   Reinforced concrete ground slabs:

6503101011A	not exceeding 150 mm thick	m³	6.00	71.72	41.90	–	113.62	159.07	227.25
6503101011B	150 – 450 mm thick	m³	8.00	95.63	55.87	–	151.50	212.10	303.00
6503101011C	exceeding 450 mm thick	m³	10.00	119.54	69.84	–	189.38	265.13	378.75

6503101021   Reinforced concrete suspended slabs and attached beams:

6503101021A	not exceeding 150 mm thick	m³	10.00	119.54	69.84	–	189.38	265.13	378.75
6503101021B	150 – 450 mm thick	m³	16.00	191.26	111.74	–	303.00	424.20	606.00
6503101021C	exceeding 450 mm thick	m³	19.00	227.13	132.69	–	359.82	503.74	719.62

6503101031   Reinforced concrete isolated beams:

6503101031A	not exceeding 0.1 m² sectional area 150 mm thick	m³	12.00	143.45	83.80	–	227.25	318.15	454.50
6503101031B	0.1 – 0.25 m² sectional area	m³	14.00	167.36	97.77	–	265.13	371.18	530.25
6503101031C	exceeding 0.25 m² sectional area	m³	19.00	227.13	132.69	–	359.82	503.74	719.62

Property – Refurbishment

Rail 2003 Ref		Unit	Labour Hours	Labour Cost £	Plant Cost £	Materials Cost £	Unit Rate Base Cost £	Unit Rate Green Zone £	Unit Rate Red Zone £
6503	**CONCRETE WORK**								
65031010	Demolition of concrete structural elements including removal from site								
6503101041	Reinforced concrete isolated columns:								
6503101041A	not exceeding 0.1 m² sectional area 150 mm thick	m³	10.00	119.54	69.84	–	189.38	265.13	378.75
6503101041B	0.1 – 0.25 m² sectional area	m³	13.00	155.40	90.79	–	246.19	344.66	492.37
6503101041C	exceeding 0.25 m² sectional area	m³	18.00	215.17	125.70	–	340.87	477.22	681.75
650320	**IN SITU CONCRETE**								
65032010	Plain in situ concrete; mix C10P								
6503201010	Foundations; poured on or against earth or unblinded hardcore:								
6503201010A	generally	m³	0.90	10.76	–	43.51	54.27	75.97	108.53
6503201020	Ground beams; poured on or against earth or unblinded hardcore:								
6503201020A	generally	m³	1.90	22.71	–	42.32	65.03	91.05	130.07
6503201030	Isolated foundations; poured on or against earth or unblinded hardcore:								
6503201030A	generally	m³	1.90	22.71	–	42.32	65.03	91.05	130.07
6503201040	Beds; poured on or against earth or unblinded hardcore:								
6503201040A	not exceeding 150 mm thick	m³	1.05	12.55	–	41.53	54.08	75.71	108.16
6503201040B	150 – 450 mm thick	m³	0.70	8.37	–	41.53	49.90	69.86	99.80
6503201040C	exceeding 450 mm thick	m³	0.50	5.98	–	41.53	47.51	66.51	95.01
6503201050	Filling hollow walls:								
6503201050A	not exceeding 150 mm thick	m³	2.70	32.28	–	41.53	73.81	103.33	147.61

Property – Refurbishment

Rail 2003 Ref		Unit	Labour Hours	Labour Cost £	Plant Cost £	Materials Cost £	Unit Rate Base Cost £	Unit Rate Green Zone £	Unit Rate Red Zone £

6503 **CONCRETE WORK**

65032020 Plain in situ concrete; mix C15P

6503202010 Foundations; poured on or against earth or unblinded hardcore:

| 6503202010A | generally | m³ | 0.90 | 10.76 | – | 49.10 | 59.86 | 83.80 | 119.72 |

6503202020 Ground beams; poured on or against earth or unblinded hardcore:

| 6503202020A | generally | m³ | 1.90 | 22.71 | – | 47.76 | 70.47 | 98.67 | 140.96 |

6503202030 Isolated foundations; poured on or against earth or unblinded hardcore:

| 6503202030A | generally | m³ | 1.05 | 12.55 | – | 46.87 | 59.42 | 83.19 | 118.84 |

6503202040 Beds; poured on or against earth or unblinded hardcore:

6503202040A	not exceeding 150 mm thick	m³	1.05	12.55	–	46.87	59.42	83.19	118.84
6503202040B	150 – 450 mm thick	m³	0.70	8.37	–	46.87	55.24	77.34	110.48
6503202040C	exceeding 450 mm thick	m³	0.50	5.98	–	46.87	52.85	73.99	105.69

65032050 Reinforced in situ concrete; mix C20P

6503205010 Foundations:

| 6503205010A | generally | m³ | 1.10 | 13.30 | 2.24 | 52.14 | 67.68 | 94.74 | 135.36 |

6503205020 Ground beams:

| 6503205020A | generally | m³ | 2.15 | 25.99 | 4.34 | 51.65 | 81.98 | 114.77 | 163.96 |

6503205030 Isolated foundations:

| 6503205030A | generally | m³ | 2.15 | 25.99 | 4.34 | 51.65 | 81.98 | 114.77 | 163.96 |

6503205040 Beds:

6503205040A	not exceeding 150 mm thick	m³	1.40	16.93	2.79	51.16	70.88	99.23	141.75
6503205040B	150 – 450 mm thick	m³	1.00	12.09	1.95	51.16	65.20	91.29	130.41
6503205040C	exceeding 450 mm thick	m³	0.70	8.46	1.39	51.16	61.01	85.43	122.05

Property – Refurbishment

Rail 2003 Ref		Unit	Labour Hours	Labour Cost £	Plant Cost £	Materials Cost £	Unit Rate Base Cost £	Unit Rate Green Zone £	Unit Rate Red Zone £

6503 CONCRETE WORK

65032050 Reinforced in situ concrete; mix C20P

6503205050 Slabs:

6503205050A	not exceeding 150 mm thick	m³	2.10	25.39	4.18	51.16	80.73	113.02	161.47
6503205050B	150 – 450 mm thick	m³	1.55	18.74	3.07	51.16	72.97	102.16	145.94
6503205050C	exceeding 450 mm thick	m³	1.05	12.70	2.09	51.16	65.95	92.33	131.90

6503205060 Walls:

6503205060A	not exceeding 150 mm thick	m³	2.31	27.93	4.62	51.16	83.71	117.19	167.42
6503205060B	150 – 450 mm thick	m³	2.05	24.79	4.06	51.16	80.01	112.01	160.01
6503205060C	exceeding 450 mm thick	m³	1.40	16.93	2.80	51.16	70.89	99.25	141.78

6503205070 Beams:

6503205070A	generally	m³	2.80	33.85	5.60	51.65	91.10	127.53	182.19

6503205080 Columns:

6503205080A	generally	m³	4.00	48.36	8.04	51.65	108.05	151.27	216.11

65032055 Reinforced in situ concrete; mix C30P

6503205510 Foundations:

6503205510A	generally	m³	1.10	13.30	2.24	58.66	74.20	103.87	148.39

6503205520 Ground beams:

6503205520A	generally	m³	2.15	25.99	4.34	58.11	88.44	123.81	176.88

6503205530 Isolated foundations:

6503205530A	generally	m³	2.15	25.99	4.34	58.11	88.44	123.81	176.88

6503205540 Beds:

6503205540A	not exceeding 150 mm thick	m³	1.40	16.93	2.79	57.56	77.28	108.19	154.54
6503205540B	150 – 450 mm thick	m³	1.00	12.09	1.95	57.56	71.60	100.25	143.20
6503205540C	exceeding 450 mm thick	m³	0.70	8.46	1.39	57.56	67.41	94.39	134.84

Property – Refurbishment

Rail 2003 Ref		Unit	Labour Hours	Labour Cost £	Plant Cost £	Materials Cost £	Unit Rate Base Cost £	Unit Rate Green Zone £	Unit Rate Red Zone £
6503	**CONCRETE WORK**								
65032055	**Reinforced in situ concrete; mix C30P**								
6503205550	**Slabs:**								
6503205550A	not exceeding 150 mm thick	m³	2.10	25.39	4.18	57.56	87.13	121.98	174.26
6503205550B	150 – 450 mm thick	m³	1.55	18.74	3.07	57.56	79.37	111.12	158.73
6503205550C	exceeding 450 mm thick	m³	1.05	12.70	2.09	57.56	72.35	101.29	144.69
6503205560	**Walls:**								
6503205560A	not exceeding 150 mm thick	m³	2.31	27.93	4.62	57.56	90.11	126.15	180.21
6503205560B	150 – 450 mm thick	m³	2.05	24.79	4.06	57.56	86.41	120.97	172.80
6503205560C	exceeding 450 mm thick	m³	1.40	16.93	2.80	57.56	77.29	108.21	154.57
6503205570	**Beams:**								
6503205570A	generally	m³	2.80	33.85	5.60	58.11	97.56	136.57	195.11
6503205580	**Columns:**								
6503205580A	generally	m³	4.00	48.36	8.04	58.11	114.51	160.31	229.03
65032060	**Formwork to general finish**								
6503206010	**Sides of foundations:**								
6503206010A	not exceeding 250 mm high	m	0.76	10.43	–	2.23	12.66	17.72	25.32
6503206010B	250 – 500 mm high	m	1.15	15.78	–	3.71	19.49	27.30	38.99
6503206010C	500 – 1000 mm high	m	1.15	15.78	–	3.71	19.49	27.30	38.99
6503206010D	exceeding 1000 mm high	m²	1.86	25.53	–	7.01	32.54	45.55	65.06
6503206020	**Sides of ground beams and edges of beds:**								
6503206020A	not exceeding 250 mm high	m	0.76	10.43	–	2.23	12.66	17.72	25.32
6503206020B	250 – 500 mm high	m	1.15	15.78	–	3.71	19.49	27.30	38.99
6503206020C	500 – 1000 mm high	m	2.04	28.00	–	6.66	34.66	48.53	69.32
6503206020D	exceeding 1000 mm high	m²	1.86	25.53	–	28.25	53.78	75.29	107.55
6503206030	**Edges of suspended slabs:**								
6503206030A	not exceeding 250 mm high	m	0.71	9.74	–	1.98	11.72	16.41	23.45
6503206030B	250 – 500 mm high	m	0.90	12.35	–	3.24	15.59	21.82	31.18
6503206030C	500 – 1000 mm high	m	1.60	21.96	2.89	5.61	30.46	42.64	60.93
6503206030D	exceeding 1000 mm high	m²	1.46	20.04	5.78	6.45	32.27	45.17	64.53

Property – Refurbishment

Rail 2003 Ref		Unit	Labour Hours	Labour Cost £	Plant Cost £	Materials Cost £	Unit Rate Base Cost £	Unit Rate Green Zone £	Unit Rate Red Zone £
6503	**CONCRETE WORK**								
65032060	**Formwork to general finish**								
6503206040	Soffits; not exceeding 1.5 m above floor level:								
6503206040A	not exceeding 200 mm thick	m²	1.46	20.04	2.41	6.86	29.31	41.02	58.60
6503206040B	200 – 300 mm thick	m²	2.04	28.00	2.41	6.86	37.27	52.17	74.52
6503206050	Soffits; exceeding 1.5 m above floor level:								
6503206050A	not exceeding 200 mm thick	m²	1.79	24.57	2.41	6.86	33.84	47.36	67.66
6503206050B	200 – 300 mm thick	m²	2.50	34.31	2.41	18.61	55.33	77.45	110.65
6503206060	Walls:								
6503206060A	vertical surfaces	m²	1.78	24.43	0.94	7.51	32.88	46.04	65.77
6503206060B	curved surfaces	m²	2.00	27.45	0.94	6.57	34.96	48.95	69.93

Property – Refurbishment

Rail 2003 Ref		Unit	Labour Hours	Labour Cost £	Plant Cost £	Materials Cost £	Unit Rate Base Cost £	Unit Rate Green Zone £	Unit Rate Red Zone £
6504	**UNDERPINNING**								
650410	**EXCAVATION**								
65041010	Excavation by hand, commencing at ground level								
6504101010	Preliminary trenches; to level of existing foundations; depth:								
6504101010A	not exceeding 0.25 m	m³	4.50	53.78	–	–	53.78	75.29	107.55
6504101010B	0.25 – 1.00 m	m³	4.20	50.19	–	–	50.19	70.27	100.38
6504101010C	1.00 – 2.00 m	m³	4.80	57.36	–	–	57.36	80.30	114.72
6504101010D	exceeding 2.00 m	m³	5.60	66.92	–	–	66.92	93.69	133.84
65041020	Excavation by hand, below level of existing foundation base								
6504102010	Trenches:								
6504102010A	generally	m³	5.60	66.92	–	–	66.92	93.69	133.84
6504102020	Extra over excavations for:								
6504102020A	working below normal ground water level	m³	6.90	82.46	–	–	82.46	115.44	164.91
65041030	Cutting away projecting foundations								
6504103010	Brick:								
6504103010A	half brick thick, three courses	m	0.40	4.78	0.31	–	5.09	7.12	10.18
6504103010T	two brick thick; five courses	m	0.60	7.17	0.46	–	7.63	10.69	15.26
6504103020	Concrete:								
6504103020A	150 – 300 mm thick; 150 – 300 mm wide	m	0.55	6.57	0.43	–	7.00	9.80	14.01
6504103020I	300 – 450 mm thick; 300 – 450 mm wide	m	0.95	11.35	0.74	–	12.09	16.93	24.19

Property – Refurbishment

Rail 2003 Ref		Unit	Labour Hours	Labour Cost £	Plant Cost £	Materials Cost £	Unit Rate Base Cost £	Unit Rate Green Zone £	Unit Rate Red Zone £
6504	**UNDERPINNING**								
65041035	**Preparing underside of existing work; wedging and grouting up**								
6504103510	Slates in cement mortar (1:3):								
6504103510A	not exceeding 150 mm wide	m	0.75	8.96	–	6.49	15.45	21.63	30.90
6504103510B	150 – 300 mm wide	m	1.25	14.94	–	9.39	24.33	34.06	48.66
6504103510C	300 – 450 mm wide	m	1.50	17.93	–	12.30	30.23	42.32	60.45
6504103520	Neat cement slurry:								
6504103520A	not exceeding 150 mm wide	m	0.70	9.07	–	1.31	10.38	14.54	20.77
6504103520B	150 – 300 mm wide	m	0.75	9.67	–	1.31	10.98	15.38	21.96
6504103520C	300 – 450 mm wide	m	0.90	11.46	–	1.31	12.77	17.89	25.55
6504103530	Epoxy cement non-shrinking grout:								
6504103530A	not exceeding 150 mm wide	m	0.70	9.07	–	0.94	10.01	14.01	20.03
6504103530B	150 – 300 mm wide	m	0.75	9.67	–	1.52	11.19	15.67	22.39
6504103530C	300 – 450 mm wide	m	0.90	11.46	–	2.13	13.59	19.03	27.18
65041040	**Breaking up pavings and obstructions with hand held mechanical tools**								
6504104010	Extra over excavation for breaking out and removing from site:								
6504104010A	soft rock or brickwork	m^3	3.60	43.02	25.14	–	68.16	95.43	136.32
6504104010B	hard rock	m^3	6.40	76.48	44.69	–	121.17	169.64	242.35
6504104010C	plain concrete	m^3	6.40	76.48	44.69	–	121.17	169.64	242.35
6504104010D	reinforced concrete	m^3	9.60	114.72	67.04	–	181.76	254.47	363.52
6504104020	Excavation in tarmacadam paving and removing from site:								
6504104020A	not exceeding 150 mm thick	m^2	0.40	4.78	1.40	–	6.18	8.65	12.35
6504104020B	150 – 300 mm thick	m^2	0.70	8.37	2.79	–	11.16	15.62	22.32
6504104020C	exceeding 300 mm thick	m^2	1.00	11.95	4.19	–	16.14	22.60	32.28

Property – Refurbishment

Rail 2003 Ref		Unit	Labour Hours	Labour Cost £	Plant Cost £	Materials Cost £	Unit Rate Base Cost £	Unit Rate Green Zone £	Unit Rate Red Zone £
6504	**UNDERPINNING**								
65041040	Breaking up pavings and obstructions with hand held mechanical tools								
6504104030	Excavation in plain concrete paving and removing from site:								
6504104030A	not exceeding 150 mm thick	m²	0.60	7.17	2.79	–	9.96	13.95	19.93
6504104030B	150 – 300 mm thick	m²	0.90	10.76	4.19	–	14.95	20.93	29.89
6504104030C	exceeding 300 mm thick	m²	1.20	14.34	5.59	–	19.93	27.90	39.85
6504104040	Excavation in reinforced concrete paving and removing from site:								
6504104040A	not exceeding 150 mm thick	m²	0.80	9.56	4.48	–	14.04	19.66	28.09
6504104040B	150 – 300 mm thick	m²	1.30	15.54	7.28	–	22.82	31.94	45.63
6504104040C	exceeding 300 mm thick	m²	2.00	23.90	11.47	–	35.37	49.52	70.74
65041050	Timber earthwork support to firm ground								
6504105010	To sides of excavation not exceeding 2.00 m apart; depth not exceeding:								
6504105010A	1.00 m	m²	1.00	11.95	–	0.95	12.90	18.07	25.81
6504105010B	2.00 m	m²	1.20	14.35	–	1.27	15.62	21.85	31.22
6504105010C	4.00 m	m²	1.60	19.13	–	1.90	21.03	29.44	42.05
6504105010D	6.00 m	m²	2.40	28.69	–	2.53	31.22	43.71	62.44
6504105020	To sides of excavation 2.00 – 4.00 m apart; depth not exceeding:								
6504105020A	1.00 m	m²	1.40	16.74	–	1.58	18.32	25.64	36.63
6504105020B	2.00 m	m²	1.60	19.13	–	1.90	21.03	29.44	42.05
6504105020C	4.00 m	m²	2.00	23.91	–	2.53	26.44	37.01	52.88
6504105020D	6.00 m	m²	2.80	33.47	–	3.17	36.64	51.30	73.28

Property – Refurbishment

Rail 2003 Ref		Unit	Labour Hours	Labour Cost £	Plant Cost £	Materials Cost £	Unit Rate Base Cost £	Unit Rate Green Zone £	Unit Rate Red Zone £
6504	**UNDERPINNING**								
65041060	Timber earthwork support to loose ground								
6504106010	To sides of excavation not exceeding 2.00 m apart; depth not exceeding:								
6504106010A	1.00 m	m²	1.32	15.78	–	3.82	19.60	27.43	39.19
6504106010B	2.00 m	m²	1.56	18.65	–	4.13	22.78	31.89	45.56
6504106010C	4.00 m	m²	1.92	22.95	–	4.76	27.71	38.80	55.43
6504106010D	6.00 m	m²	2.76	32.99	–	5.40	38.39	53.74	76.78
6504106020	To sides of excavation 2.00 – 4.00 m apart; depth not exceeding:								
6504106020A	1.00 m	m²	2.04	24.39	–	4.44	28.83	40.36	57.66
6504106020B	2.00 m	m²	1.84	22.00	–	4.76	26.76	37.46	53.52
6504106020C	4.00 m	m²	2.32	27.73	–	5.40	33.13	46.38	66.26
6504106020D	6.00 m	m²	3.04	36.34	–	6.03	42.37	59.33	84.75
650420	**IN SITU CONCRETE IN UNDERPINNING**								
65042010	Plain in situ concrete; mix C15P								
6504201010	Foundations; poured on or against earth or unblinded hardcore:								
6504201010A	generally	m³	0.90	10.76	–	49.10	59.86	83.80	119.72
6504201020	Isolated foundations; poured on or against earth or unblinded hardcore:								
6504201020A	generally	m³	1.10	13.15	–	49.10	62.25	87.15	124.50
65042020	Plain in situ concrete; mix C20P								
6504202010	Foundations; poured on or against earth or unblinded hardcore:								
6504202010A	generally	m³	0.90	10.76	–	53.60	64.36	90.10	128.72

Property – Refurbishment

Rail 2003 Ref		Unit	Labour Hours	Labour Cost £	Plant Cost £	Materials Cost £	Unit Rate Base Cost £	Unit Rate Green Zone £	Unit Rate Red Zone £
6504	**UNDERPINNING**								
65042020	Plain in situ concrete; mix C20P								
6504202020	Isolated foundations; poured on or against earth or unblinded hardcore:								
6504202020A	generally	m³	1.90	22.71	–	53.60	76.31	106.84	152.63
65042030	Plain in situ concrete; mix C25P								
6504203010	Foundations; poured on or against earth or unblinded hardcore:								
6504203010A	generally	m³	0.90	10.76	–	56.96	67.72	94.80	135.43
6504203020	Isolated foundations; poured on or against earth or unblinded hardcore:								
6504203020A	generally	m³	1.90	22.71	–	56.96	79.67	111.54	159.34
65042040	Reinforced in situ concrete; mix C25P								
6504204010	Foundations; poured on or against earth or unblinded hardcore:								
6504204010A	generally	m³	1.10	13.30	2.24	55.40	70.94	99.32	141.89
6504204020	Isolated foundations; poured on or against earth or unblinded hardcore:								
6504204020A	generally	m³	2.15	25.99	4.34	54.89	85.22	119.30	170.43
6504204030	Walls:								
6504204030A	not exceeding 150 mm thick	m²	2.31	27.93	4.62	54.37	86.92	121.68	173.83
6504204030B	150 – 300 mm thick	m²	2.05	24.79	4.06	54.37	83.22	116.50	166.42
6504204030C	exceeding 300 mm thick	m²	1.40	16.93	2.80	54.37	74.10	103.74	148.19
6504204040	Upstands:								
6504204040A	generally	m³	2.75	33.25	5.46	55.40	94.11	131.76	188.22

Property – Refurbishment

Rail 2003 Ref		Unit	Labour Hours	Labour Cost £	Plant Cost £	Materials Cost £	Unit Rate Base Cost £	Unit Rate Green Zone £	Unit Rate Red Zone £
6504	**UNDERPINNING**								
650430	**REINFORCEMENT**								
65043010	Bar reinforcement; mild steel bars, BS 4449, delivered to site cut, bent and labelled								
6504301010	Bars, fixing with tying wire; diameter:								
6504301010A	not exceeding 12 mm	Tonne	38.00	474.24	–	387.84	862.08	1206.92	1724.16
6504301010B	12 – 25 mm	Tonne	23.00	287.04	–	342.77	629.81	881.74	1259.62
6504301010C	exceeding 25 mm	Tonne	17.00	212.16	–	330.21	542.37	759.31	1084.73
65043020	Bar reinforcement; high yield steel bars BS 4449, delivered to site cut, bent and labelled								
6504302010	Bars, fixing with tying wire; diameter:								
6504302010A	not exceeding 12 mm	Tonne	38.00	474.24	–	402.59	876.83	1227.56	1753.65
6504302010B	12 – 25 mm	Tonne	23.00	287.04	–	356.07	643.11	900.36	1286.23
6504302010C	exceeding 25 mm	Tonne	17.00	212.16	–	343.19	555.35	777.48	1110.70
650440	**FORMWORK TO UNDERPINNING**								
65044010	Formwork to general finish								
6504401010	Sides of foundations:								
6504401010A	not exceeding 250 mm high	m	0.76	10.43	–	5.41	15.84	22.17	31.67
6504401010B	250 – 500 mm high	m	1.15	15.78	–	10.12	25.90	36.26	51.80
6504401010C	500 – 1000 mm high	m	2.04	28.00	–	19.42	47.42	66.39	94.83
6504401010D	exceeding 1000 mm high	m^2	1.86	25.53	–	18.61	44.14	61.79	88.26
6504401020	Sides of upstands:								
6504401020A	not exceeding 250 mm high	m	0.71	9.74	–	1.98	11.72	16.41	23.45
6504401020B	250 – 500 mm high	m	0.90	12.35	–	3.24	15.59	21.82	31.18
6504401020C	500 – 1000 mm high	m	1.60	21.96	–	5.61	27.57	38.59	55.14
6504401020D	exceeding 1000 mm high	m^2	1.46	20.04	–	6.45	26.49	37.08	52.97
6504401030	Walls:								
6504401030A	vertical surfaces	m^2	1.78	24.43	0.94	6.45	31.82	44.55	63.65
6504401030B	curved walls	m^2	2.00	27.45	0.94	6.57	34.96	48.95	69.93
6504401030C	battering surfaces	m^2	2.13	29.23	0.94	6.76	36.93	51.71	73.88

Property – Refurbishment

Rail 2003 Ref		Unit	Labour Hours	Labour Cost	Plant Cost	Materials Cost	Unit Rate	Unit Rate	Unit Rate
				£	£	£	Base Cost £	Green Zone £	Red Zone £

6504 UNDERPINNING

65044010 Formwork to general finish

6504401040 Wall ends, steps and openings in walls:

6504401040A	not exceeding 250 mm high	m	0.50	6.86	–	1.68	8.54	11.97	17.09
6504401040B	250 – 500 mm high	m	0.69	9.47	–	3.41	12.88	18.03	25.76
6504401040C	500 – 1000 mm high	m	0.83	11.39	0.72	5.94	18.05	25.27	36.09
6504401040D	exceeding 1000 mm high	m^2	0.75	10.29	0.94	6.28	17.51	24.52	35.04

650450 BRICKWORK IN UNDERPINNING

65045010 Walls

6504501010 Common bricks in cement mortar (1:3):

6504501010A	half brick thick	m^2	0.97	19.10	–	12.29	31.39	43.95	62.79
6504501010B	one brick thick	m^2	1.83	36.03	–	24.59	60.62	84.86	121.24
6504501010C	one and a half brick thick	m^2	2.70	53.16	–	37.92	91.08	127.51	182.17
6504501010D	two brick thick	m^2	3.66	72.06	–	50.89	122.95	172.13	245.90

6504501020 Class A engineering bricks in cement mortar (1:3):

6504501020A	half brick thick	m^2	1.02	20.08	–	31.00	51.08	71.52	102.17
6504501020B	one brick thick	m^2	1.98	38.98	–	62.67	101.65	142.32	203.31
6504501020C	one and a half brick thick	m^2	2.97	58.48	–	98.61	157.09	219.93	314.17
6504501020D	two brick thick	m^2	3.96	77.97	–	130.28	208.25	291.55	416.50

6504501030 Class B engineering bricks in cement mortar (1:3):

6504501030A	half brick thick	m^2	1.09	21.46	–	21.11	42.57	59.61	85.15
6504501030B	one brick thick	m^2	1.90	37.41	–	42.90	80.31	112.43	160.61
6504501030C	one and a half brick thick	m^2	2.86	56.31	–	67.31	123.62	173.07	247.23
6504501030D	two brick thick	m^2	3.81	75.02	–	89.09	164.11	229.75	328.21

6504501040 Facing bricks, PC £250:00 per 1000, in gauged mortar (1:1:6):

6504501040A	half brick thick; stretcher bond	m^2	2.03	39.97	–	33.22	73.19	102.47	146.38
6504501040B	one brick thick; English bond	m^2	2.13	41.94	–	33.86	75.80	106.12	151.61
6504501040C	one brick thick; Flemish bond	m^2	2.20	43.32	–	33.86	77.18	108.05	154.36

Property – Refurbishment

Rail 2003 Ref		Unit	Labour Hours	Labour Cost £	Plant Cost £	Materials Cost £	Unit Rate Base Cost £	Unit Rate Green Zone £	Unit Rate Red Zone £
6504	**UNDERPINNING**								
650460	**BRICKWORK SUNDRIES**								
65046010	Extra over general brickwork fo fair faced work								
6504601010	Fair facing and flush pointing:								
6504601010A	stretcher bond	m²	0.06	1.18	–	0.32	1.50	2.10	3.00
6504601010B	English bond	m²	0.06	1.18	–	0.32	1.50	2.10	3.00
6504601010C	Flemish bond	m²	0.06	1.18	–	0.32	1.50	2.10	3.00
6504601020	Fair facing and struck or weather struck pointing:								
6504601020A	stretcher bond	m²	0.06	1.18	–	0.32	1.50	2.10	3.00
6504601020B	English bond	m²	0.06	1.18	–	0.32	1.50	2.10	3.00
6504601020C	Flemish bond	m²	0.06	1.18	–	0.32	1.50	2.10	3.00
6504601030	Fair facing and tooled or keyed pointing:								
6504601030A	stretcher bond	m²	0.06	1.18	–	0.32	1.50	2.10	3.00
6504601030B	English bond	m²	0.07	1.38	–	0.32	1.70	2.38	3.40
6504601030C	Flemish bond	m²	0.07	1.38	–	0.32	1.70	2.38	3.40
65046020	Cutting, toothing and bonding ends of new walls into existing								
6504602010	Common brickwork in alternate courses:								
6504602010A	half brick wall	m	0.52	10.24	–	2.39	12.63	17.67	25.25
6504602010B	one brick wall	m	0.84	16.54	–	4.10	20.64	28.89	41.28
6504602010C	one and a half brick wall	m	1.10	21.66	–	4.10	25.76	36.06	51.52
6504602010D	two brick wall	m	1.65	32.49	–	6.48	38.97	54.56	77.94
6504602020	Engineering brickwork in alternate courses:								
6504602020A	half brick wall	m	0.53	10.44	–	3.63	14.07	19.69	28.13
6504602020B	one brick wall	m	0.88	17.33	–	6.93	24.26	33.96	48.50
6504602020C	one and a half brick wall	m	1.15	22.64	–	7.26	29.90	41.87	59.80
6504602020D	two brick wall	m	1.69	33.27	–	10.56	43.83	61.36	87.66

Property – Refurbishment

Rail 2003 Ref		Unit	Labour Hours	Labour Cost	Plant Cost	Materials Cost	Unit Rate	Unit Rate	Unit Rate
				£	£	£	Base Cost £	Green Zone £	Red Zone £
6504	**UNDERPINNING**								
65046020	Cutting, toothing and bonding ends of new walls into existing								
6504602030	Facing brickwork in alternate courses:								
6504602030A	half brick wall	m	0.58	11.42	–	2.83	14.25	19.96	28.51
6504602030B	one brick wall	m	0.93	18.31	–	5.33	23.64	33.11	47.29
650470	**DAMP PROOF COURSES AND MEMBRANES IN UNDERPINNING**								
65047010	Membranes and compounds								
6504701010	RIW liquid asphaltic composition; two coats on horizontal concrete surfaces:								
6504701010A	not exceeding 150 mm wide	m	0.02	0.24	–	1.04	1.28	1.78	2.56
6504701010B	150 – 300 mm wide	m	0.03	0.36	–	2.08	2.44	3.41	4.87
6504701010C	exceeding 300 mm wide	m^2	0.08	0.96	–	6.98	7.94	11.11	15.86
6504701020	RIW liquid asphaltic composition; two coats on vertical concrete surfaces:								
6504701020A	not exceeding 150 mm high	m	0.02	0.24	–	1.04	1.28	1.78	2.56
6504701020B	150 – 300 mm high	m	0.03	0.36	–	2.08	2.44	3.41	4.87
6504701020C	exceeding 300 mm high	m^2	0.09	1.08	–	6.98	8.06	11.28	16.10
6504701030	RIW liquid asphaltic composition; two coats on vertical masonry surfaces:								
6504701030A	not exceeding 150 mm high	m	0.03	0.36	–	1.00	1.36	1.90	2.71
6504701030B	150 – 300 mm high	m	0.04	0.48	–	1.99	2.47	3.46	4.95
6504701030C	exceeding 300 mm high	m^2	0.10	1.20	–	6.60	7.80	10.91	15.60
6504701040	Synthaprufe waterproofing compound; two coats on horizontal concrete surfaces; final coat dusted with sharp sand:								
6504701040A	not exceeding 150 mm wide	m	0.02	0.24	–	0.77	1.01	1.41	2.03
6504701040B	150 – 300 mm wide	m	0.03	0.36	–	1.46	1.82	2.55	3.65
6504701040C	exceeding 300 mm wide	m^2	0.09	1.08	–	4.75	5.83	8.15	11.64

Property – Refurbishment

Rail 2003 Ref		Unit	Labour Hours	Labour Cost £	Plant Cost £	Materials Cost £	Unit Rate Base Cost £	Unit Rate Green Zone £	Unit Rate Red Zone £
6504	**UNDERPINNING**								
65047010	**Membranes and compounds**								
6504701050	Synthaprufe waterproofing compound; two coats on vertical concrete surfaces; final coat dusted with sharp sand:								
6504701050A	not exceeding 150 mm high	m	0.02	0.24	–	0.76	1.00	1.40	2.01
6504701050B	150 – 300 mm high	m	0.03	0.36	–	1.45	1.81	2.53	3.63
6504701050C	exceeding 300 mm high	m^2	0.10	1.20	–	4.75	5.95	8.31	11.88

Property – Refurbishment

Rail 2003 Ref	Unit	Labour Hours	Labour Cost £	Plant Cost £	Materials Cost £	Unit Rate Base Cost £	Unit Rate Green Zone £	Unit Rate Red Zone £
6505	**BRICKWORK AND BLOCKWORK**							
650510	**GENERAL REMOVALS**							
65051010	**Demolish and remove from site**							
6505101010	Brickwork walls in cement mortar:							
6505101010A	half brick thick.................... m^2	0.80	9.56	0.62	–	10.18	14.25	20.36
6505101010B	one brick thick..................... m^2	1.40	16.74	1.08	–	17.82	24.94	35.63
6505101010C	one and a half brick thick............ m^2	2.10	25.10	1.62	–	26.72	37.41	53.45
6505101010D	two brick thick..................... m^2	2.70	32.28	2.08	–	34.36	48.10	68.71
6505101020	Blockwork walls in cement mortar:							
6505101020A	not exceeding 150 mm thick.......... m^2	0.70	8.37	0.54	–	8.91	12.47	17.82
6505101020B	150 – 300 mm thick................ m^2	1.10	13.15	0.85	–	14.00	19.60	28.00
6505101020C	300 – 450 mm thick................ m^2	1.70	20.32	1.31	–	21.63	30.28	43.26
6505101030	Attached brickwork piers and chimney breasts:							
6505101030A	half brick thick..................... m^2	0.90	10.76	0.69	–	11.45	16.03	22.91
6505101030B	one brick thick..................... m^2	1.60	19.13	1.23	–	20.36	28.51	40.72
6505101030C	one and a half brick thick............ m^2	2.30	27.49	1.77	–	29.26	40.97	58.53
6505101030D	two brick thick..................... m^2	2.80	33.47	2.16	–	35.63	49.88	71.25
6505101040	Isolated brickwork piers and chimney stacks:							
6505101040A	brickwork; internally................ m^2	9.60	114.76	3.70	–	118.46	165.84	236.92
6505101040B	brickwork; externally .... m^2	8.70	104.00	3.39	–	107.39	150.35	214.78
65051020	**Carefully take down and set aside salvaged materials for re-use**							
6505102010	Brickwork walls in soft mortar:							
6505102010A	half brick thick..................... m^2	1.20	14.35	–	–	14.35	20.08	28.69
6505102010B	one brick thick..................... m^2	2.10	25.10	–	–	25.10	35.14	50.21
6505102010C	one and a half brick thick............ m^2	3.15	37.66	–	–	37.66	52.72	75.31
6505102010D	two brick thick..................... m^2	4.10	49.01	–	–	49.01	68.62	98.02

Property – Refurbishment

Rail 2003 Ref		Unit	Labour Hours	Labour Cost £	Plant Cost £	Materials Cost £	Unit Rate Base Cost £	Unit Rate Green Zone £	Unit Rate Red Zone £
6505	**BRICKWORK AND BLOCKWORK**								
65051020	Carefully take down and set aside salvaged materials for re-use								
6505102020	Extra over for selecting, cleaning and stacking salvaged materials:								
6505102020A	general bricks	m²	14.00	167.36	–	–	167.36	234.30	334.71
65051030	Take down and remove from site								
6505103010	Chimney pots:								
6505103010A	any size	m²	1.00	11.95	–	–	11.95	16.74	23.91
6505103020	Tiled fireplace units:								
6505103020A	including hearth	m²	5.15	61.56	–	–	61.56	86.19	123.13
650520	**NEW WORK**								
65052010	Brick and block walling								
6505201010	Common bricks in cement mortar (1:3):								
6505201010A	half brick thick	m²	0.68	13.39	–	11.62	25.01	35.01	50.03
6505201010B	one brick thick	m²	1.36	26.78	–	24.59	51.37	71.91	102.73
6505201010C	one and a half brick thick	m²	2.03	39.97	–	37.92	77.89	109.05	155.79
6505201010D	two brick thick	m²	2.71	53.36	–	50.89	104.25	145.94	208.49
6505201020	Class A engineerinrg bricks in cement mortar (1:3):								
6505201020A	half brick thick	m²	0.73	14.37	–	31.00	45.37	63.52	90.75
6505201020B	one brick thick	m²	1.47	28.94	–	62.67	91.61	128.26	183.23
6505201020C	one and a half brick thick	m²	2.20	43.32	–	98.61	141.93	198.70	283.85
6505201020D	two brick thick	m²	2.94	57.89	–	130.28	188.17	263.43	376.33
6505201030	Class B engineerinrg bricks in cement mortar (1:3):								
6505201030A	half brick thick	m²	0.71	13.98	–	21.11	35.09	49.13	70.19
6505201030B	one brick thick	m²	1.41	27.76	–	42.90	70.66	98.93	141.31
6505201030C	one and a half brick thick	m²	2.12	41.74	–	67.31	109.05	152.67	218.09
6505201030D	two brick thick	m²	2.82	55.52	–	89.09	144.61	202.46	289.23

Property – Refurbishment

Rail 2003 Ref		Unit	Labour Hours	Labour Cost £	Plant Cost £	Materials Cost £	Unit Rate Base Cost £	Unit Rate Green Zone £	Unit Rate Red Zone £
6505	**BRICKWORK AND BLOCKWORK**								
65052010	**Brick and block walling**								
6505201040	Facing bricks, PC £250:00 per 1000, in gauged mortar (1:1:6):								
6505201040A	half brick thick; stretcher bond	m^2	0.79	15.55	–	16.34	31.89	44.66	63.79
6505201040B	one brick thick; English bond	m^2	1.58	31.11	–	34.02	65.13	91.19	130.27
6505201040C	one brick thick; Flemish bond	m^2	1.63	32.09	–	33.35	65.44	91.63	130.90
6505201050	Precast concrete blockwork, BS 6073, 7 N/mm^2, in cement mortar (1:3):								
6505201050A	100 mm solid blocks	m^2	0.57	11.22	–	7.27	18.49	25.88	36.98
6505201050B	150 mm solid blocks	m^2	0.63	12.40	–	10.56	22.96	32.16	45.93
6505201050C	200 mm solid blocks	m^2	0.71	13.98	–	14.41	28.39	39.74	56.77
6505201050D	215 mm solid blocks	m^2	0.72	14.18	–	15.85	30.03	42.05	60.06
6505201050E	215 mm hollow blocks	m^2	0.66	13.00	–	15.85	28.85	40.38	57.69
6505201060	Thermalite blocks in gauged mortar (1:1:6):								
6505201060A	100 mm Shield blocks	m^2	0.48	9.45	–	8.78	18.23	25.52	36.46
6505201060B	150 mm Shield blocks	m^2	0.52	10.24	–	12.85	23.09	32.31	46.17
6505201060C	200 mm Shield blocks	m^2	0.59	11.62	–	16.84	28.46	39.83	56.90
6505201060D	215 mm Shield blocks	m^2	0.60	11.81	–	18.12	29.93	41.90	59.86
6505201070	Extra over general brickwork for fair facing and flush pointing:								
6505201070A	stretcher bond	m^2	0.06	1.18	–	0.32	1.50	2.10	3.00
6505201070B	English bond	m^2	0.06	1.18	–	0.32	1.50	2.10	3.00
6505201070C	Flemish bond	m^2	0.06	1.18	–	0.32	1.50	2.10	3.00
6505201080	Extra over general blockwork for fair facing and flush pointing:								
6505201080A	concrete blocks	m^2	0.02	0.39	–	0.32	0.71	1.00	1.43
6505201080B	Thermalite blocks	m^2	0.03	0.59	–	0.32	0.91	1.28	1.82

Property – Refurbishment

Rail 2003 Ref		Unit	Labour Hours	Labour Cost £	Plant Cost £	Materials Cost £	Unit Rate Base Cost £	Unit Rate Green Zone £	Unit Rate Red Zone £
6505	**BRICKWORK AND BLOCKWORK**								
65052020	**Sills and copings**								
6505202010	Sills and copings in facing bricks, PC £250:00 per 1000, in gauged mortar (1:1:6); flush pointing all exposed edges:								
6505202010A	brick-on-edge sills; flush; flat top; half brick wide (snapped headers)........	m²	0.24	4.73	–	2.82	7.55	10.57	15.09
6505202010B	brick-on-edge sills; flush; flat top; one brick wide	m²	0.37	7.29	–	3.14	10.43	14.60	20.86
6505202010C	brick-on-edge sills; 50 mm projection; weathered; half brick wide (snapped headers).........................	m²	0.47	9.25	–	2.82	12.07	16.91	24.15
6505202010D	brick-on-edge sills; 50 mm projection; weathered; one brick wide...........	m²	0.53	10.44	–	5.64	16.08	22.51	32.16
6505202010E	brick-on-edge copings; one brick wide	m²	0.37	7.29	–	3.14	10.43	14.60	20.86
6505202010F	brick-on-edge copings; half brick wide	m²	0.56	11.03	–	3.14	14.17	19.84	28.34
65052025	**Tile sills and creasings**								
6505202510	Sills; 265 × 165 mm plain concrete tiles in gauged mortar (1:1:6); double course breaking joint; 50 mm projection; weathered; flush pointing all exposed edges:								
6505202510A	152 mm wide	m²	0.25	4.92	–	4.68	9.60	13.45	19.21
6505202510B	265 mm wide	m²	0.28	5.51	–	4.68	10.19	14.28	20.40
6505202520	Creasings; 265 × 165 mm plain concrete tiles in gauged mortar (1:1:6); single course; 50 mm projection one side; flush pointing all exposed edges:								
6505202520A	152 mm wide	m²	0.22	4.33	–	4.68	9.01	12.62	18.03
6505202520B	265 mm wide	m²	0.23	4.53	–	4.68	9.21	12.90	18.43
6505202530	Creasings; 265 × 165 mm plain concrete tiles in gauged mortar (1:1:6); double course; 50 mm projection both sides; flush pointing all exposed edges:								
6505202530A	202 mm wide	m²	0.22	4.33	–	4.68	9.01	12.62	18.03
6505202530B	315 mm wide	m²	0.28	5.51	–	4.68	10.19	14.28	20.40

Property – Refurbishment

Rail 2003 Ref		Unit	Labour Hours	Labour Cost £	Plant Cost £	Materials Cost £	Unit Rate Base Cost £	Unit Rate Green Zone £	Unit Rate Red Zone £
6505	**BRICKWORK AND BLOCKWORK**								
65052025	Tile sills and creasings								
6505202540	Creasings; 265 × 165 mm plain concrete tiles in gauged mortar (1:1:6); double course; 50 mm projection one side; flush pointing all exposed edges								
6505202540A	152 mm wide	m²	0.38	7.48	–	4.68	12.16	17.03	24.33
6505202540B	265 mm wide	m²	0.42	8.27	–	5.00	13.27	18.59	26.55
6505202550	Creasings; 265 × 165 mm plain concrete tiles in gauged mortar (1:1:6); double course; 50 mm projection both sides; flush pointing all exposed edges:								
6505202550A	202 mm wide	m²	0.47	9.25	–	4.68	13.93	19.52	27.88
6505202550B	315 mm wide	m²	0.61	12.01	–	13.73	25.74	36.03	51.47
65052030	Forming cavities								
6505203010	Form cavity in hollow walls:								
6505203010A	25 mm wide	m²	0.02	0.39	–	–	0.39	0.55	0.79
6505203010B	50 mm wide	m²	0.02	0.39	–	–	0.39	0.55	0.79
6505203010C	75 mm wide	m²	0.02	0.39	–	–	0.39	0.55	0.79
6505203020	Build in wall ties; 5 per m²:								
6505203020A	3 mm galvanised steel wire butterfly ties, 200 mm long	m²	0.15	2.95	–	0.94	3.89	5.44	7.79
6505203020B	3 mm stainless steel wire butterfly ties, 200 mm long	m²	0.15	2.95	–	0.76	3.71	5.19	7.42
6505203020C	3 × 19 mm galvanised steel twisted ties; 200 mm long	m²	0.15	2.95	–	1.41	4.36	6.11	8.74
6505203020D	3 × 19 mm stainless steel twisted ties; 200 mm long	m²	0.15	2.95	–	1.88	4.83	6.76	9.66

Property – Refurbishment

Rail 2003 Ref		Unit	Labour Hours	Labour Cost £	Plant Cost £	Materials Cost £	Unit Rate Base Cost £	Unit Rate Green Zone £	Unit Rate Red Zone £
6505	**BRICKWORK AND BLOCKWORK**								
65052035	Cavity wall insulation								
6505203510	Jablite expanded polystyrene; fitting between wall ties:								
6505203510A	25 mm thick	m²	0.15	2.95	–	2.66	5.61	7.85	11.23
6505203510B	50 mm thick	m²	0.16	3.15	–	4.70	7.85	10.98	15.69
6505203510C	75 mm thick	m²	0.18	3.54	–	6.57	10.11	14.15	20.22
6505203520	Dritherm glassfibre cavity wall insulation slabs; fitting between wall ties:								
6505203520A	50 mm thick	m²	0.16	3.15	–	3.23	6.38	8.93	12.75
6505203520B	75 mm thick	m²	0.18	3.54	–	4.23	7.77	10.88	15.54
65052040	Damp proof courses, cavity trays and membranes								
6505204010	Hyload pitch polymer d.p.c.; bedded in cement mortar; horizontal:								
6505204010A	100 mm wide	m	0.04	0.79	–	1.13	1.92	2.68	3.84
6505204010B	225 mm wide	m	0.08	1.58	–	2.37	3.95	5.52	7.88
6505204010C	exceeding 225 mm wide	m²	0.35	6.89	–	11.32	18.21	25.50	36.42
6505204020	Hyload pitch polymer d.p.c.; bedded in cement mortar; vertical:								
6505204020A	150 mm wide	m	0.06	1.18	–	1.71	2.89	4.05	5.79
6505204030	Hyload pitch polymer d.p.c.; bedded in cement mortar; forming cavity trays:								
6505204030A	exceeding 300 mm girth	m²	0.56	11.03	–	11.32	22.35	31.29	44.69
6505204040	Bituthene bitumen coated 1000 gauge polythene sheeting; sticking to masonry surfaces primed with bituthene primer:								
6505204040A	horizontal; over 300 mm wide	m²	0.10	1.20	–	8.77	9.97	13.95	19.93
6505204040B	vertical; over 300 mm wide	m²	0.13	1.55	–	9.18	10.73	15.04	21.48

Property – Refurbishment

Rail 2003 Ref		Unit	Labour Hours	Labour Cost	Plant Cost	Materials Cost	Unit Rate	Unit Rate	Unit Rate
				£	£	£	Base Cost £	Green Zone £	Red Zone £
6505	**BRICKWORK AND BLOCKWORK**								
65052040	**Damp proof courses, cavity trays and membranes**								
6505204050	Waterproof building paper; BS 1521:								
6505204050A	horizontal; over 300 mm wide	m²	0.03	0.36	–	1.00	1.36	1.89	2.71
6505204050B	vertical; fixing with staples; over 300 mm wide	m²	0.04	0.48	–	1.00	1.48	2.06	2.95
6505204060	Newtonite lathing to concrete or masonry backgrounds:								
6505204060A	walls, reveals, recesses	m²	0.12	1.43	–	9.00	10.43	14.61	20.88
6505204060B	walls; not exceeding 300 mm wide	m²	0.18	2.15	–	9.00	11.15	15.61	22.31
65052050	**Expanded metal reinforcement**								
6505205010	24 gauge galvanised mild steel expanded brick reinforcement:								
6505205010A	half brick wide	m	0.15	2.95	–	0.41	3.36	4.70	6.72
6505205010B	one brick wide	m	0.18	3.54	–	0.71	4.25	5.96	8.51
6505205010C	one and a half brick wide	m	0.19	3.74	–	1.09	4.83	6.77	9.67
6505205010D	two brick wide	m	0.21	4.14	–	1.42	5.56	7.78	11.12
6505205020	24 gauge stainless steel expanded brick reinforcement:								
6505205020A	half brick wide	m	0.15	2.95	–	0.69	3.64	5.09	7.28
6505205020B	one brick wide	m	0.17	3.35	–	1.07	4.42	6.19	8.83
6505205020C	one and a half brick wide	m	0.19	3.74	–	1.64	5.38	7.53	10.76
6505205020D	two brick wide	m	0.21	4.14	–	2.14	6.28	8.78	12.54
65052060	**Expansion joints and the like**								
6505206010	19 mm Flexcell compressible joint filler; fixing in place in brickwork or blockwork:								
6505206010A	half brick wide	m	0.13	2.56	–	1.46	4.02	5.63	8.05
6505206010B	one brick wide	m	0.15	2.95	–	2.91	5.86	8.21	11.73
6505206010C	one and a half brick wide	m	0.17	3.35	–	4.38	7.73	10.82	15.44
6505206010D	two brick wide	m	0.19	3.74	–	5.83	9.57	13.40	19.13
6505206010E	100 mm wide	m	0.13	2.56	–	1.32	3.88	5.43	7.77
6505206010F	150 mm wide	m	0.14	2.76	–	1.99	4.75	6.65	9.49
6505206010G	200 mm wide	m	0.15	2.95	–	2.65	5.60	7.84	11.21

Property – Refurbishment

Rail 2003 Ref		Unit	Labour Hours	Labour Cost £	Plant Cost £	Materials Cost £	Unit Rate Base Cost £	Unit Rate Green Zone £	Unit Rate Red Zone £
6505	**BRICKWORK AND BLOCKWORK**								
65052060	Expansion joints and the like								
6505206010	19 mm Flexcell compressible joint filler; fixing in place in brickwork or blockwork:								
6505206010H	215 mm wide	m	0.15	2.95	–	2.85	5.80	8.12	11.61
6505206020	Gun grade polysulphide rubber compound:								
6505206020A	19 × 19 mm	m	0.14	2.76	–	1.22	3.98	5.56	7.94

Property – Refurbishment

Rail 2003 Ref		Unit	Labour Hours	Labour Cost £	Plant Cost £	Materials Cost £	Unit Rate Base Cost £	Unit Rate Green Zone £	Unit Rate Red Zone £
6506	**ASPHALT WORK**								
650610	**GENERAL REMOVALS**								
65061010	**Pavings and the like**								
6506101010	Hack up defective areas of existing paving, tanking and coverings, load into skips; horizontal:								
6506101010A	on loose underlay	m^2	0.12	1.43	–	–	1.43	2.01	2.87
6506101010B	keyed to substrate	m^2	0.45	5.38	–	–	5.38	7.53	10.76
6506101020	Hack up defective areas of existing roof coverings, load into skips; horizontal or sloping:								
6506101020B	on loose underlay	m^2	0.15	1.79	–	–	1.79	2.51	3.59
6506101030	Hack up defective areas of existing skirtings, load into skips; vertical upstands:								
6506101030A	keyed; not exceeding 150 mm high	m	0.18	2.15	–	–	2.15	3.01	4.30
6506101030B	keyed; exceeding 150 mm high	m	0.27	3.23	–	–	3.23	4.52	6.46

Property – Refurbishment

Rail 2003 Ref		Unit	Specialist Cost £	Unit Rate Base Cost £	Unit Rate Green Zone £	Unit Rate Red Zone £

6506 ASPHALT WORK

650620 NEW WORK

65062010 Mastic asphalt tanking; BS 6925

6506201010 13 mm one coat horizontal covering on concrete:

6506201010A	not exceeding 150 mm wide	m	–	–	–	8.99	8.99	12.59	17.98
6506201010B	150 – 300 mm wide	m	–	–	–	13.68	13.68	19.15	27.36
6506201010C	exceeding 300 mm wide	m²	–	–	–	31.31	31.31	43.83	62.62

6506201020 20 mm two coat horizontal covering on concrete:

6506201020A	not exceeding 150 mm wide	m	–	–	–	11.24	11.24	15.74	22.48
6506201020B	150 – 300 mm wide	m	–	–	–	17.09	17.09	23.93	34.18
6506201020C	exceeding 300 mm wide	m²	–	–	–	39.19	39.19	54.87	78.38

6506201030 30 mm three coat horizontal covering on concrete:

6506201030A	not exceeding 150 mm wide	m	–	–	–	18.94	18.94	26.52	37.88
6506201030B	150 – 300 mm wide	m	–	–	–	29.07	29.07	40.70	58.14
6506201030C	exceeding 300 mm wide	m²	–	–	–	66.56	66.56	93.18	133.12

6506201040 13 mm two coat vertical covering on concrete or brickwork:

6506201040A	not exceeding 150 mm wide	m	–	–	–	24.20	24.20	33.88	48.40
6506201040B	150 – 300 mm wide	m	–	–	–	37.31	37.31	52.23	74.62
6506201040C	exceeding 300 mm wide	m²	–	–	–	85.68	85.68	119.95	171.36

6506201050 20 mm three coat vertical covering on concrete or brickwork:

6506201050A	not exceeding 150 mm wide	m	–	–	–	32.07	32.07	44.90	64.14
6506201050B	150 – 300 mm wide	m	–	–	–	49.12	49.12	68.77	98.24
6506201050C	exceeding 300 mm wide	m²	–	–	–	112.51	112.51	157.51	225.02

6506201080 General labours:

6506201080A	internal angle fillets	m	–	–	–	12.00	12.00	16.80	24.00
6506201080B	turning nibs into grooves	m	–	–	–	6.76	6.76	9.46	13.52
6506201080C	working into outlets	Nr	–	–	–	71.24	71.24	99.74	142.48
6506201080D	collars to small pipes	Nr	–	–	–	57.93	57.93	81.10	115.86
6506201080E	collars to large pipes	Nr	–	–	–	84.56	84.56	118.38	169.12

Property – Refurbishment

Rail 2003 Ref	Unit	Specialist Cost £	Unit Rate Base Cost £	Unit Rate Green Zone £	Unit Rate Red Zone £
6506	**ASPHALT WORK**				
65062020	**Mastic asphalt tanking; BS 6577**				
6506202010	**13 mm one coat horizontal covering on concrete:**				
6506202010A	not exceeding 150 mm wide .... m	–	–	–	13.84	13.84	19.38	27.68
6506202010B	150 – 300 mm wide .... m	–	–	–	21.18	21.18	29.65	42.36
6506202010C	exceeding 300 mm wide .... m^2	–	–	–	48.01	48.01	67.21	96.02
6506202020	**20 mm two coat horizontal covering on concrete:**				
6506202020A	not exceeding 150 mm wide .... m	–	–	–	16.88	16.88	23.63	33.76
6506202020B	150 – 300 mm wide .... m	–	–	–	25.68	25.68	35.95	51.36
6506202020C	exceeding 300 mm wide .... m^2	–	–	–	59.26	59.26	82.96	118.52
6506202030	**30 mm three coat horizontal covering on concrete:**				
6506202030A	not exceeding 150 mm wide .... m	–	–	–	27.74	27.74	38.84	55.48
6506202030B	150 – 300 mm wide .... m	–	–	–	42.38	42.38	59.33	84.76
6506202030C	exceeding 300 mm wide .... m^2	–	–	–	97.13	97.13	135.98	194.26
6506202040	**13 mm two coat vertical covering on concrete or brickwork:**				
6506202040A	not exceeding 150 mm wide .... m	–	–	–	30.04	30.04	42.06	60.08
6506202040B	150 – 300 mm wide .... m	–	–	–	46.12	46.12	64.57	92.24
6506202040C	exceeding 300 mm wide .... m^2	–	–	–	104.45	104.45	146.23	208.90
6506202050	**20 mm three coat vertical covering on concrete or brickwork:**				
6506202050A	not exceeding 150 mm wide .... m	–	–	–	40.13	40.13	56.18	80.26
6506202050B	150 – 300 mm wide .... m	–	–	–	61.12	61.12	85.57	122.24
6506202050C	exceeding 300 mm wide .... m^2	–	–	–	140.43	140.43	196.60	280.86
6506202080	**General labours:**				
6506202080A	internal angle fillets .... m	–	–	–	12.00	12.00	16.80	24.00
6506202080B	turning nibs into grooves .... m	–	–	–	6.76	6.76	9.46	13.52
6506202080C	working into outlets .... Nr	–	–	–	71.24	71.24	99.74	142.48
6506202080D	collars to small pipes .... Nr	–	–	–	57.93	57.93	81.10	115.86
6506202080E	collars to large pipes .... Nr	–	–	–	84.56	84.56	118.38	169.12

Property – Refurbishment

Rail 2003 Ref	Unit	Specialist Cost £	Unit Rate Base Cost £	Unit Rate Green Zone £	Unit Rate Red Zone £

6506 ASPHALT WORK

65062030 Mastic asphalt flooring; BS 6925

6506203010 15 mm one coat light duty flooring on and including isolating membrane:

6506203010A	not exceeding 150 mm wide .... m	– – –	11.26	11.26	15.76	22.52
6506203010B	150 – 300 mm wide .... m	– – –	17.07	17.07	23.90	34.14
6506203010C	exceeding 300 mm wide .... m^2	– – –	39.38	39.38	55.13	78.76

6506203020 20 mm one coat medium duty flooring on and including isolating membrane:

6506203020A	not exceeding 150 mm wide .... m	– – –	12.00	12.00	16.80	24.00
6506203020B	150 – 300 mm wide .... m	– – –	18.38	18.38	25.73	36.76
6506203020C	exceeding 300 mm wide .... m^2	– – –	42.19	42.19	59.07	84.38

6506203030 30 mm one coat medium duty flooring on and including isolating membrane:

6506203030A	not exceeding 150 mm wide .... m	– – –	16.50	16.50	23.10	33.00
6506203030B	150 – 300 mm wide .... m	– – –	25.12	25.12	35.17	50.24
6506203030C	150 – 300 mm wide .... m^2	– – –	57.75	57.75	80.85	115.50

6506203040 General labours:

6506203040A	working against metal frames........ m	– – –	3.18	3.18	4.45	6.36
6506203040B	extra for working into recessed covers; not exceeding 1.00 m^2 .... m	– – –	26.25	26.25	36.75	52.50

6506203050 Skirtings; 13 mm two coat; fair edge, angles, coved angle fillet and nib turned into groove:

6506203050A	150 mm high..................... m	– – –	25.51	25.51	35.71	51.02

Property – Refurbishment

Rail 2003 Ref		Unit	Specialist Cost £	Unit Rate Base Cost £	Unit Rate Green Zone £	Unit Rate Red Zone £
6506	**ASPHALT WORK**					
65062040	Coloured mastic asphalt flooring; BS 6925					
6506204010	15 mm one coat light duty brown flooring on and including isolating membrane:					
6506204010A	not exceeding 150 mm wide	m	–	–	–	13.51	13.51	18.91	27.02
6506204010B	150 – 300 mm wide	m	–	–	–	20.64	20.64	28.90	41.28
6506204010C	exceeding 300 mm wide	m²	–	–	–	47.64	47.64	66.70	95.28
6506204050	Skirtings; 13 mm two coat; brown with fair edge, angles, coved angle fillet and nib turned into groove:					
6506204050A	150 mm high	m	–	–	–	27.94	27.94	39.12	55.88
65062080	**Sundries**					
6506208010	13 mm layer of limestone chippings; bedding to bitumen compound:					
6506208010A	exceeding 300 mm wide	m²	–	–	–	4.61	4.61	6.45	9.22
6506208020	Prepare for, prime and apply one coat solar reflective paint to bitumen covering:					
6506208020A	exceeding 300 mm wide	m²	–	–	–	5.13	5.13	7.18	10.26
650630	**REPAIRS AND REMEDIAL WORKS**					
65063010	Remedial work to existing coverings					
6506301010	Cut out crack to existing asphalt paving and make good in matching materials:					
6506301010A	not exceeding 150 mm wide	m	–	–	–	18.71	18.71	26.19	37.42
6506301010B	150 – 300 mm wide	m	–	–	–	25.22	25.22	35.31	50.44

Rail 2003 Ref		Unit				Specialist Cost	Unit Rate	Unit Rate	Unit Rate
						£	Base Cost £	Green Zone £	Red Zone £
6506	**ASPHALT WORK**								
65063010	**Remedial work to existing coverings**								
6506301020	**Cut out blister to existing asphalt paving and make good in matching materials:**								
6506301020A	not exceeding 0.25 m^2	Nr	–	–	–	12.94	12.94	18.12	25.88
6506301020B	0.25 – 0.50 m^2	Nr	–	–	–	20.00	20.00	28.00	40.00
6506301020C	exceeding 0.50 m^2	m^2	–	–	–	63.98	63.98	89.57	127.96
6506301030	**Cut out isolated areas of defective asphalt paving and make good in matching materials:**								
6506301030A	not exceeding 1.0 m^2	Nr	–	–	–	71.62	71.62	100.27	143.24
6506301030B	1.0 – 1.5 m^2	Nr	–	–	–	94.78	94.78	132.69	189.56
6506301030C	exceeding 1.5 m^2	m^2	–	–	–	65.09	65.09	91.13	130.18

Property – Refurbishment

Rail 2003 Ref		Unit	Labour Hours	Labour Cost	Plant Cost	Materials Cost	Unit Rate	Unit Rate	Unit Rate
				£	£	£	Base Cost £	Green Zone £	Red Zone £
6507	**ROOFING**								
650710	**WORK TO EXISTING**								
65071010	**Removal of existing roof finishes**								
6507101010	Strip existing roof finishes and load into skips for disposal:								
6507101010A	slates	m²	0.25	2.99	–	–	2.99	4.18	5.98
6507101010B	tiles	m²	0.23	2.75	–	–	2.75	3.85	5.50
6507101010C	corrugated metal sheeting	m²	0.21	2.51	–	–	2.51	3.51	5.02
6507101010D	underfelt	m²	0.07	0.84	–	–	0.84	1.17	1.67
6507101010E	felt in three layers	m²	0.25	2.99	–	–	2.99	4.18	5.98
6507101010F	sheet metal finishings	m²	0.33	3.95	–	–	3.95	5.52	7.89
6507101020	Remove metal flashings and load into skips:								
6507101020A	horizontal	m	0.10	1.20	–	–	1.20	1.67	2.39
6507101020B	stepped	m	0.13	1.55	–	–	1.55	2.18	3.11
6507101030	Remove battens and load into skips:								
6507101030A	tiles or slates	m²	0.16	1.91	–	–	1.91	2.68	3.83
6507101040	Remove coverings, carefully handling and disposing of by an approved method toxic waste or other special waste:								
6507101040A	asbetos cement sheeting	m²	0.58	6.93	–	–	6.93	9.71	13.87
6507101050	Clean and set aside sound materials for re-use:								
6507101050A	slates	m²	0.40	4.78	–	–	4.78	6.69	9.56
6507101050B	tiles	m²	0.37	4.42	–	–	4.42	6.19	8.85
6507101060	Take up roof finish substrate and load into skips:								
6507101060A	chipboard	m²	0.35	4.18	–	–	4.18	5.86	8.37
6507101060B	plywood	m²	0.33	3.95	–	–	3.95	5.52	7.89
6507101060C	woodwool	m²	0.31	3.71	–	–	3.71	5.19	7.41

Property – Refurbishment

Rail 2003 Ref		Unit	Labour Hours	Labour Cost £	Plant Cost £	Materials Cost £	Unit Rate Base Cost £	Unit Rate Green Zone £	Unit Rate Red Zone £

6507 ROOFING

650720 NEW WORKS

65072010 Slate roofing

6507201010 Welsh slates; size 510 × 255 mm; 75 mm lap; 50 × 25 mm treated softwood battens reinforced slaters underlining felt type 1F:

6507201010A	sloping	m²	0.50	9.08	–	476.43	485.51	679.71	971.02
6507201010B	Extra for double eaves course	m	0.45	8.17	–	49.88	58.05	81.27	116.09
6507201010C	Extra for mitred hips; both sides measured	m	2.70	49.02	–	74.82	123.84	173.37	247.68
6507201010D	Extra for cutting to valleys; both sides measured	m	1.75	31.78	–	99.75	131.53	184.15	263.06

6507201020 Artificial slates; asbestos free slates; blue/black; Eternit 2000; 600 × 300 mm; 75 mm lap; 50 × 25 mm treated sawn softwood battens; reinforced slaters underlining felt type 1F:

6507201020E	sloping	m²	0.50	9.08	–	21.66	30.74	43.03	61.47
6507201020F	Extra for double eaves course	m	0.45	8.17	–	7.75	15.92	22.29	31.84
6507201020G	Extra for mitred hips or cutting to valleys; both sides measured:	m	1.45	26.33	–	9.69	36.02	50.43	72.04

6507201030 Hardrow slates; 457 × 305 mm; 75 mm lap; 25 × 50 mm treated softwood battens; reinforced slaters underlining felt type 1F:

6507201030A	sloping	m²	0.50	9.08	–	23.59	32.67	45.74	65.35
6507201030B	Extra for eaves course	m	0.20	3.63	–	4.99	8.62	12.07	17.24
6507201030C	Extra for ridge slates	m	0.30	5.45	–	15.40	20.85	29.19	41.68
6507201030D	Extra for mitred hips or cutting to valleys; both sides measured	m	1.45	26.33	–	9.98	36.31	50.84	72.63

Property – Refurbishment

Rail 2003 Ref		Unit	Labour Hours	Labour Cost £	Plant Cost £	Materials Cost £	Unit Rate Base Cost £	Unit Rate Green Zone £	Unit Rate Red Zone £

6507   ROOFING

65072020   Tile roofing

6507202010   Clay plain tiling; machine made; smooth red; 265 × 165 mm; 64 mm lap; 19 × 38 mm treated sawn softwood battens; reinforced slaters felt type 1F:

6507202010A	sloping	m²	1.40	25.42	–	42.96	68.38	95.73	136.76
6507202010B	Extra for verges	m²	0.20	3.63	–	0.67	4.30	6.02	8.60
6507202010C	Extra for double eaves course	m²	0.20	3.63	–	6.70	10.33	14.45	20.65
6507202010D	Extra for half round ridge tiles	m²	0.85	15.43	–	15.60	31.03	43.45	62.07

6507202020   Interlocking tiles; concrete; granular finish; 418 × 330 mm; 75 mm lap; 19 × 38 mm treated sawn softwood battens; reinforced slaters underlining felt type 1F:

6507202020A	sloping	m²	0.22	4.00	–	11.44	15.44	21.60	30.86
6507202020B	Extra for verges; 150 mm fibre reinforced cement strip undercloak	m²	0.35	6.36	–	5.60	11.96	16.74	23.90
6507202020C	Extra for ridge tiles	m²	0.64	11.62	–	7.30	18.92	26.49	37.84

Property – Refurbishment

Rail 2003 Ref		Unit	Labour Hours	Labour Cost £	Plant Cost £	Materials Cost £	Unit Rate Base Cost £	Unit Rate Green Zone £	Unit Rate Red Zone £
6509	**IRONMONGERY**								
650910	**REMOVALS**								
65091010	General removals; load debris into skips								
6509101010	Remove door ironmongery and piece in:								
6509101010A	100 mm steel butt hinges	Nr	0.75	10.29	–	–	10.29	14.41	20.58
6509101010B	rim lock and furniture	Nr	0.80	10.98	–	–	10.98	15.37	21.95
6509101010C	mortice latch and furniture	Nr	1.20	16.46	–	–	16.46	23.05	32.93
6509101010D	mortice lock and furniture	Nr	1.55	21.27	–	–	21.27	29.77	42.53
650920	**REPAIRS AND ALTERATIONS**								
65092010	General renewals; prepare for and make good on completion								
6509201010	Take off and renew door ironmongery fixed to softwood:								
6509201010A	75 mm steel butt hinges	Nr	1.33	18.25	–	0.48	18.73	26.22	37.46
6509201010B	100 mm steel butt hinges	Pr	1.40	19.21	–	0.83	20.04	28.06	40.09
6509201010C	305 mm steel tee hinges	Pr	1.13	15.50	–	2.10	17.60	24.65	35.21
6509201010D	457 mm steel tee hinges	Pr	1.22	16.74	–	4.14	20.88	29.22	41.76
6509201010E	rim lock and furniture	Nr	1.58	21.68	–	27.46	49.14	68.79	98.27
6509201010F	mortice latch and furniture	Nr	1.75	24.01	–	20.09	44.10	61.73	88.19
6509201010G	mortice lock and furniture	Nr	1.84	25.25	–	20.46	45.71	63.98	91.40
6509201010H	Suffolk latch	Nr	1.33	18.25	–	7.75	26.00	36.40	52.00
6509201010I	150 mm bolt	Nr	0.95	13.03	–	2.64	15.67	21.95	31.35
6509201010J	225 mm bolt	Nr	1.05	14.41	–	3.39	17.80	24.92	35.59
6509201020	Take off and renew window furniture fixed to softwood:								
6509201020A	250 mm casement stay and pins	Nr	0.62	8.51	–	4.07	12.58	17.60	25.14
6509201020B	300 mm casement stay and pins	Nr	0.62	8.51	–	7.22	15.73	22.02	31.46
6509201020C	casement fastener	Nr	0.70	9.60	–	3.91	13.51	18.92	27.02
6509201020D	sash fastener	Nr	1.47	20.17	–	3.46	23.63	33.09	47.27
6509201020E	sash lift	Nr	0.47	6.45	–	2.86	9.31	13.04	18.62

Property – Refurbishment

Rail 2003 Ref		Unit	Labour Hours	Labour Cost £	Plant Cost £	Materials Cost £	Unit Rate Base Cost £	Unit Rate Green Zone £	Unit Rate Red Zone £
6509	**IRONMONGERY**								
650930	**NEW WORKS**								
65093010	**Door furniture**								
6509301010	Light pattern pressed steel butts and labour hanging door:								
6509301010A	50 mm	Pr	1.65	22.64	–	0.44	23.08	32.31	46.17
6509301010B	75 mm	Pr	1.70	23.32	–	0.48	23.80	33.32	47.61
6509301010C	100 mm	Pr	1.85	25.38	–	0.83	26.21	36.70	52.43
6509301015	Strong pattern pressed steel butts and labour hanging door:								
6509301015A	50 mm	Pr	1.70	23.32	–	0.44	23.76	33.27	47.54
6509301015B	75 mm	Pr	1.76	24.15	–	0.48	24.63	34.48	49.25
6509301015C	100 mm	Pr	1.76	24.15	–	0.83	24.98	34.98	49.96
6509301020	Steel rising butts and labour hanging door:								
6509301020A	75 × 70 mm	Pr	1.45	19.89	–	2.56	22.45	31.43	44.90
6509301020B	100 × 81 mm	Pr	1.70	23.32	–	3.61	26.93	37.71	53.87
6509301025	Steel washered brass butts and labour hanging door:								
6509301025A	50 mm	Pr	1.70	23.32	–	3.07	26.39	36.95	52.79
6509301025B	65 mm	Pr	1.75	24.01	–	3.33	27.34	38.27	54.68
6509301025C	75 mm	Pr	1.80	24.70	–	4.61	29.31	41.02	58.60
6509301025D	100 mm	Pr	1.90	26.07	–	5.84	31.91	44.67	63.81
6509301030	Steel tee hinges and labour hanging door:								
6509301030A	305 mm	Pr	1.35	18.52	–	2.10	20.62	28.87	41.24
6509301030B	457 mm	Pr	1.55	21.27	–	4.14	25.41	35.56	50.81
6509301035	Interior straight sliding door gear; top track with wheel hangers; door guides; stops; finger pulls; steel pelmet and labour hanging door:								
6509301035A	35-44 mm thick single door	Nr	2.45	33.61	–	15.42	49.03	68.65	98.07

Property – Refurbishment

Rail 2003 Ref		Unit	Labour Hours	Labour Cost	Plant Cost	Materials Cost	Unit Rate	Unit Rate	Unit Rate
				£	£	£	Base Cost £	Green Zone £	Red Zone £

6509 IRONMONGERY

65093010 Door furniture

6509301040 Locks and latches:

6509301040A	rim lock and furniture	Nr	1.58	21.68	–	18.19	39.87	55.82	79.75
6509301040B	mortice latch and furniture	Nr	1.75	24.01	–	20.09	44.10	61.73	88.19
6509301040C	cylinder night latch	Nr	1.84	25.25	–	20.09	45.34	63.46	90.66
6509301040D	cylinder night latch	Nr	2.20	30.18	–	27.07	57.25	80.16	114.52
6509301040E	Suffolk latch	Nr	1.20	16.46	–	7.75	24.21	33.90	48.43

6509301045 General door furniture:

6509301045A	escutcheon	Nr	0.50	6.86	–	0.29	7.15	10.01	14.31
6509301045B	Bales catch	Nr	0.85	11.66	–	3.76	15.42	21.59	30.84
6509301045C	cupboard catch	Nr	0.90	12.35	–	5.50	17.85	24.99	35.70
6509301045D	cupboard or drawer lock	Nr	1.10	15.09	–	7.25	22.34	31.28	44.68

6509301050 Door closers:

6509301050A	overhead closer; surface fixing	Nr	1.75	24.01	–	11.43	35.44	49.61	70.87
6509301050B	Perko closer	Nr	1.75	24.01	–	11.66	35.67	49.93	71.34
6509301050C	coil gate spring	Nr	0.40	5.49	–	6.00	11.49	16.08	22.98

6509301055 Bolts:

6509301055A	100 mm barrel bolt; straight	Nr	0.40	5.49	–	2.64	8.13	11.38	16.26
6509301055B	150 mm barrel bolt; straight	Nr	0.45	6.17	–	3.39	9.56	13.39	19.13
6509301055C	200 mm barrel bolt; straight	Nr	0.50	6.86	–	4.14	11.00	15.39	22.00
6509301055D	100 mm barrel bolt; cranked	Nr	0.40	5.49	–	9.03	14.52	20.32	29.04
6509301055E	150 mm barrel bolt; cranked	Nr	0.45	6.17	–	12.79	18.96	26.54	37.93
6509301055F	200 mm barrel bolt; cranked	Nr	0.50	6.86	–	15.04	21.90	30.65	43.79
6509301055G	455 mm monkey tail	Nr	0.55	7.55	–	18.43	25.98	36.36	51.95
6509301055H	225 mm flush	Nr	0.85	11.66	–	15.04	26.70	37.38	53.39
6509301055I	indicator	Nr	1.20	16.46	–	10.07	26.53	37.15	53.08
6509301055J	single door panic	Nr	1.75	24.01	–	99.26	123.27	172.57	246.54
6509301055K	double door panic	Nr	2.30	31.56	–	117.32	148.88	208.42	297.74

6509301060 Handles and pulls:

6509301060A	door handle	Nr	0.30	4.12	–	4.14	8.26	11.55	16.51
6509301060B	drawer pull	Nr	0.20	2.74	–	6.50	9.24	12.94	18.49
6509301060C	cupboard knob	Nr	0.20	2.74	–	7.15	9.89	13.85	19.79

Property – Refurbishment

Rail 2003 Ref		Unit	Labour Hours	Labour Cost £	Plant Cost £	Materials Cost £	Unit Rate Base Cost £	Unit Rate Green Zone £	Unit Rate Red Zone £
6509	**IRONMONGERY**								
65093010	**Door furniture**								
6509301065	Plates:								
6509301065A	door push plate	Nr	0.40	5.49	–	5.42	10.91	15.26	21.81
6509301065B	letter plate including opening in door	Nr	1.75	24.01	–	6.17	30.18	42.24	60.35
65093020	**Window furniture**								
6509302010	Window fittings:								
6509302010A	250 mm casement stay with pins	Nr	0.40	5.49	–	4.07	9.56	13.37	19.11
6509302010B	300 mm casement stay with pins	Nr	0.40	5.49	–	1.00	6.49	9.08	12.98
6509302010C	casement fastener	Nr	0.50	6.86	–	4.75	11.61	16.25	23.22
6509302010D	sash lift	Nr	0.30	4.12	–	3.50	7.62	10.66	15.23
6509302010E	sash fastener	Nr	1.30	17.84	–	8.60	26.44	37.01	52.87
6509302010F	quadrant stay	Nr	0.40	5.49	–	10.50	15.99	22.38	31.98
65093050	**General ironmongery**								
6509305010	Sundry items:								
6509305010A	hat and coat hooks	Nr	0.23	3.16	–	5.00	8.16	11.42	16.31
6509305010B	cabin hook and eye	Nr	0.30	4.12	–	8.50	12.62	17.66	25.23
6509305010C	padlock hasp and staple	Nr	0.30	4.12	–	3.00	7.12	9.96	14.23
6509305010D	swivel locking bar	Nr	0.40	5.49	–	35.00	40.49	56.68	80.98
6509305010E	rubber door stop	Nr	0.20	2.74	–	0.30	3.04	4.26	6.09
6509305010F	door buffer	Nr	0.30	4.12	–	6.50	10.62	14.86	21.23
6509305010G	shelf bracket	Nr	0.40	5.49	–	3.25	8.74	12.23	17.48
6509305010H	numerals	Nr	0.20	2.74	–	2.65	5.39	7.55	10.79

Property – Refurbishment

Rail 2003 Ref		Unit	Labour Hours	Labour Cost £	Plant Cost £	Materials Cost £	Unit Rate Base Cost £	Unit Rate Green Zone £	Unit Rate Red Zone £

6510 STRUCTURAL STEELWORK

651010 WORKS TO EXISTING

65101010 Removal of steelwork elements

6510101010 Steel beams, joists and lintels:

6510101010A	not exceeding 10 Kg/m	Tonne	18.00	318.58	–	–	318.58	446.01	637.16
6510101010B	10 – 20 Kg/m......................	Tonne	17.00	300.88	–	–	300.88	421.24	601.77
6510101010C	20 – 50 Kg/m......................	Tonne	16.25	287.61	–	–	287.61	402.65	575.22

6510101020 Steel columns and stanchions:

6510101020A	not exceeding 10 Kg/m	Tonne	18.00	318.58	–	–	318.58	446.01	637.16
6510101020B	10 – 20 Kg/m......................	Tonne	17.00	300.88	–	–	300.88	421.24	601.77
6510101020C	20 – 50 Kg/m......................	Tonne	16.25	287.61	–	–	287.61	402.65	575.22

6510101030 Steel rails and purlins:

6510101030A	not exceeding 10 Kg/m	Tonne	22.00	389.38	–	–	389.38	545.13	778.76
6510101030B	10 – 20 Kg/m......................	Tonne	21.00	371.68	–	–	371.68	520.35	743.36
6510101030C	20 – 50 Kg/m......................	Tonne	19.25	340.71	–	–	340.71	476.99	681.41

651020 NEW WORKS

65102010 Structural steel framing

6510201010 Universal columns; BS 4360; shot blasted and primed at works; fixed at not exceeding 3.00 m above ground level:

6510201010A	356 × 406 mm × 235 Kg/m	Tonne	13.00	230.09	56.58	606.59	893.26	1250.56	1786.51
6510201010B	305 × 305 mm × 283 Kg/m	Tonne	12.00	212.39	52.23	606.59	871.21	1219.69	1742.41
6510201010C	254 × 254 mm × 167 Kg/m	Tonne	15.00	265.49	65.28	586.37	917.14	1284.00	1834.28
6510201010D	203 × 203 mm × 52 Kg/m	Tonne	18.00	318.58	78.34	566.15	963.07	1348.30	1926.14

6510201020 Universal columns; BS 4360; shot blasted and primed at works; fixed at exceeding 3.00 m but not exceeding 6.00 m above ground level:

6510201020A	356 × 406 mm × 235 Kg/m	Tonne	18.00	318.58	56.58	606.59	981.75	1374.45	1963.50
6510201020B	305 × 305 mm × 283 Kg/m	Tonne	17.00	300.88	52.23	606.59	959.70	1343.59	1919.40
6510201020C	254 × 254 mm × 167 Kg/m	Tonne	21.00	371.68	65.28	586.37	1023.33	1432.67	2046.67
6510201020D	203 × 203 mm × 52 Kg/m	Tonne	25.00	442.48	78.34	566.15	1086.97	1521.76	2173.93

Property – Refurbishment

Rail 2003 Ref		Unit	Labour Hours	Labour Cost £	Plant Cost £	Materials Cost £	Unit Rate Base Cost £	Unit Rate Green Zone £	Unit Rate Red Zone £
6510	**STRUCTURAL STEELWORK**								
65102010	Structural steel framing								
6510201030	Universal beams; BS 4360; shot blasted and primed at works; fixed at not exceeding 3.00 m above ground level:								
6510201030A	610 × 305 mm × 238 Kg/m	Tonne	13.00	230.09	56.58	606.59	893.26	1250.56	1786.51
6510201030B	457 × 191 mm × 98 Kg/m	Tonne	15.00	265.49	65.28	586.37	917.14	1284.00	1834.28
6510201030C	406 × 178 mm × 74 Kg/m	Tonne	16.00	283.18	69.64	566.15	918.97	1286.56	1837.94
6510201030D	305 × 165 mm × 54 Kg/m	Tonne	18.00	318.58	78.34	566.15	963.07	1348.30	1926.14
6510201030E	254 × 146 mm × 43 Kg/m	Tonne	19.00	336.28	82.69	566.15	985.12	1379.17	1970.25
6510201030F	203 × 133 mm × 30 Kg/m	Tonne	20.00	353.98	87.05	545.93	986.96	1381.73	1973.91
6510201040	Universal beams; BS 4360; shot blasted and primed at works; fixed at exceeding 3.00 m but not exceeding 6.00 m above ground level:								
6510201040A	610 × 305 mm × 238 Kg/m	Tonne	18.00	318.58	56.58	606.59	981.75	1374.45	1963.50
6510201040B	457 × 191 mm × 98 Kg/m	Tonne	21.00	371.68	65.28	586.37	1023.33	1432.67	2046.67
6510201040C	406 × 178 mm × 74 Kg/m	Tonne	22.00	389.38	69.64	566.15	1025.17	1435.23	2050.33
6510201040D	305 × 165 mm × 54 Kg/m	Tonne	25.00	442.48	78.34	566.15	1086.97	1521.76	2173.93
6510201040E	254 × 146 mm × 43 Kg/m	Tonne	26.00	460.17	82.69	566.15	1109.01	1552.62	2218.04
6510201040F	203 × 133 mm × 30 Kg/m	Tonne	27.00	477.87	87.05	545.93	1110.85	1555.18	2221.70
65102020	Support steelwork and bracing								
6510202010	Equal angles; BS 4360; shot blasted and primed at works:								
6510202010A	200 × 200 × 16 mm × 48.50 Kg/m	Tonne	15.00	265.49	7.85	545.93	819.27	1146.97	1638.53
6510202010B	150 × 150 × 10 mm × 23.00 Kg/m	Tonne	17.00	300.88	8.90	535.82	845.60	1183.84	1691.20
6510202010C	100 × 100 × 15 mm × 21.90 Kg/m	Tonne	18.00	318.58	9.42	535.82	863.82	1209.35	1727.64
6510202010D	80 × 80 × 10 mm × 11.90 Kg/m	Tonne	24.00	424.78	12.56	525.71	963.05	1348.26	1926.09
6510202010E	50 × 50 × 8 mm × 5.82 Kg/m	Tonne	26.00	460.17	13.61	525.71	999.49	1399.28	1998.98
6510202020	Unequal angles; BS 4360; shot blasted and primed at works:								
6510202020A	200 × 150 × 18 mm × 47.10 Kg/m	Tonne	15.00	265.49	7.85	545.93	819.27	1146.97	1638.53
6510202020B	200 × 100 × 15 mm × 33.70 Kg/m	Tonne	16.00	283.18	8.37	545.93	837.48	1172.48	1674.97
6510202020C	150 × 75 × 15 mm × 24.80 Kg/m	Tonne	17.00	300.88	8.90	545.93	855.71	1197.99	1711.42
6510202020D	100 × 75 × 12 mm × 15.40 Kg/m	Tonne	22.00	389.38	11.51	545.93	946.82	1325.55	1893.64
6510202020E	75 × 50 × 8 mm × 7.39 Kg/m	Tonne	26.00	460.17	13.61	525.71	999.49	1399.28	1998.98
6510202020F	60 × 30 × 6 mm × 3.99 Kg/m	Tonne	28.00	495.57	14.65	525.71	1035.93	1450.30	2071.86

Property – Refurbishment

Rail 2003 Ref		Unit	Labour Hours	Labour Cost	Plant Cost	Materials Cost	Unit Rate	Unit Rate	Unit Rate
				£	£	£	Base Cost £	Green Zone £	Red Zone £

6510 STRUCTURAL STEELWORK

65102030 Fixings

6510203010 Bolts; BS 4190; high strength friction grip (HSFG) black metric hexagon head with nut and washer:

6510203010A	M6 × 25 mm	Nr	0.05	0.88	–	0.12	1.00	1.40	2.00
6510203010B	M6 × 50 mm	Nr	0.06	1.06	–	0.13	1.19	1.67	2.38
6510203010C	M6 × 75 mm	Nr	0.07	1.24	–	0.18	1.42	1.98	2.84
6510203010D	M8 × 25 mm	Nr	0.07	1.24	–	0.18	1.42	1.98	2.84
6510203010E	M8 × 50 mm	Nr	0.09	1.59	–	0.20	1.79	2.52	3.60
6510203010F	M8 × 75 mm	Nr	0.10	1.77	–	0.23	2.00	2.80	4.00
6510203010G	M8 × 100 mm	Nr	0.11	1.95	–	0.31	2.26	3.16	4.51
6510203010H	M10 × 50 mm	Nr	0.11	1.95	–	0.31	2.26	3.16	4.51
6510203010I	M10 × 75 mm	Nr	0.12	2.12	–	0.36	2.48	3.47	4.97
6510203010J	M10 × 100 mm	Nr	0.13	2.30	–	0.53	2.83	3.96	5.66
6510203010K	M12 × 50 mm	Nr	0.13	2.30	–	0.45	2.75	3.85	5.50
6510203010L	M12 × 75 mm	Nr	0.14	2.48	–	0.53	3.01	4.21	6.02
6510203010M	M12 × 100 mm	Nr	0.15	2.66	–	0.62	3.28	4.58	6.54
6510203010P	M16 × 100 mm	Nr	0.19	3.36	–	1.07	4.43	6.21	8.87

6510203020 Drill steelwork for bolts:

6510203020A	6 mm steel for M6 bolt	Nr	0.15	2.66	–	–	2.66	3.72	5.31
6510203020B	6 mm steel for M8 bolt	Nr	0.16	2.83	–	–	2.83	3.96	5.66
6510203020C	6 mm steel for M10 bolt	Nr	0.18	3.19	–	–	3.19	4.46	6.37
6510203020D	6 mm steel for M12 bolt	Nr	0.20	3.54	–	–	3.54	4.96	7.08
6510203020E	6 mm steel for M16 bolt	Nr	0.22	3.89	–	–	3.89	5.45	7.79
6510203020F	8 mm steel for M6 bolt	Nr	0.18	3.19	–	–	3.19	4.46	6.37
6510203020G	8 mm steel for M8 bolt	Nr	0.19	3.36	–	–	3.36	4.71	6.73
6510203020H	8 mm steel for M10 bolt	Nr	0.20	3.54	–	–	3.54	4.96	7.08
6510203020I	8 mm steel for M12 bolt	Nr	0.22	3.89	–	–	3.89	5.45	7.79
6510203020J	8 mm steel for M16 bolt	Nr	0.24	4.25	–	–	4.25	5.95	8.50
6510203020K	10 mm steel for M6 bolt	Nr	0.19	3.36	–	–	3.36	4.71	6.73
6510203020L	10 mm steel for M8 bolt	Nr	0.21	3.72	–	–	3.72	5.20	7.43
6510203020M	10 mm steel for M10 bolt	Nr	0.23	4.07	–	–	4.07	5.70	8.14
6510203020N	10 mm steel for M12 bolt	Nr	0.24	4.25	–	–	4.25	5.95	8.50
6510203020O	10 mm steel for M16 bolt	Nr	0.25	4.42	–	–	4.42	6.20	8.85

Property – Refurbishment

Rail 2003 Ref		Unit	Labour Hours	Labour Cost £	Plant Cost £	Materials Cost £	Unit Rate Base Cost £	Unit Rate Green Zone £	Unit Rate Red Zone £
6510	**STRUCTURAL STEELWORK**								
65102050	**Wedging and grouting bases**								
6510205010	Wedging up with steel shims and grouting under bases with cement mortar (1:3); 25 mm thick:								
6510205010A	not exceeding 0.10 m^2	Nr	0.50	5.98	–	0.34	6.32	8.84	12.62
6510205010B	0.10 – 0.25 m^2	Nr	0.60	7.17	–	0.67	7.84	10.98	15.68
6510205010C	0.25 – 0.50 m^2	Nr	0.75	8.97	–	0.67	9.64	13.49	19.27
6510205010D	exceeding 0.50 m^2	m^2	0.95	11.36	–	1.34	12.70	17.78	25.39
6510205020	Wedging up with steel shims and grouting under bases with cement mortar (1:3); 50 mm thick:								
6510205020A	not exceeding 0.10 m^2	Nr	0.55	6.58	–	0.67	7.25	10.15	14.49
6510205020B	0.10 – 0.25 m^2	Nr	0.66	7.89	–	0.67	8.56	11.99	17.12
6510205020C	0.25 – 0.50 m^2	Nr	0.83	9.92	–	2.01	11.93	16.71	23.87
6510205020D	exceeding 0.50 m^2	m^2	1.05	12.55	–	2.01	14.56	20.39	29.13
65102060	**Holding down assemblies**								
6510206010	Holding down bolts and assemblies; M20 bolt with 75 × 75 mm tacked plate washer; nut and washer:								
6510206010A	300 mm long	Nr	1.25	14.94	–	12.00	26.94	37.72	53.89
6510206010B	375 mm long	Nr	1.50	17.93	–	12.60	30.53	42.74	61.06
6510206010C	450 mm long	Nr	1.65	19.72	–	14.25	33.97	47.56	67.95
6510206020	Holding down bolts and assemblies; M16 indented bolt with nut and washers:								
6510206020A	120 mm long	Nr	1.25	14.94	–	8.24	23.18	32.46	46.37
6510206020B	160 mm long	Nr	1.65	19.72	–	9.05	28.77	40.28	57.55
6510206020C	200 mm long	Nr	1.65	19.72	–	11.19	30.91	43.28	61.83

Property – Refurbishment

Rail 2003 Ref		Unit	Labour Hours	Labour Cost £	Plant Cost £	Materials Cost £	Unit Rate Base Cost £	Unit Rate Green Zone £	Unit Rate Red Zone £
6511	**METALWORK**								
651110	**WORKS TO EXISTING**								
65111010	**Removal of metalwork items**								
6511101010	Take down stairs, steps, balustrades and handrails including disposal:								
6511101010A	stair or step unit, straight flight	Nr	4.00	47.82	–	–	47.82	66.94	95.63
6511101010B	stair or step unit, two or more flights	Nr	6.70	80.09	–	–	80.09	112.13	160.18
6511101010C	balustrading	m	0.60	7.17	–	–	7.17	10.04	14.34
6511101010D	handrail	m	0.35	4.18	–	–	4.18	5.86	8.37
6511101020	Signage, including disposal:								
6511101020A	road signs complete with posts	Nr	1.35	16.14	–	–	16.14	22.59	32.28
6511101020B	general signage complete with fixing brackets	Nr	1.10	13.15	–	–	13.15	18.41	26.30
6511101030	Signage, setting aside for re-use:								
6511101030A	road signs complete with posts	Nr	1.65	19.72	–	–	19.72	27.61	39.45
6511101030B	general signage complete with fixing brackets	Nr	1.40	16.74	–	–	16.74	23.43	33.47
6511101040	Take down and remove metal fencing and barriers complete with posts, including disposal:								
6511101040A	weldmesh fabrication	m^2	2.10	25.10	–	–	25.10	35.14	50.21
6511101040B	balustrade type	m^2	2.35	28.09	–	–	28.09	39.33	56.18
6511101050	Remove general metalwork, including disposal:								
6511101050A	window frame; small	Nr	0.75	8.97	–	–	8.97	12.55	17.93
6511101050B	window frame; medium	Nr	0.90	10.76	–	–	10.76	15.06	21.52
6511101050C	window frame; large	Nr	1.15	13.75	–	–	13.75	19.25	27.49
6511101050D	door frame; single	Nr	0.95	11.36	–	–	11.36	15.90	22.71
6511101050E	door frame; double	Nr	1.25	14.94	–	–	14.94	20.92	29.89
6511101050F	fixing bracket	Nr	0.30	3.59	–	–	3.59	5.02	7.17

Property – Refurbishment

Rail 2003 Ref		Unit	Labour Hours	Labour Cost £	Plant Cost £	Materials Cost £	Unit Rate Base Cost £	Unit Rate Green Zone £	Unit Rate Red Zone £
6511	METALWORK								
651120	NEW WORK								
65112010	General fixtures and equipment								
6511201010	Matwell frames; fabricated, welded or brazed construction with lugs; galvanised mild steel angle section; 32 × 32 × 6 mm; size:								
6511201010A	600 × 450 mm	Nr	0.80	14.16	–	52.86	67.02	93.83	134.04
6511201010B	750 × 450 mm	Nr	0.90	15.93	–	55.87	71.80	100.51	143.59
6511201010C	750 × 600 mm	Nr	1.00	17.70	–	61.45	79.15	110.81	158.30
6511201020	Matwell frames; fabricated welded or brazed construction with lugs; polished brass angle section; 32 × 32 × 6 mm; size:								
6511201020A	600 × 450 mm	Nr	1.20	21.24	–	126.75	147.99	207.18	295.98
6511201020B	750 × 450 mm	Nr	1.30	23.01	–	143.57	166.58	233.21	333.16
6511201020C	750 × 600 mm	Nr	1.40	24.78	–	158.29	183.07	256.29	366.13
6511201030	Matwell frames; fabricated, welded or brazed construction with lugs; aluminium angle section; 32 × 32 × 6 mm; size:								
6511201030A	600 × 450 mm	Nr	1.10	19.47	–	63.86	83.33	116.66	166.65
6511201030B	750 × 450 mm	Nr	1.20	21.24	–	71.06	92.30	129.22	184.61
6511201030C	750 × 600 mm	Nr	1.30	23.01	–	79.59	102.60	143.64	205.21

Property – Refurbishment

Rail 2003 Ref		Unit	Labour Hours	Labour Cost £	Plant Cost £	Materials Cost £	Unit Rate Base Cost £	Unit Rate Green Zone £	Unit Rate Red Zone £
6514	**INTERNAL FINISHINGS**								
651410	**REMOVAL OF OLD WORK**								
65141010	**Removal of surface finishes**								
6514101010	**Floors:**								
6514101010A	linoleum	m²	0.12	1.43	–	–	1.43	2.01	2.87
6514101010B	carpet and underlay	m²	0.13	1.55	–	–	1.55	2.18	3.11
6514101010C	screed	m²	0.50	5.98	–	–	5.98	8.37	11.95
6514101010D	granolithic screed	m²	0.67	8.01	–	–	8.01	11.21	16.01
6514101010E	terrazzo or ceramic tiles; screed	m²	1.10	13.15	–	–	13.15	18.40	26.29
6514101020	**Walls:**								
6514101020A	plasterboard	m²	0.45	5.38	–	–	5.38	7.53	10.76
6514101020B	plaster	m²	0.22	2.63	–	–	2.63	3.68	5.26
6514101020C	cement rendering	m²	0.45	5.38	–	–	5.38	7.53	10.76
6514101020D	tiling and screed	m²	0.55	6.57	–	–	6.57	9.20	13.15
6514101030	**Ceilings:**								
6514101030A	plasterboard and skim; including withdrawing nails	m²	0.33	3.94	–	–	3.94	5.52	7.89
6514101030B	wood lath and plaster; including withdrawing nails	m²	0.55	6.57	–	–	6.57	9.20	13.15
6514101030C	suspended	m²	0.83	9.92	–	–	9.92	13.89	19.84
6514101030D	plaster moulded cornice; per 25 mm	m	0.17	2.03	–	–	2.03	2.84	4.06
65141020	**Preparation of surfaces**								
6514102010	**Prepare surface, apply Unibond universal pva adhesive and sealer to receive plaster or cement rendering; walls:**								
6514102010A	existing cement and sand base	m²	0.30	5.91	–	0.59	6.50	9.10	13.00
6514102010B	existing glazed tile base	m²	0.24	4.73	–	0.40	5.13	7.18	10.26
6514102010C	existing paint base	m²	0.28	5.52	–	0.46	5.98	8.36	11.95
6514102010D	existing concrete base	m²	0.30	5.91	–	0.46	6.37	8.91	12.74

Property – Refurbishment

Rail 2003 Ref		Unit	Labour Hours	Labour Cost £	Plant Cost £	Materials Cost £	Unit Rate Base Cost £	Unit Rate Green Zone £	Unit Rate Red Zone £
6514	**INTERNAL FINISHINGS**								
65141020	**Preparation of surfaces**								
6514102020	**Prepare surface, apply Unibond universal pva adhesive and sealer to receive plaster or cement rendering; ceilings:**								
6514102020A	existing cement and sand base	m²	0.37	7.29	–	0.59	7.88	11.03	15.76
6514102020B	existing painted base	m²	0.34	6.70	–	0.46	7.16	10.02	14.32
6514102020C	existing concrete base	m²	0.37	7.29	–	0.50	7.79	10.90	15.58
6514102030	**Hack down defective ceiling plaster and laths:**								
6514102030A	remove nails	m²	0.85	16.75	–	–	16.75	23.44	33.49
6514102040	**Take down temporary boarded linings and clean joists:**								
6514102040A	to ceilings	m²	0.40	7.88	–	–	7.88	11.03	15.76
651440	**NEW WORK**								
65144010	**In situ finishings**								
6514401010	**Carlite plaster; 8 mm bonding; 2 mm finish; to concrete or plasterboard base:**								
6514401010A	walls; over 300 mm wide	m²	0.52	10.24	–	2.22	12.46	17.44	24.92
6514401010B	walls; not exceeding 300 mm wide	m²	0.78	15.37	–	2.22	17.59	24.61	35.16
6514401010C	ceilings; over 300 mm wide	m²	0.65	12.81	–	2.22	15.03	21.03	30.04
6514401010D	ceilings; not exceeding 300 mm wide	m²	0.97	19.11	–	2.22	21.33	29.85	42.65
6514401020	**Carlite plaster; 11 mm browning; 2 mm finish; to brick or block base:**								
6514401020A	walls; over 300 mm wide	m²	0.53	10.44	–	2.26	12.70	17.79	25.40
6514401020B	walls; not exceeding 300 mm wide	m²	0.79	15.56	–	2.26	17.82	24.96	35.65

Property – Refurbishment

Rail 2003 Ref		Unit	Labour Hours	Labour Cost £	Plant Cost £	Materials Cost £	Unit Rate Base Cost £	Unit Rate Green Zone £	Unit Rate Red Zone £
6514	**INTERNAL FINISHINGS**								
65144010	**In situ finishings**								
6514401030	Carlite plaster; 11 mm metal lathing undercoat; 2 mm finish; to metal lathing base:								
6514401030A	walls; over 300 mm wide............	m²	0.53	10.44	–	3.74	14.18	19.86	28.37
6514401030B	walls; not exceeding 300 mm wide ...	m²	0.79	15.56	–	3.74	19.30	27.03	38.62
6514401030C	ceilings; over 300 mm wide	m²	0.68	13.40	–	3.74	17.14	23.99	34.28
6514401030D	ceilings; not exceeding 300 mm wide .	m²	1.02	20.09	–	3.74	23.83	33.37	47.68
6514401040	Plaster; Universal one coat; 10 mm work; concrete base:								
6514401040A	walls; over 300 mm wide............	m²	0.40	7.57	–	1.71	9.28	13.00	18.56
6514401040B	walls; not exceeding 300 mm wide ...	m²	0.60	11.36	–	1.71	13.07	18.30	26.13
6514401040C	ceilings; over 300 mm wide	m²	0.52	9.84	–	1.71	11.55	16.18	23.11
6514401040D	ceilings; not exceeding 300 mm wide .	m²	0.77	14.58	–	1.71	16.29	22.81	32.57
6514401050	Plaster; Universal one coat; 13 mm work; brick or block base:								
6514401050A	walls; over 300 mm wide............	m²	0.42	7.95	–	1.71	9.66	13.53	19.32
6514401050B	walls; not exceeding 300 mm wide ...	m²	0.62	11.74	–	1.71	13.45	18.83	26.89
6514401060	Plaster; Universal one coat; 5 mm work; plasterboard base:								
6514401060A	walls; over 300 mm wide............	m²	0.16	3.03	–	1.71	4.74	6.64	9.48
6514401060B	walls; not exceeding 300 mm wide ...	m²	0.24	4.54	–	1.71	6.25	8.76	12.51
6514401060C	ceilings; over 300 mm wide	m²	0.18	3.41	–	1.71	5.12	7.17	10.23
6514401060D	ceilings; not exceeding 300 mm wide .	m²	0.27	5.11	–	1.71	6.82	9.56	13.64
6514401070	Thistle; 3 mm one coat board finish; plasterboard base:								
6514401070A	walls; over 300 mm wide............	m²	0.37	7.00	–	0.17	7.17	10.05	14.35
6514401070B	walls; not exceeding 300 mm wide ...	m²	0.57	10.79	–	0.17	10.96	15.35	21.92
6514401070C	ceilings; over 300 mm wide	m²	0.49	9.28	–	0.17	9.45	13.23	18.89
6514401070D	ceilings; not exceeding 300 mm wide .	m²	0.74	14.01	–	0.17	14.18	19.85	28.35

Property – Refurbishment

Rail 2003 Ref		Unit	Labour Hours	Labour Cost £	Plant Cost £	Materials Cost £	Unit Rate Base Cost £	Unit Rate Green Zone £	Unit Rate Red Zone £

6514 INTERNAL FINISHINGS

65144010 In situ finishings

6514401080 Thistle; 10 mm cement and sand (1:3); 3 mm finish; concrete, brick or block base:

| 6514401080A | walls; over 300 mm wide | m² | 0.53 | 10.03 | – | 1.44 | 11.47 | 16.06 | 22.94 |
| 6514401080B | walls; not exceeding 300 mm wide | m² | 0.78 | 14.77 | – | 1.44 | 16.21 | 22.69 | 32.41 |

65144030 Labours on plastering

6514403010 Rounded internal angle:

| 6514403010A | not exceeding 10 mm radius | m | 0.06 | 1.14 | – | – | 1.14 | 1.59 | 2.27 |
| 6514403010B | exceeding 10 mm radius | m | 0.08 | 1.51 | – | – | 1.51 | 2.12 | 3.03 |

6514403020 Rounded external angle:

| 6514403020A | not exceeding 10 mm radius | m | 0.07 | 1.33 | – | – | 1.33 | 1.86 | 2.65 |
| 6514403020B | exceeding 10 mm radius | m | 0.10 | 1.89 | – | – | 1.89 | 2.65 | 3.79 |

65144035 Plaster beads and the like

6514403510 Catnic steel beads; fixed with plaster dabs:

6514403510A	standard angle bead	m	0.04	0.76	–	0.84	1.60	2.23	3.19
6514403510B	dry wall angle bead	m	0.04	0.76	–	0.93	1.69	2.36	3.37
6514403510C	dry wall stop bead; 3 mm	m	0.04	0.76	–	1.19	1.95	2.72	3.88
6514403510D	dry wall stop bead; 6 mm	m	0.04	0.76	–	1.28	2.04	2.85	4.06
6514403510E	renderstop	m	0.04	0.76	–	1.12	1.88	2.63	3.75
6514403510F	plaster stop; 12 mm	m	0.04	0.76	–	1.10	1.86	2.59	3.70
6514403510G	plaster stop; 15 mm	m	0.04	0.76	–	1.48	2.24	3.14	4.48
6514403510H	plaster stop; 18 mm	m	0.04	0.76	–	1.41	2.17	3.03	4.32

Property – Refurbishment

Rail 2003 Ref		Unit	Labour Hours	Labour Cost £	Plant Cost £	Materials Cost £	Unit Rate Base Cost £	Unit Rate Green Zone £	Unit Rate Red Zone £
6514	**INTERNAL FINISHINGS**								
65144040	**Portland cement finishes**								
6514404010	Render; cement and sand (1:3); 6 mm; dubbing out; brick or block base:								
6514404010A	walls; over 300 mm wide	m²	0.14	2.65	–	0.67	3.32	4.65	6.64
6514404010B	wallls; not exceeding 300 mm wide	m	0.06	1.14	–	0.20	1.34	1.87	2.67
6514404020	Render; cement and sand (1:3); 13 mm; dubbing out; brick or block base:								
6514404020A	walls; over 300 mm wide	m²	0.18	3.41	–	0.67	4.08	5.71	8.15
6514404020B	walls; not exceeding 300 mm wide	m	0.07	1.33	–	0.20	1.53	2.14	3.05
6514404030	Render; cement and sand (1:3); 19 mm; dubbing out; brick or block base:								
6514404030A	walls; over 300 mm wide	m²	0.27	5.11	–	1.34	6.45	9.04	12.90
6514404030B	walls; not exceeding 300 mm wide	m	0.11	2.08	–	0.67	2.75	3.85	5.50
6514404040	Render; cement and sand (1:3); 25 mm; dubbing out; brick or block base:								
6514404040A	walls; over 300 mm wide	m²	0.36	6.81	–	2.01	8.82	12.36	17.66
6514404040B	walls; not exceeding 300 mm wide	m	0.15	2.84	–	0.67	3.51	4.91	7.02
65144045	**Labours on rendering**								
6514404510	Rounded internal angle:								
6514404510A	not exceeding 10 mm radius	m	0.06	1.14	–	–	1.14	1.59	2.27
6514404510B	exceeding 10 mm radius	m	0.08	1.51	–	–	1.51	2.12	3.03
6514404520	Rounded external angle:								
6514404520A	not exceeding 10 mm radius	m	0.08	1.51	–	–	1.51	2.12	3.03
6514404520B	exceeding 10 mm radius	m	0.10	1.89	–	–	1.89	2.65	3.79
6514404530	Expamet steel beads; stainless steel:								
6514404530A	angle bead	m	0.04	0.76	–	2.63	3.39	4.74	6.77
6514404530B	movement bead	m	0.04	0.76	–	6.50	7.26	10.16	14.51
6514404530C	stop bead; 10 mm	m	0.04	0.76	–	2.35	3.11	4.35	6.21
6514404530D	stop bead; 13 mm	m	0.04	0.76	–	2.35	3.11	4.35	6.21
6514404530E	stop bead; 16 mm	m	0.04	0.76	–	2.35	3.11	4.35	6.21
6514404530F	stop bead; 19 mm	m	0.04	0.76	–	1.74	2.50	3.50	4.99

Property – Refurbishment

Rail 2003 Ref		Unit	Labour Hours	Labour Cost £	Plant Cost £	Materials Cost £	Unit Rate Base Cost £	Unit Rate Green Zone £	Unit Rate Red Zone £

6514 INTERNAL FINISHINGS

65144045 Labours on rendering

6514404530 Expamet steel beads; stainless steel:

6514404530G	external render stop	m	0.04	0.76	–	1.74	2.50	3.50	4.99

65144050 Cement screeds

6514405010 Screed; cement and sand (1:3); floors; level and to falls:

6514405010A	25 mm thick	m²	0.25	4.73	–	2.01	6.74	9.44	13.49
6514405010B	32 mm thick	m²	0.27	5.11	–	2.68	7.79	10.92	15.59
6514405010C	38 mm thick	m²	0.29	5.49	–	2.68	8.17	11.44	16.35
6514405010D	50 mm thick	m²	0.32	6.06	–	4.03	10.09	14.12	20.16
6514405010E	65 mm thick	m²	0.36	6.81	–	4.70	11.51	16.11	23.02

6514405020 Skirtings; cement and sand (1:3); 150 mm high; fair edge; ends and the like:

6514405020A	straight	m	0.20	3.79	–	0.13	3.92	5.49	7.84
6514405020B	curved	m	0.33	6.25	–	0.07	6.32	8.85	12.64

65144055 Granolithic screeds

6514405510 Screed; cement and granite chippings (2:5); floors; level and to falls:

6514405510A	25 mm thick	m²	0.27	5.11	–	2.34	7.45	10.44	14.90
6514405510B	32 mm thick	m²	0.29	5.49	–	2.34	7.83	10.96	15.66
6514405510C	38 mm thick	m²	0.32	6.06	–	3.34	9.40	13.16	18.79
6514405510D	50 mm thick	m²	0.36	6.81	–	4.01	10.82	15.16	21.65
6514405510E	65 mm thick	m²	0.42	7.95	–	4.68	12.63	17.69	25.27

6514405520 Skirtings; rounded top edge; coved junction to paving; ends, angles and the like; 150 mm high:

6514405520A	13 mm	m	0.42	7.95	–	0.63	8.58	12.02	17.17
6514405520B	19 mm	m	0.49	9.28	–	0.63	9.91	13.88	19.82

Property – Refurbishment

Rail 2003 Ref		Unit	Labour Hours	Labour Cost £	Plant Cost £	Materials Cost £	Unit Rate Base Cost £	Unit Rate Green Zone £	Unit Rate Red Zone £
6514	**INTERNAL FINISHINGS**								
65144055	Granolithic screeds								
6514405530	Treads; rounded nosing; ends, angles and the like; 275 mm wide:								
6514405530A	25 mm thick	m	0.28	5.30	–	0.84	6.14	8.59	12.27
6514405530B	32 mm thick	m	0.31	5.87	–	0.84	6.71	9.39	13.41
6514405530C	38 mm thick	m	0.36	6.81	–	1.00	7.81	10.94	15.63
6514405530D	50 mm thick	m	0.44	8.33	–	1.67	10.00	14.00	20.00
6514405530E	65 mm thick	m	0.53	10.03	–	1.67	11.70	16.38	23.40
6514405540	Risers; rounded top edge; coved juction to tread; undercut; 150 mm high:								
6514405540A	13 mm	m	0.31	5.87	–	1.00	6.87	9.62	13.74
6514405540B	19 mm	m	0.36	6.81	–	1.67	8.48	11.88	16.97
6514405550	Strings and aprons; rounded top edge; ends, angles and the like; 275 mm high:								
6514405550A	13 mm	m	0.28	5.30	–	0.90	6.20	8.68	12.40
6514405550B	19 mm	m	0.34	6.44	–	0.90	7.34	10.27	14.67
6514405560	Carborundum surface dressing:								
6514405560A	1 Kg per m^2	m	0.06	1.14	–	0.33	1.47	2.05	2.93
65144060	Glazed wall tiling								
6514406010	Ceramic glazed wall tiles; fixed with adhesive; flush pointing with white grout; to walls:								
6514406010A	102 × 102 × 5.5 mm	m^2	1.63	30.85	–	20.47	51.32	71.86	102.65
6514406010B	152 × 152 × 5.5 mm	m^2	1.22	23.09	–	21.03	44.12	61.77	88.25
6514406010C	straight cutting	m	0.33	6.25	–	–	6.25	8.75	12.49
6514406010D	raking cutting	m	0.45	8.52	–	–	8.52	11.93	17.04

Property – Refurbishment

Rail 2003 Ref		Unit	Labour Hours	Labour Cost	Plant Cost	Materials Cost	Unit Rate	Unit Rate	Unit Rate
				£	£	£	Base Cost £	Green Zone £	Red Zone £
6514	**INTERNAL FINISHINGS**								
65144065	**Clay floor tiling**								
6514406510	Heather brown quarry tiles; bedded and jointed in cement mortar (1:3); flush pointed with grout:								
6514406510A	150 × 150 × 19 mm; over 300 mm wide	m^2	1.03	19.50	–	22.60	42.10	58.94	84.19
6514406510B	150 × 150 × 19 mm; not exceeding 300 mm wide	m	0.42	7.95	–	6.81	14.76	20.66	29.52
6514406550	**Labours on tiling:**								
6514406550A	straight cutting	m	0.85	16.09	–	–	16.09	22.53	32.18
6514406550B	raking cutting	m	1.22	23.09	–	–	23.09	32.33	46.19
6514406550C	curved cutting	m	1.55	29.34	–	–	29.34	41.08	58.68

Rail 2003 Ref		Unit	Specialist Cost	Unit Rate	Unit Rate	Unit Rate
			£	Base Cost £	Green Zone £	Red Zone £
65144070	**Vinyl floor finishes**					
6514407010	Marley vinyl tiles; 300 × 300 mm:					
6514407010A	Econoflex; 2 mm	m	26.67	26.67	37.34	53.34
6514407010B	Marleyflex Interational; 2 mm	m	28.99	28.99	40.59	57.98
6514407010C	Marleyflex International; 2.5 mm	m	33.24	33.24	46.54	66.48
6514407020	**Marley vinyl sheeting:**					
6514407020A	Vynatred felt backed	m	27.83	27.83	38.96	55.66
6514407020B	Safetred Universal	m	34.01	34.01	47.61	68.02
6514407020C	HD series 2, 2 mm	m	36.72	36.72	51.41	73.44
6514407020D	HD acoustic foam backed	m	41.36	41.36	57.90	82.72

Property – Refurbishment

Rail 2003 Ref		Unit	Labour Hours	Labour Cost £	Plant Cost £	Materials Cost £	Unit Rate Base Cost £	Unit Rate Green Zone £	Unit Rate Red Zone £
6514	**INTERNAL FINISHINGS**								
651450	**REPAIRS AND ALTERATIONS**								
65145010	Ceiling repairs								
6514501010	Expanded metal lathing and 13 mm Carlite plaster:								
6514501010A	to ceiling joists	m²	1.70	32.18	–	10.10	42.28	59.19	84.56
6514501020	Thistle baseboard, scrim and 3 mm Thistle finish:								
6514501020A	to ceilings	m²	1.00	18.93	–	4.38	23.31	32.63	46.62
6514501030	Hack down defective ceiling plaster and fix Thistle baseboard, scrim and 3 mm Thistle finish including jointing to existing:								
6514501030A	area not exceeding 1 m²	m²	2.00	37.86	–	3.30	41.16	57.62	82.32
6514501030B	area 1 – 4 m²	m²	1.40	26.50	–	3.30	29.80	41.72	59.60
65145020	Plasterwork to walls repairs								
6514502010	Make good at intersection of wall and ceiling plaster after replastering:								
6514502010A	wall or ceiling	m	0.55	10.41	–	0.17	10.58	14.82	21.16
6514502020	Make good cracks in:								
6514502020A	ceiling plaster	m	0.45	8.52	–	0.17	8.69	12.17	17.38
6514502030	Dub out uneven walls to receive:								
6514502030A	new plaster	m²	0.40	7.57	–	0.88	8.45	11.83	16.89
6514502040	13 mm Thistle plaster on:								
6514502040A	brick or block walls	m²	0.85	16.09	–	2.08	18.17	25.44	36.34

Property – Refurbishment

Rail 2003 Ref		Unit	Labour Hours	Labour Cost £	Plant Cost £	Materials Cost £	Unit Rate Base Cost £	Unit Rate Green Zone £	Unit Rate Red Zone £

6514	**INTERNAL FINISHINGS**								
65145020	Plasterwork to walls repairs								
6514502050	Hack down defective wall plaster in small quantities; apply 13mm Thistle plaster including jointing to existing:								
6514502050A	area not exceeding 1 m^2	m^2	1.75	33.13	–	2.71	35.84	50.17	71.66
6514502050B	area 1 – 4 m^2	m^2	0.90	17.04	–	2.71	19.75	27.64	39.48
6514502060	Make good cracks in plaster:								
6514502060A	walls	m	0.45	8.52	–	0.17	8.69	12.17	17.38
6514502060B	around door and window frames and repoint	m	0.40	7.57	–	0.17	7.74	10.84	15.48
65145030	Ceramic tiling repairs								
6514503010	Glazed ceramic wall tiles; fixed with adhesive, pointed in grout; in small quantities in repairs:								
6514503010A	102 × 102 × 6.5 mm	m^2	2.13	40.32	–	20.47	60.79	85.11	121.58
6514503010B	150 × 150 × 5.5 mm	m^2	1.80	34.07	–	21.49	55.56	77.78	111.12
6514503020	Refixing salvaged tiles:								
6514503020A	with adhesive	m^2	2.20	41.64	–	2.77	44.41	62.18	88.83

Property – Refurbishment

Rail 2003 Ref		Unit	Labour Hours	Labour Cost £	Plant Cost £	Materials Cost £	Unit Rate Base Cost £	Unit Rate Green Zone £	Unit Rate Red Zone £

6515 GLAZING

651510 WORKS OF REMOVAL

65151010 General glazing

6515101010 Hack out old glazing, clean out rebates:

| 6515101010A | prepare for new glazing | m | 0.22 | 2.79 | – | – | 2.79 | 3.91 | 5.58 |

6515101020 Hack out old glazing, remove glazing beads, clean out rebates:

| 6515101020A | prepare for new glazing | m | 0.28 | 3.55 | – | – | 3.55 | 4.97 | 7.10 |

6515101030 Take out patent glazing; clear debris to rubbish skip:

| 6515101030A | generally | m^2 | 0.85 | 10.78 | – | – | 10.78 | 15.09 | 21.56 |

651550 NEW WORKS

65155010 Plain glazing

6515501010 Clear sheet or float glass; to wood with putty:

6515501010A	3 mm	m^2	0.60	7.61	–	24.01	31.62	44.26	63.23
6515501010B	4 mm	m^2	0.60	7.61	–	27.18	34.79	48.70	69.58
6515501010C	5 mm	m^2	0.70	8.88	–	39.95	48.83	68.36	97.65
6515501010D	6 mm	m^2	0.70	8.88	–	44.94	53.82	75.35	107.64

6515501015 Clear sheet or float glass; to wood with beads:

6515501015A	3 mm	m^2	0.70	8.88	–	24.01	32.89	46.04	65.77
6515501015B	4 mm	m^2	0.70	8.88	–	28.15	37.03	51.83	74.05
6515501015C	5 mm	m^2	0.80	10.15	–	40.92	51.07	71.49	102.12
6515501015D	6 mm	m^2	0.80	10.15	–	45.91	56.06	78.48	112.11

6515501020 Georgian wired glass; to wood with putty:

| 6515501020A | 6 mm polished plate | m^2 | 0.70 | 8.88 | – | 77.08 | 85.96 | 120.35 | 171.93 |
| 6515501020B | 7 mm cast glass | m^2 | 0.75 | 9.51 | – | 31.04 | 40.55 | 56.77 | 81.10 |

Property – Refurbishment

Rail 2003 Ref		Unit	Labour Hours	Labour Cost	Plant Cost	Materials Cost	Unit Rate	Unit Rate	Unit Rate
				£	£	£	Base Cost £	Green Zone £	Red Zone £
6515	**GLAZING**								
65155010	Plain glazing								
6515501025	Georgian wired glass; to wood with beads:								
6515501025A	6 mm polished plate	m²	0.80	10.15	–	75.93	86.08	120.52	172.16
6515501025B	7 mm cast glass	m²	0.86	10.91	–	32.00	42.91	60.07	85.83
6515501030	Toughened safety glass; to wood with putty:								
6515501030A	4 mm	m²	0.60	7.61	–	48.49	56.10	78.53	112.20
6515501030B	5 mm	m²	0.60	7.61	–	54.86	62.47	87.45	124.94
6515501030C	6 mm	m²	0.70	8.88	–	59.68	68.56	95.99	137.13
6515501030D	10 mm	m²	0.90	11.42	–	118.06	129.48	181.27	258.95
6515501040	Extra over clear sheet or float glass for:								
6515501040A	4 mm patterned glass	m²	–	–	–	3.35	3.35	4.69	6.70
6515501040B	6 mm patterned glass	m²	–	–	–	6.90	6.90	9.66	13.80
6515501045	Extra over toughened safety glass for:								
6515501045A	4 mm tempered safety glass	m²	–	–	–	3.64	3.64	5.10	7.28
6515501045B	6 mm tempered safety glass	m²	–	–	–	4.75	4.75	6.65	9.50
6515501045C	4 mm white patterned safety glass	m²	–	–	–	3.64	3.64	5.10	7.28
6515501045D									9.50
	6 mm white patterned safety glass	m²	–	–	–	4.75	4.75	6.65	

Rail 2003 Ref		Unit				Specialist Cost	Unit Rate	Unit Rate	Unit Rate
						£	Base Cost £	Green Zone £	Red Zone £
65155050	Special glazing								
6515505010	Hermetically sealed toughened double glazing units; to wood with beads:								
6515505010A	not exceeding 0.5 m²	m²	–	–	–	74.99	74.99	104.99	149.98
6515505010B	0.5 – 1.0 m²	m²	–	–	–	72.28	72.28	101.19	144.56
6515505010C	exceeding 1.0 m²	m²	–	–	–	69.57	69.57	97.40	139.14

Property – Refurbishment

Rail 2003 Ref		Unit	Labour Hours	Labour Cost £	Plant Cost £	Materials Cost £	Unit Rate Base Cost £	Unit Rate Green Zone £	Unit Rate Red Zone £
6516	**PAINTING AND DECORATING**								
651610	**NEW WORKS**								
65161010	**Emulsion paint**								
6516101010	Prepare, apply one mist coat and two full coats; plaster backgrounds:								
6516101010A	walls	m²	0.18	2.47	–	1.59	4.06	5.69	8.13
6516101010B	walls; not exceeding 300 mm girth	m	0.07	0.96	–	0.53	1.49	2.09	2.98
6516101010C	ceilings	m²	0.20	2.75	–	1.59	4.34	6.07	8.68
6516101020	Prepare; apply one mist coat and two full caots; brickwork backgrounds:								
6516101020A	walls	m²	0.25	3.43	–	2.39	5.82	8.15	11.64
6516101020B	walls; not exceeding 300 mm girth	m	0.10	1.37	–	0.80	2.17	3.04	4.33
6516101030	Prepare; apply one mist coat and two full coats; concrete backgrounds:								
6516101030A	walls	m²	0.20	2.75	–	1.86	4.61	6.44	9.21
6516101030B	walls; not exceeding 300 mm girth	m	0.09	1.24	–	0.53	1.77	2.47	3.53
6516101030C	ceilings	m²	0.21	2.88	–	1.86	4.74	6.63	9.48
65161020	**Masonry paint**								
6516102010	Prepare, apply masonry base coat sealer and two coats of masonry paint finish; rendered backgrounds:								
6516102010A	walls	m²	0.27	3.71	–	2.63	6.34	8.87	12.67
6516102010B	walls; not exceeding 300 mm girth	m	0.11	1.51	–	3.10	4.61	6.44	9.21
6516102010C	ceilings	m²	0.28	3.84	–	2.63	6.47	9.06	12.95
6516102020	Prepare, apply masonry base coat sealer and two coats of masonry paint finish; concrete backgrounds:								
6516102020A	walls	m²	0.02	0.27	–	2.36	2.63	3.69	5.28
6516102020B	walls; not exceeding 300 mm girth	m	0.10	1.37	–	0.79	2.16	3.02	4.31
6516102020C	ceiling	m²	0.26	3.57	–	2.36	5.93	8.31	11.87

Property – Refurbishment

Rail 2003 Ref		Unit	Labour Hours	Labour Cost £	Plant Cost £	Materials Cost £	Unit Rate Base Cost £	Unit Rate Green Zone £	Unit Rate Red Zone £
6516	**PAINTING AND DECORATING**								
65161020	**Masonry paint**								
6516102030	**Prepare, apply masonry base coat sealer and two coats of masonry paint finish; brickwork backgrounds:**								
6516102030A	walls............................	m²	0.30	4.12	–	2.63	6.75	9.44	13.49
6516102030B	walls; not exceeding 300 mm girth ...	m	0.12	1.65	–	0.79	2.44	3.41	4.86

Level Crossings

Rail 2003 Ref		Unit	Composite Cost £	Unit Rate Base Cost £	Unit Rate Green Zone £	Unit Rate Red Zone £
7	**LEVEL CROSSINGS**					
730	**AUTOMATIC HALF BARRIER (AHB)** Typical elemental cost analyses					
73010	**CONTRACTOR'S PRELIMINARIES**					
7301010	**General Preliminaries**					
730101010	Generally:					
730101010A	per crossing	Item	–	–	–	73439.75 73439.75 102815.65 146879.50
73020	**CONTRACTOR'S DESIGN**					
7302010	**Design Works**					
730201010	Generally:					
730201010A	per crossing	Item	–	–	–	76145.43 76145.43 106603.60 152290.86
73030	**CIVIL WORKS**					
7303030	**Foundations**					
730303010	Generally:					
730303010A	per crossing	Item	–	–	–	6570.93 6570.93 9199.30 13141.86
7303040	**Road Works**					
730304010	Generally:					
730304010A	per crossing	Item	–	–	–	1546.10 1546.10 2164.54 3092.20
7303050	**Vehicle Parking**					
730305010	Generally:					
730305010A	per crossing	Item	–	–	–	4174.47 4174.47 5844.26 8348.94

Level Crossings

Rail 2003 Ref		Unit	Composite Cost £	Unit Rate Base Cost £	Unit Rate Green Zone £	Unit Rate Red Zone £
730	**AUTOMATIC HALF BARRIER (AHB)** Typical elemental cost analyses					
7303060	Signage					
730306010	Generally:					
730306010A	per crossing	Item	– – –	4792.91	4792.91	6710.07	9585.82
7303070	Griddage					
730307010	Generally:					
730307010A	per crossing	Item	– – –	2319.15	2319.15	3246.81	4638.30
73035	**CROSSING DECKS**					
7303530	Supply and Install					
730353010	Generally:					
730353010A	per crossing	Item	– – –	36797.18	36797.18	51516.05	73594.36
73040	**BARRIER EQUIPMENT**					
7304030	Barriers					
730403010	Generally:					
730403010A	per crossing	Item	– – –	11904.97	11904.97	16666.96	23809.94
7304040	Traffic Lights					
730404010	Generally:					
730404010A	per crossing	Item	– – –	9972.35	9972.35	13961.29	19944.70

Level Crossings

Rail 2003 Ref		Unit	Composite Cost £	Unit Rate Base Cost £	Unit Rate Green Zone £	Unit Rate Red Zone £
730	**AUTOMATIC HALF BARRIER (AHB)** Typical elemental cost analyses					
7304050	**GATSO Equipment**					
730405010	Generally:					
730405010A	per crossing	Item	– – –	75449.68	105629.55	150899.36
			75449.68			
7304060	**Apparatus Cases**					
730406010	Generally:					
730406010A	per crossing	Item	– – –	9431.21	13203.69	18862.42
			9431.21			
73045	**REBs**					
7304530	**REB Supply**					
730453010	Generally:					
730453010A	per crossing	Item	– – –	18939.73	26515.62	37879.46
			18939.73			
7304540	**REB Racking**					
730454010	Generally:					
730454010A	per crossing	Item	– – –	5720.57	8008.80	11441.14
			5720.57			
7304550	**REB Batteries**					
730455010	Generally:					
730455010A	per crossing	Item	– – –	4251.78	5952.49	8503.56
			4251.78			
7304560	**REB Relays**					
730456010	Generally:					
730456010A	per crossing	Item	– – –	19326.25	27056.75	38652.50
			19326.25			

Level Crossings

Rail 2003 Ref	Unit		Composite Cost £	Unit Rate Base Cost £	Unit Rate Green Zone £	Unit Rate Red Zone £
730	**AUTOMATIC HALF BARRIER (AHB)**					
	Typical elemental cost analyses					
7304570	REB Transformers					
730457010	Generally:					
730457010A	per crossing	Item	– – – 7189.37	7189.37	10065.12	14378.74
73050	**POWER SUPPLIES**					
7305030	Level Crossing Equipment					
730503010	Generally:					
730503010A	per crossing	Item	– – – 5565.96	5565.96	7792.34	11131.92
73055	**SIGNALS**					
7305530	Track Circuits					
730553010	Generally:					
730553010A	per crossing	Item	– – – 29221.29	29221.29	40909.81	58442.58
7305540	Local Control Units					
730554010	Generally:					
730554010A	per crossing	Item	– – – 2473.76	2473.76	3463.26	4947.52
7305550	Recorder Equipment					
730555010	Generally:					
730555010A	per crossing	Item	– – – 15924.83	15924.83	22294.76	31849.66

Rail 2003 Ref		Unit				Composite Cost £	Unit Rate Base Cost £	Unit Rate Green Zone £	Unit Rate Red Zone £
730	**AUTOMATIC HALF BARRIER (AHB) Typical elemental cost analyses**								
7305560	**Treadle Equipment**								
730556010	**Generally:**								
730556010A	per crossing	Item	–	–	–	8348.94	8348.94	11688.52	16697.88
73060	**TELECOMS**								
7306030	**Public Emergency Telecom Equipment**								
730603010	**Generally:**								
730603010A	per crossing	Item	–	–	–	21645.40	21645.40	30303.56	43290.80
7306060	**REB Telephone**								
730606010	**Generally:**								
730606010A	per crossing	Item	–	–	–	463.83	463.83	649.36	927.66
73070	**CCTV**								
7307030	**CCTV Equipment**								
730703010	**Generally:**								
730703010A	per crossing	Item	–	–	–	20679.09	20679.09	28950.73	41358.18
73080	**CONTRACTOR'S TESTING AND COMMISSIONING**								
7308030	**GATSO Installation**								
730803010	**Generally:**								
730803010A	per crossing	Item	–	–	–	12523.41	12523.41	17532.77	25046.82

Level Crossings

Rail 2003 Ref		Unit	Composite Cost £	Unit Rate Base Cost £	Unit Rate Green Zone £	Unit Rate Red Zone £
730	**AUTOMATIC HALF BARRIER (AHB)** Typical elemental cost analyses					
7308060	**PET System**					
730806010	**Generally:**					
730806010A	per crossing	Item	– – – 2319.15	2319.15	3246.81	4638.30
73090	**RECOVERIES**					
7309010	**Stripping Out and Recoveries**					
730901010	**Generally:**					
730901010A	per crossing	item	– – – 19326.25	19326.25	27056.75	38652.50

Level Crossings

Rail 2003 Ref		Unit	Composite Cost £	Unit Rate Base Cost £	Unit Rate Green Zone £	Unit Rate Red Zone £
740	**AUTOMATIC OPEN CROSSINGS REMOTELY MONITORED (AOCR)** Typical elemental cost analyses					
74010	**CONTRACTOR'S PRELIMINARIES**					
7401010	**General Preliminaries**					
740101010	Generally:					
740101010A	per crossing	Item	– – –	83025.57	83025.57 116235.80	166051.14
74020	**CONTRACTOR'S DESIGN**					
7402010	**Design Works**					
740201010	Generally:					
740201010A	per crossing	Item	– – –	76531.95	76531.95 107144.73	153063.90
74030	**CIVIL WORKS**					
7403030	**Foundations**					
740303010	Generally:					
740303010A	per crossing	Item	– – –	7343.98	7343.98 10281.57	14687.96
7403040	**Road Works**					
740304010	Generally:					
740304010A	per crossing	Item	– – –	7421.28	7421.28 10389.79	14842.56
7403050	**Vehicle Parking**					
740305010	Generally:					
740305010A	per crossing	Item	– – –	4947.52	4947.52 6926.53	9895.04

Level Crossings

Rail 2003 Ref		Unit	Composite Cost £	Unit Rate Base Cost £	Unit Rate Green Zone £	Unit Rate Red Zone £
740	**AUTOMATIC OPEN CROSSINGS REMOTELY MONITORED (AOCR)** Typical elemental cost analyses					
7403060	**Signage**					
740306010	Generally:					
740306010A	per crossing	Item	– – – 6493.62	6493.62	9091.07	12987.24
7403070	**Griddage**					
740307010	Generally:					
740307010A	per crossing	Item	– – – 2319.15	2319.15	3246.81	4638.30
74035	**CROSSING DECKS**					
7403530	**Supply and Install**					
740353010	Generally:					
740353010A	per crossing	Item	– – – 39425.55	39425.55	55195.77	78851.10
74040	**BARRIER EQUIPMENT**					
7404030	**Barriers**					
740403010	Generally:					
740403010A	per crossing	Item	– – – 12678.02	12678.02	17749.23	25356.04
7404040	**Traffic Lights**					
740404010	Generally:					
740404010A	per crossing	Item	– – – 9972.35	9972.35	13961.29	19944.70

Level Crossings

Rail 2003 Ref		Unit	Composite Cost £	Unit Rate Base Cost £	Unit Rate Green Zone £	Unit Rate Red Zone £
740	**AUTOMATIC OPEN CROSSINGS REMOTELY MONITORED (AOCR)** Typical elemental cost analyses					
7404050	**GATSO Equipment**					
740405010	Generally:					
740405010A	per crossing	Item	– – –	76918.48	107685.87	153836.96
7404060	**Apparatus Cases**					
740406010	Generally:					
740406010A	per crossing	Item	– – –	9431.21	13203.69	18862.42
74045	**REBs**					
7404530	**REB Supply**					
740453010	Generally:					
740453010A	per crossing	Item	– – –	22727.67	31818.74	45455.34
7404540	**REB Racking**					
740454010	Generally:					
740454010A	per crossing	Item	– – –	6029.79	8441.71	12059.58
7404550	**REB Batteries**					
740455010	Generally:					
740455010A	per crossing	Item	– – –	5024.83	7034.76	10049.66
7404560	**REB Relays**					
740456010	Generally:					
740456010A	per crossing	Item	– – –	21645.40	30303.56	43290.80

Level Crossings

Rail 2003 Ref		Unit	Composite Cost £	Unit Rate Base Cost £	Unit Rate Green Zone £	Unit Rate Red Zone £
740	**AUTOMATIC OPEN CROSSINGS REMOTELY MONITORED (AOCR)** Typical elemental cost analyses					
7404570	**REB Transformers**					
740457010	Generally:					
740457010A	per crossing	Item	– – – 7730.50	7730.50	10822.70	15461.00
74050	**POWER SUPPLIES**					
7405030	**Level Crossing Equipment**					
740503010	Generally:					
740503010A	per crossing	Item	– – – 6957.45	6957.45	9740.43	13914.90
74055	**SIGNALS**					
7405530	**Track Circuits**					
740553010	Generally:					
740553010A	per crossing	Item	– – – 31076.61	31076.61	43507.25	62153.22
7405540	**Local Control Units**					
740554010	Generally					
740554010A	per crossing	Item	– – – 2782.98	2782.98	3896.17	5565.96
7405550	**Recorder Equipment**					
740555010	Generally:					
740555010A	per crossing	Item	– – – 18707.81	18707.81	26190.93	37415.62

Level Crossings

Rail 2003 Ref		Unit	Composite Cost £	Unit Rate Base Cost £	Unit Rate Green Zone £	Unit Rate Red Zone £
740	**AUTOMATIC OPEN CROSSINGS REMOTELY MONITORED (AOCR)** Typical elemental cost analyses					
7405560	Treadle Equipment					
740556010	Generally:					
740556010A	per crossing	Item	– – –	9199.30	9199.30	12879.02	18398.60
74060	TELECOMS					
7406030	Public Emergency Telephone Equipment					
740603010	Generally:					
740603010A	per crossing	Item	– – –	32468.10	32468.10	45455.34	64936.20
7406060	REB Telephone					
740606010	Generally:					
740606010A	per crossing	Item	– – –	850.36	850.36	1190.50	1700.72
74070	CCTV					
7407030	CCTV Equipment					
740703010	Generally:					
740703010A	per crossing	Item	– – –	25974.48	25974.48	36364.27	51948.96
74080	CONTRACTOR'S TESTING AND COMMISSIONING					
7408030	GATSO Installation					
740803010	Generally:					
740803010A	per crossing	Item	– – –	14301.43	14301.43	20022.00	28602.86

Level Crossings

Rail 2003 Ref		Unit	Composite Cost £	Unit Rate Base Cost £	Unit Rate Green Zone £	Unit Rate Red Zone £
740	**AUTOMATIC OPEN CROSSINGS REMOTELY MONITORED (AOCR)** Typical elemental cost analyses					
7408060	**PET System**					
740806010	Generally:					
740806010A	per crossing	Item	– – – 3053.55	3053.55	4274.97	6107.10
74090	**RECOVERIES**					
7409010	**Stripping Out and Recoveries**					
740901010	Generally:					
740901010A	per crossing	Item	– – – 16234.05	16234.05	22727.67	32468.10

Level Crossings

Rail 2003 Ref		Unit	Composite Cost £	Unit Rate Base Cost £	Unit Rate Green Zone £	Unit Rate Red Zone £
750	**AUTOMATIC OPEN CROSSINGS LOCALLY MONITORED (AOCL)** Typical elemental cost analyses					
75010	**CONTRACTOR'S PRELIMINARIES**					
7501010	General Preliminaries					
750101010	Generally:					
750101010A	per crossing	Item	–	–	–	34787.25	34787.25	48702.15	69574.50
75020	**CONTRACTOR'S DESIGN**					
7502010	Design Works					
750201010	Generally:					
750201010A	per crossing	Item	–	–	–	49475.20	49475.20	69265.28	98950.40
75030	**CIVIL WORKS**					
7503030	Foundations					
750303010	Generally:					
750303010A	per crossing	Item	–	–	–	9663.13	9663.13	13528.38	19326.26
7503040	Road Works					
750304010	Generally:					
750304010A	per crossing	Item	–	–	–	6957.45	6957.45	9740.43	13914.90
7503050	General Works					
750305010	Generally:					
750305010A	per crossing	Item	–	–	–	4174.47	4174.47	5844.26	8348.94

Level Crossings

Rail 2003 Ref		Unit	Composite Cost £	Unit Rate Base Cost £	Unit Rate Green Zone £	Unit Rate Red Zone £
750	**AUTOMATIC OPEN CROSSINGS LOCALLY MONITORED (AOCL)** Typical elemental cost analyses					
7503060	Signage					
750306010	Generally:					
750306010A	per crossing	Item	– – –	6957.45	6957.45	9740.43	13914.90
7503070	Griddage					
750307010	Generally:					
750307010A	per crossing	Item	– – –	2319.15	2319.15	3246.81	4638.30
75035	**CROSSING DECKS**					
7503530	Supply and Install					
750353010	Generally:					
750353010A	per crossing	Item	– – –	41435.48	41435.48	58009.67	82870.96
75040	**BARRIER EQUIPMENT**					
7504030	Barriers					
750403010	Generally:					
750403010A	per crossing	Item	– – –	11904.97	11904.97	16666.96	23809.94
7504040	Traffic Lights					
750404010	Generally:					
750404010A	per crossing	Item	– – –	9972.35	9972.35	13961.29	19944.70

Level Crossings

Rail 2003 Ref		Unit			Composite Cost £	Unit Rate Base Cost £	Unit Rate Green Zone £	Unit Rate Red Zone £
750	**AUTOMATIC OPEN CROSSINGS LOCALLY MONITORED (AOCL)** Typical elemental cost analyses							
7504050	**GATSO Equipment**							
750405010	Generally:							
750405010A	per crossing	Item	–	–	–	56741.87	56741.87	79438.62	113483.74
7504060	**Apparatus Cases**							
750406010	Generally:							
750406010A	per crossing	Item	–	–	–	1546.10	1546.10	2164.54	3092.20
75045	**REBs**							
7504530	**REB Supply**							
750453010	Generally:							
750453010A	per crossing	Item	–	–	–	18939.73	18939.73	26515.62	37879.46
7504540	**REB Racking**							
750454010	Generally:							
750454010A	per crossing	Item	–	–	–	3865.25	3865.25	5411.35	7730.50
7504550	**REB Batteries**							
750455010	Generally:							
750455010A	per crossing	Item	–	–	–	4251.78	4251.78	5952.49	8503.56
7504560	**REB Relays**							
750456010	Generally:							
750456010A	per crossing	Item	–	–	–	12368.80	12368.80	17316.32	24737.60

Level Crossings

Rail 2003 Ref		Unit	Composite Cost £	Unit Rate Base Cost £	Unit Rate Green Zone £	Unit Rate Red Zone £
750	**AUTOMATIC OPEN CROSSINGS LOCALLY MONITORED (AOCL)** Typical elemental cost analyses					
7504570	REB Transformers					
750457010	Generally:					
750457010A	per crossing	Item	– – – 7189.37	7189.37	10065.12	14378.74
75050	POWER SUPPLIES					
7505030	Level Crossing Equipment					
750503010	Generally:					
750503010A	per crossing	Item	– – – 46383.00	46383.00	64936.20	92766.00
75055	SIGNALS					
7505530	Track Circuits					
750553010	Generally:					
750553010A	per crossing	Item	– – – 23655.33	23655.33	33117.46	47310.66
7505540	Local Control Units					
750554010	Generally:					
750554010A	per crossing	Item	– – – 2473.76	2473.76	3463.26	4947.52
7505550	Recorder Equipment					
750555010	Generally:					
750555010A	per crossing	Item	– – – 15924.83	15924.83	22294.76	31849.66

Level Crossings

Rail 2003 Ref		Unit	Composite Cost £	Unit Rate Base Cost £	Unit Rate Green Zone £	Unit Rate Red Zone £
750	**AUTOMATIC OPEN CROSSINGS LOCALLY MONITORED (AOCL)** Typical elemental cost analyses					
7505560	Treadle Equipment					
750556010	Generally:					
750556010A	per crossing	Item	– – – 8348.94	8348.94	11688.52	16697.88
75060	**TELECOMS**					
7506030	Public Emergency Telephone Equipment					
750603010	Generally:					
750603010A	per crossing	Item	– – – 10822.70	10822.70	15151.78	21645.40
7506060	REB Telephone					
750606010	Generally:					
750606010A	per crossing	Item	– – – 463.83	463.83	649.36	927.66
75070	**CCTV**					
7507030	CCTV Equipment					
750703010	Generally:					
750703010A	per crossing	Item	– – – 16002.14	16002.14	22403.00	32004.28
75080	**CONTRACTOR'S TESTING AND COMMISSIONING**					
7508030	GATSO Installation					
750803010	Generally:					
750803010A	per crossing	Item	– – – 10358.87	10358.87	14502.42	20717.74

Level Crossings

Rail 2003 Ref		Unit	Composite Cost £	Unit Rate Base Cost £	Unit Rate Green Zone £	Unit Rate Red Zone £
750	**AUTOMATIC OPEN CROSSINGS LOCALLY MONITORED (AOCL)** Typical elemental cost analyses					
7508060	**PET System**					
750806010	Generally:					
750806010A	per crossing	Item	– – –	1855.32	1855.32	2597.45	3710.64
75090	**RECOVERIES**					
7509010	Stripping Out and Recoveries					
750901010	Generally:					
750901010A	per crossing	Item	– – –	14687.95	14687.95	20563.13	29375.90

Bridges

Rail 2003 Ref		Unit	Labour Hours	Labour Cost £	Plant Cost £	Materials Cost £	Unit Rate Base Cost £	Unit Rate Green Zone £	Unit Rate Red Zone £

8 BRIDGES

820 DEMOLITIONS AND REMOVALS

82010 DEMOLITION OF STRUCTURES

8201010 Masonry structural elements

820101010 Brick walls and partitions:

Ref	Description	Unit	Hours	Labour	Plant	Mat.	Base	Green	Red
820101010A	half brick thick	m^2	0.80	9.56	0.62	–	10.18	14.25	20.36
820101010B	one brick thick	m^2	1.40	16.74	1.08	–	17.82	24.94	35.63
820101010C	one and a half brick thick	m^2	2.10	25.10	1.62	–	26.72	37.41	53.45
820101010D	two brick thick	m^2	2.70	32.28	2.08	–	34.36	48.10	68.71

820101015 Block walls and partitions:

Ref	Description	Unit	Hours	Labour	Plant	Mat.	Base	Green	Red
820101015A	100 mm thick	m^2	0.30	3.59	0.93	–	4.52	6.32	9.02
820101015B	125 mm thick	m^2	0.33	3.95	1.00	–	4.95	6.92	9.89
820101015C	150 mm thick	m^2	0.35	4.18	1.08	–	5.26	7.37	10.53
820101015D	200 mm thick	m^2	0.40	4.78	1.23	–	6.01	8.42	12.03
820101015E	225 mm thick	m^2	0.55	6.58	1.70	–	8.28	11.58	16.54

820101018 Attached piers and buttresses:

Ref	Description	Unit	Hours	Labour	Plant	Mat.	Base	Green	Red
820101018A	half brick	m^2	0.45	5.38	1.39	–	6.77	9.47	13.53
820101018B	one brick	m^2	0.80	9.56	2.47	–	12.03	16.84	24.06
820101018C	one and a half brick	m^2	1.15	13.75	3.54	–	17.29	24.21	34.58
820101018D	two brick	m^2	1.40	16.74	4.32	–	21.06	29.47	42.10

8201020 Concrete structural elements

820102020 Reinforced walls and attached columns:

Ref	Description	Unit	Hours	Labour	Plant	Mat.	Base	Green	Red
820102020A	not exceeding 150 mm thick	m^3	12.00	143.45	83.80	–	227.25	318.15	454.50
820102020B	150 – 300 mm thick	m^3	14.00	167.36	97.77	–	265.13	371.18	530.25
820102020C	over 300 mm thick	m^3	18.00	215.17	125.70	–	340.87	477.22	681.75

820102022 Reinforced concrete ground slabs:

Ref	Description	Unit	Hours	Labour	Plant	Mat.	Base	Green	Red
820102022A	not exceeding 150 mm thick	m^3	6.00	71.72	41.90	–	113.62	159.07	227.25
820102022B	150 – 300 mm thick	m^3	8.00	95.63	55.87	–	151.50	212.10	303.00
820102022C	over 300 mm thick	m^3	10.00	119.54	69.84	–	189.38	265.13	378.75

Bridges

Rail 2003 Ref		Unit	Labour Hours	Labour Cost £	Plant Cost £	Materials Cost £	Unit Rate Base Cost £	Unit Rate Green Zone £	Unit Rate Red Zone £

820	**DEMOLITIONS AND REMOVALS**								
8201020	Concrete structural elements								
820102025	Reinforced concrete suspended slabs and attached beams:								
820102025A	not exceeding 150 mm thick	m³	10.00	119.54	69.84	–	189.38	265.13	378.75
820102025B	150 – 300 mm thick	m³	16.00	191.26	111.74	–	303.00	424.20	606.00
820102025C	over 300 mm thick	m³	19.00	227.13	132.69	–	359.82	503.74	719.62
820102027	Reinforced isolated columns; sectional area:								
820102027A	not exceeding 0.1 m²	m³	10.00	119.54	69.84	–	189.38	265.13	378.75
820102027B	0.10 – 0.25 m²	m³	13.00	155.40	90.79	–	246.19	344.66	492.37
820102027C	over 0.25 m²	m³	18.00	215.17	125.70	–	340.87	477.22	681.75
820102029	Reinforced isolated piers; sectional area:								
820102029A	not exceeding 0.1 m²	m³	12.00	143.45	83.80	–	227.25	318.15	454.50
820102029B	0.10 – 0.25 m²	m³	14.00	167.36	97.77	–	265.13	371.18	530.25
820102029C	over 0.25 m²	m³	19.00	227.13	132.69	–	359.82	503.74	719.62
8201030	Steelwork structural elements								
820103010	Steel beams, joists and lintels:								
820103010A	not exceeding 10 kg/m	Tonne	50.00	597.70	22.99	–	620.69	868.97	1241.38
820103010B	10 – 20 kg/m	Tonne	46.00	549.88	20.90	–	570.78	799.10	1141.57
820103010C	20 – 50 kg/m	Tonne	40.00	478.16	18.81	–	496.97	695.75	993.94
820103015	Steel columns and stanchions:								
820103015A	not exceeding 10 kg/m	Tonne	60.00	717.24	22.99	–	740.23	1036.33	1480.46
820103015B	10 – 20 kg/m	Tonne	56.00	669.42	20.90	–	690.32	966.45	1380.65
820103015C	20 – 50 kg/m	Tonne	50.00	597.70	18.81	–	616.51	863.11	1233.02
820103018	Steel rails and purlins:								
820103018A	not exceeding 10 kg/m	Tonne	50.00	597.70	20.90	–	618.60	866.04	1237.20
820103018B	10 – 20 kg/m	Tonne	46.00	549.88	18.81	–	568.69	796.17	1137.39

Bridges

Rail 2003 Ref		Unit	Labour Hours	Labour Cost £	Plant Cost £	Materials Cost £	Unit Rate Base Cost £	Unit Rate Green Zone £	Unit Rate Red Zone £
820	**DEMOLITIONS AND REMOVALS**								
82020	**REMOVAL OF OLD WORK**								
8202010	General metalwork								
820201010	Stairs, steps, balustrades and handrails; including disposal:								
820201010A	stair or step unit; straight flight	Nr	3.50	41.84	18.78	–	60.62	84.86	121.24
820201010B	stair or step unit; two or more flights.	Nr	5.00	59.77	23.48	–	83.25	116.55	166.49
820201010C	balustrading	m	0.80	9.56	7.04	–	16.60	23.25	33.22
820201010D	handrail	m	0.30	3.59	2.35	–	5.94	8.31	11.87
820201020	Signage; including disposal:								
820201020A	road signs complete with posts	Nr	0.70	8.37	9.39	–	17.76	24.87	35.52
820201020B	general signage complete with fixing brackets	Nr	0.55	6.58	9.39	–	15.97	22.36	31.93
820201025	Signage; setting aside for re-use:								
820201025A	road signs complete with posts	Nr	0.75	8.97	–	–	8.97	12.55	17.93
820201025B	general signage complete with fixing brackets	Nr	1.00	11.95	–	–	11.95	16.74	23.91
8202020	General elements								
820202010	Timber decking; setting aside for re-use:								
820202010A	not exceeding 50 m^3	m^2	0.75	8.97	–	–	8.97	12.55	17.93
820202010B	over 50 m^3	m^2	0.68	8.13	–	–	8.13	11.38	16.26
820202015	Timber decking; including disposal:								
820202015A	not exceeding 50 m^3	m^2	0.33	3.95	2.35	–	6.30	8.81	12.59
820202015B	over 50 m^3	m^2	0.30	3.59	2.35	–	5.94	8.31	11.87

Bridges

Rail 2003 Ref		Unit	Labour Hours	Labour Cost £	Plant Cost £	Materials Cost £	Unit Rate Base Cost £	Unit Rate Green Zone £	Unit Rate Red Zone £
820	**DEMOLITIONS AND REMOVALS**								
8202030	**Fencing**								
820203010	Open post and rail; including disposal:								
820203010A	not exceeding 1.2 m high	m	0.28	3.35	1.88	–	5.23	7.32	10.45
820203010B	1.2 – 1.8 m high	m	0.31	3.71	2.82	–	6.53	9.13	13.04
820203010C	over 1.8 m high	m	0.34	4.06	3.76	–	7.82	10.95	15.64
820203020	Post and wire; including disposal:								
820203020A	not exceeding 1.2 m high	m	0.22	2.63	1.41	–	4.04	5.65	8.08
820203020B	1.2 – 1.8 m high	m	0.24	2.87	2.07	–	4.94	6.91	9.87
820203020C	over 1.8 m high	m	0.26	3.11	3.29	–	6.40	8.95	12.79
820203030	Post and chain link; including disposal:								
820203030A	not exceeding 1.2 m high	m	0.30	3.59	1.69	–	5.28	7.39	10.55
820203030B	1.2 – 1.8 m high	m	0.33	3.95	2.44	–	6.39	8.94	12.77
820203030C	over 1.8 m high	m	0.36	4.30	3.57	–	7.87	11.02	15.75
820203040	Post and palisade; including disposal:								
820203040A	not exceeding 1.2 m high	m	0.60	7.17	3.10	–	10.27	14.38	20.54
820203040B	1.2 – 1.8 m high	m	0.66	7.89	4.51	–	12.40	17.36	24.79
820203040C	over 1.8 m high	m	0.73	8.73	5.54	–	14.27	19.98	28.53
820203050	Post and weldmesh; including disposal:								
820203050A	not exceeding 1.2 m high	m	0.45	5.38	3.10	–	8.48	11.87	16.96
820203050B	1.2 – 1.8 m high	m	0.50	5.98	4.51	–	10.49	14.68	20.96
820203050C	over 1.8 m high	m	0.55	6.58	5.54	–	12.12	16.97	24.23
8202040	**Removal of hard access constructions**								
820204010	Break out and remove hard surfacings; dispose of off-site:								
820204010A	asphalt wearing and base courses	m^2	0.40	4.78	1.40	–	6.18	8.65	12.35
820204010B	macadam wearing and base courses	m^2	0.35	4.18	1.40	–	5.58	7.82	11.16
820204010C	reinforced concrete pavings	m^2	1.30	15.54	7.28	–	22.82	31.94	45.63
820204010D	plain concrete pavings	m^2	1.30	15.54	6.98	–	22.52	31.53	45.04

Bridges

Rail 2003 Ref		Unit	Labour Hours	Labour Cost	Plant Cost	Materials Cost	Unit Rate	Unit Rate	Unit Rate
				£	£	£	Base Cost £	Green Zone £	Red Zone £
820	**DEMOLITIONS AND REMOVALS**								
8202040	Removal of hard access constructions								
820204020	Take up kerbs and edgings complete with foundations; dispose of off-site; size:								
820204020A	125 × 225 mm	m	0.56	6.69	0.94	–	7.63	10.68	15.26
820204020B	150 × 305 mm	m	0.80	9.56	1.32	–	10.88	15.22	21.75
820204020C	50 × 150 mm	m	0.20	2.39	0.38	–	2.77	3.88	5.53
820204020D	50 × 250 mm	m	0.30	3.59	0.56	–	4.15	5.81	8.30
8202050	Pipelines and ducts								
820205010	General drainage and duct pipework; disposal; size:								
820205010A	not exceeding 300 mm diameter	m	0.30	3.59	0.56	–	4.15	5.81	8.30
820205010B	300 – 500 mm diameter	m	0.30	3.59	0.56	–	4.15	5.81	8.30
820205010C	over 500 mm	m	0.30	3.59	0.56	–	4.15	5.81	8.30

Bridges

Rail 2003 Ref		Unit	Labour Hours	Labour Cost £	Plant Cost £	Materials Cost £	Unit Rate Base Cost £	Unit Rate Green Zone £	Unit Rate Red Zone £
825	**SITE CLEARANCE**								
82510	**GENERAL SITE CLEARANCE**								
8251010	Trees, stumps and general vegetation								
825101010	Trees; including disposal; girth:								
825101010A	600 mm – 1.5 m	Nr	2.40	28.68	25.35	–	54.03	75.65	108.07
825101010B	1.5 – 3.0 m	Nr	9.60	114.72	101.42	–	216.14	302.59	432.27
825101010C	3.0 – 4.5 m	Nr	14.35	171.48	150.20	–	321.68	450.35	643.36
825101020	Stumps; including disposal; backfilling with imported materials; diameter:								
825101020A	not exceeding 600 mm	Nr	5.50	65.73	125.06	–	190.79	267.10	381.57
825101020B	600 mm – 1 m	Nr	9.00	107.55	196.34	–	303.89	425.45	607.78
825101020C	over 1.0 m	Nr	14.00	167.30	294.51	–	461.81	646.53	923.62
825101030	Vegetation; undergrowth, bushes and the like; remove from site:								
825101030A	generally	m^2	0.02	0.24	0.81	–	1.05	1.46	2.09
825101040	Cut down hedges; grub up roots and remove from site; height:								
825101040A	not exceeding 2.0 m	m	1.00	11.95	31.59	–	43.54	60.95	87.07
825101040B	2.0 – 3.0 m	m	2.50	29.88	76.62	–	106.50	149.09	212.98
825101040C	3.0 – 4.0 m	m	4.00	47.80	116.95	–	164.75	230.65	329.50
825101040D	4.0 – 5.0 m	m	5.00	59.75	157.93	–	217.68	304.75	435.35
825101040E	over 5.0 m	m	7.00	83.65	207.01	–	290.66	406.92	581.32

Bridges

Rail 2003 Ref		Unit	Labour Hours	Labour Cost	Plant Cost	Materials Cost	Unit Rate	Unit Rate	Unit Rate
				£	£	£	Base Cost £	Green Zone £	Red Zone £

830 EARTHWORKS

83010 GENERAL EXCAVATIONS

8301010 Excavation by machine in material other than rock

830101010 Oversite excavation to remove topsoil; average depth:

830101010A	150 mm	m³	–	–	0.35	–	0.35	0.49	0.70
830101010B	300 mm	m³	–	–	0.53	–	0.53	0.74	1.05

830101015 Excavation to reduce levels; depth:

830101015A	not exceeding 0.25 m	m³	0.10	1.20	2.00	–	3.20	4.47	6.39
830101015B	0.25 – 0.5 m	m³	0.08	0.96	1.60	–	2.56	3.58	5.11
830101015C	0.5 – 1.0 m	m³	0.10	1.20	2.00	–	3.20	4.47	6.39
830101015D	1.0 – 2.0 m	m³	0.12	1.43	2.40	–	3.83	5.37	7.67
830101015E	2.0 – 5.0 m	m³	0.14	1.67	2.80	–	4.47	6.26	8.95
830101015F	5.0 – 10.0 m	m³	0.15	1.79	3.00	–	4.79	6.71	9.59
830101015G	over 10.0 m	m³	0.16	1.91	3.20	–	5.11	7.16	10.23

830101020 Excavation in cuttings; depth:

830101020A	not exceeding 0.25 m	m³	0.45	5.38	9.00	–	14.38	20.13	28.76
830101020B	0.25 – 0.5 m	m³	0.43	5.14	8.60	–	13.74	19.24	27.48
830101020C	0.5 – 1.0 m	m³	0.45	5.38	9.00	–	14.38	20.13	28.76
830101020D	1.0 – 2.0 m	m³	0.47	5.62	9.40	–	15.02	21.03	30.04
830101020E	2.0 – 5.0 m	m³	0.49	5.86	9.80	–	15.66	21.92	31.31
830101020F	5.0 – 10.0 m	m³	0.50	5.98	10.00	–	15.98	22.37	31.95
830101020G	over 10.0 m	m³	0.53	6.34	10.60	–	16.94	23.71	33.87

830101025 Excavation for foundations; depth:

830101025A	not exceeding 0.25 m	m³	0.30	3.59	5.25	–	8.84	12.37	17.67
830101025B	0.25 – 0.5 m	m³	0.28	3.35	4.90	–	8.25	11.55	16.49
830101025C	0.5 – 1.0 m	m³	0.28	3.35	4.90	–	8.25	11.55	16.49
830101025D	1.0 – 2.0 m	m³	0.30	3.59	5.25	–	8.84	12.37	17.67
830101025E	2.0 – 5.0 m	m³	0.31	3.71	5.43	–	9.14	12.79	18.26
830101025F	5.0 – 10.0 m	m³	0.33	3.95	5.78	–	9.73	13.61	19.44
830101025G	over 10.0 m	m³	0.35	4.18	6.13	–	10.31	14.44	20.62

Bridges

Rail 2003 Ref		Unit	Labour Hours	Labour Cost £	Plant Cost £	Materials Cost £	Unit Rate Base Cost £	Unit Rate Green Zone £	Unit Rate Red Zone £
830	**EARTHWORKS**								
8301010	Excavation by machine in material other than rock								
830101030	General excavations; depth:								
830101030A	not exceeding 0.25 m	m³	0.65	7.77	11.37	–	19.14	26.81	38.29
830101030B	0.25 – 0.5 m	m³	0.63	7.53	11.03	–	18.56	25.98	37.11
830101030C	0.5 – 1.0 m	m³	0.65	7.77	11.37	–	19.14	26.81	38.29
830101030D	1.0 – 2.0 m	m³	0.65	7.77	11.37	–	19.14	26.81	38.29
830101030E	2.0 – 5.0 m	m³	0.66	7.89	11.55	–	19.44	27.22	38.88
830101030F	5.0 – 10.0 m	m³	0.68	8.13	11.90	–	20.03	28.04	40.06
830101030G	over 10.0 m	m³	0.70	8.37	12.25	–	20.62	28.87	41.24
8301020	Excavation by hand in material other than topsoil, rock or other artificial material								
830102010	Oversite excavation to remove topsoil; average depth:								
830102010A	150 mm	m²	0.18	2.15	–	–	2.15	3.01	4.30
830102010B	300 mm	m²	0.32	3.82	–	–	3.82	5.35	7.65
830102020	Excavation to reduce levels; depth:								
830102020A	not exceeding 0.25 m	m³	1.50	17.93	–	–	17.93	25.10	35.85
830102020B	0.25 – 1.0 m	m³	2.00	23.90	–	–	23.90	33.46	47.80
830102020C	1.0 – 2.0 m	m³	2.30	27.49	–	–	27.49	38.48	54.97
830102030	Excavation in cuttings; depth:								
830102030A	not exceeding 0.25 m	m³	1.80	21.51	–	–	21.51	30.11	43.02
830102030B	0.25 – 1.0 m	m³	2.20	26.29	–	–	26.29	36.81	52.58
830102030C	1.0 – 2.0 m	m³	2.45	29.28	–	–	29.28	40.99	58.56
830102040	Excavation for foundations; depth:								
830102040A	not exceeding 0.25 m	m³	2.30	27.49	–	–	27.49	38.48	54.97
830102040B	0.25 – 1.0 m	m³	2.85	34.06	–	–	34.06	47.68	68.12
830102040C	1.0 – 2.0 m	m³	3.33	39.79	–	–	39.79	55.71	79.59

Bridges

Rail 2003 Ref		Unit	Labour Hours	Labour Cost £	Plant Cost £	Materials Cost £	Unit Rate Base Cost £	Unit Rate Green Zone £	Unit Rate Red Zone £
830	**EARTHWORKS**								
8301020	Excavation by hand in material other than topsoil, rock or other artificial material								
830102050	General excavations; depth:								
830102050A	not exceeding 0.25 m	m³	3.25	38.84	–	–	38.84	54.37	77.68
830102050B	0.25 – 1.0 m	m³	4.33	51.74	–	–	51.74	72.44	103.49
830102050C	1.0 – 2.0 m	m³	4.98	59.51	–	–	59.51	83.32	119.02
83030	**EXCAVATION ANCILLARIES**								
8303010	Trimming of excavated surfaces								
830301010	Topsoil:								
830301010A	horizontal	m²	0.01	0.12	0.18	–	0.30	0.42	0.59
830301010B	sloping; not exceeding 45 degrees to horizontal	m²	0.01	0.12	0.18	–	0.30	0.42	0.59
830301020	Natural material other than top soil or rock:								
830301020A	horizontal	m²	0.02	0.24	0.35	–	0.59	0.82	1.18
830301020B	sloping; not exceeding 45 degrees to horizontal	m²	0.02	0.24	0.35	–	0.59	0.82	1.18
830301020C	sloping; over 45 degrees to horizontal	m²	0.03	0.36	0.53	–	0.89	1.24	1.77
830301020D	vertical	m²	0.04	0.48	0.70	–	1.18	1.65	2.36
830301030	Rock:								
830301030A	horizontal	m²	0.22	2.63	1.54	–	4.17	5.83	8.33
830301030B	sloping; not exceeding 45 degrees to horizontal	m²	0.24	2.87	1.68	–	4.55	6.37	9.09
830301030C	sloping; over 45 degrees to horizontal	m²	0.32	3.82	2.24	–	6.06	8.48	12.12
830301030D	vertical	m²	0.42	5.02	2.93	–	7.95	11.14	15.91

Bridges

Rail 2003 Ref		Unit	Labour Hours	Labour Cost £	Plant Cost £	Materials Cost £	Unit Rate Base Cost £	Unit Rate Green Zone £	Unit Rate Red Zone £
830	**EARTHWORKS**								
8303020	Offsite disposal of materials arising from earthworks								
830302010	Removed including providing a suitable site:								
830302010A	hand loading	m³	1.37	16.37	7.05	–	23.42	32.79	46.84
830302010B	machine loading	m³	–	–	8.28	–	8.28	11.59	16.55
83050	**FILLING TO EXCAVATIONS**								
8305010	Filling to excavations with imported materials								
830501010	Filling to excavations; by machine; compacting in layers:								
830501010A	sand	m³	0.30	3.59	3.75	10.25	17.59	24.61	35.16
830501010B	hardcore	m³	0.33	3.94	6.40	14.18	24.52	34.34	49.06
830501010C	topsoil	m³	0.32	3.82	4.64	7.45	15.91	22.28	31.83
830501010D	MOT type 1	m³	0.33	3.94	6.40	13.62	23.96	33.54	47.92
830501010E	MOT type 2	m³	0.33	3.94	6.40	13.20	23.54	32.96	47.09
830501020	Filling to excavations; by hand; compacting in layers:								
830501020A	sand	m³	1.03	12.31	0.95	10.25	23.51	32.90	47.01
830501020B	hardcore	m³	1.66	19.84	0.95	14.18	34.97	48.96	69.94
830501020C	hoggin	m³	1.13	13.50	0.95	11.47	25.92	36.30	51.86
830501020D	stone rejects	m³	1.66	19.84	0.95	13.71	34.50	48.29	68.99
830501020E	granite scalpings	m³	1.66	19.84	0.95	12.36	33.15	46.40	66.28
830501020F	MOT type 1	m³	1.60	19.12	0.95	13.62	33.69	47.16	67.37
830501020G	MOT type 2	m³	1.60	19.12	0.95	13.20	33.27	46.58	66.54
830501030	Filling in making up levels; by machine; compacting in layers:								
830501030A	sand	m³	0.17	2.03	2.70	10.25	14.98	20.96	29.95
830501030B	hardcore	m³	0.17	2.03	3.58	14.18	19.79	27.71	39.58
830501030C	hoggin	m³	0.17	2.03	3.23	11.47	16.73	23.42	33.46
830501030D	stone rejects	m³	0.17	2.03	3.58	13.71	19.32	27.04	38.63
830501030E	granite scalpings	m³	0.17	2.03	3.58	12.36	17.97	25.15	35.92
830501030F	MOT type 1	m³	0.17	2.03	3.23	13.62	18.88	26.42	37.74
830501030G	MOT type 2	m³	0.17	2.03	3.23	13.20	18.46	25.84	36.91

Bridges

Rail 2003 Ref		Unit	Labour Hours	Labour Cost £	Plant Cost £	Materials Cost £	Unit Rate Base Cost £	Unit Rate Green Zone £	Unit Rate Red Zone £

830 EARTHWORKS

8305010 Filling to excavations with imported materials

830501040 Filling in making up levels; by hand; compacting in layers

Ref	Material	Unit	Lab Hrs	Lab Cost	Plant	Materials	Base	Green	Red
830501040A	sand	m³	0.83	9.92	0.73	10.25	20.90	29.25	41.78
830501040B	hardcore	m³	1.46	17.45	0.73	14.18	32.36	45.31	64.71
830501040C	hoggin	m³	0.93	11.11	0.73	11.47	23.31	32.64	46.63
830501040D	stone rejects	m³	1.46	17.45	0.73	13.71	31.89	44.64	63.76
830501040E	granite scalpings	m³	1.46	17.45	0.73	12.36	30.54	42.75	61.05
830501040F	MOT type 1	m³	1.40	16.73	0.73	13.62	31.08	43.50	62.14
830501040G	MOT type 2	m³	1.40	16.73	0.73	13.20	30.66	42.92	61.31

830501050 Filling into oversite to make up levels; by machine; compacting in layers; average 150 mm thick:

Ref	Material	Unit	Lab Hrs	Lab Cost	Plant	Materials	Base	Green	Red
830501050A	sand	m³	0.13	1.55	1.60	10.25	13.40	18.76	26.80
830501050B	hardcore	m³	0.13	1.55	2.48	14.18	18.21	25.51	36.43
830501050C	hoggin	m³	0.13	1.55	2.13	11.47	15.15	21.22	30.31
830501050D	stone rejects	m³	0.13	1.55	2.48	13.71	17.74	24.84	35.48
830501050E	granite scalpings	m³	0.13	1.55	2.48	12.36	16.39	22.95	32.77
830501050F	MOT type 1	m³	0.13	1.55	2.13	13.62	17.30	24.22	34.59
830501050G	MOT type 2	m³	0.13	1.55	2.13	13.20	16.88	23.64	33.76

830501060 Filling into oversite to make up levels; by hand; compacting in layers; average 150 mm thick;

Ref	Material	Unit	Lab Hrs	Lab Cost	Plant	Materials	Base	Green	Red
830501060A	sand	m³	1.03	12.31	0.95	10.25	23.51	32.90	47.01
830501060B	hardcore	m³	1.66	19.84	0.95	14.18	34.97	48.96	69.94
830501060C	hoggin	m³	1.13	13.50	0.95	11.47	25.92	36.30	51.86
830501060D	stone rejects	m³	1.66	19.84	0.95	13.71	34.50	48.29	68.99
830501060E	granite scalpings	m³	1.66	19.84	0.95	12.36	33.15	46.40	66.28
830501060F	MOT type 1	m³	1.60	19.12	0.95	13.62	33.69	47.16	67.37
830501060G	MOT type 2	m³	1.60	19.12	0.95	13.20	33.27	46.58	66.54

830501070 Grading filled oversite to contours, embankments and the like; by machine:

Ref	Description	Unit	Lab Hrs	Lab Cost	Plant	Materials	Base	Green	Red
830501070A	to falls	m³	0.03	0.36	0.53	–	0.89	1.24	1.77
830501070B	to falls and crossfalls	m³	0.05	0.60	0.88	–	1.48	2.07	2.95
830501070C	to slopes	m³	0.02	0.24	0.35	–	0.59	0.82	1.18

Bridges

Rail 2003 Ref		Unit	Labour Hours	Labour Cost £	Plant Cost £	Materials Cost £	Unit Rate Base Cost £	Unit Rate Green Zone £	Unit Rate Red Zone £
830	**EARTHWORKS**								
8305010	Filling to excavations with imported materials								
830501080	Grading filled oversite to contours, embankments and the like; by hand:								
830501080A	to falls	m³	0.10	1.20	–	–	1.20	1.67	2.39
830501080B	to falls and crossfalls	m³	0.17	2.03	–	–	2.03	2.84	4.06
830501080C	to slopes	m³	0.09	1.08	–	–	1.08	1.51	2.15
830501090	Blinding to hardcore:								
830501090A	25 mm sand	m³	0.04	0.48	–	0.25	0.73	1.01	1.45
830501090B	50 mm sand	m³	0.06	0.72	–	0.49	1.21	1.69	2.41

Bridges

Rail 2003 Ref		Unit	Labour Hours	Labour Cost £	Plant Cost £	Materials Cost £	Unit Rate Base Cost £	Unit Rate Green Zone £	Unit Rate Red Zone £
840	**CONCRETE WORK**								
84010	**IN SITU CONCRETE**								
8401010	Plain in situ concrete; grade C15; 20 mm aggregate								
840101010	Blinding; thickness:								
840101010A	not exceeding 150 mm	m³	1.25	14.94	–	49.10	64.04	89.66	128.09
840101010B	150 – 450 mm	m³	1.10	13.15	–	49.10	62.25	87.15	124.50
840101010C	450 – 600 mm	m³	0.95	11.36	–	49.10	60.46	84.64	120.91
840101010D	over 600 mm	m³	0.90	10.76	–	49.10	59.86	83.80	119.72
840101020	Base, footings, pile caps and ground slabs; thickness:								
840101020A	not exceeding 150 mm	m³	1.25	14.94	–	49.10	64.04	89.66	128.09
840101020B	150 – 450 mm	m³	1.10	13.15	–	49.10	62.25	87.15	124.50
840101020C	450 – 600 mm	m³	0.95	11.36	–	49.10	60.46	84.64	120.91
840101020D	over 600 mm	m³	0.90	10.76	–	49.10	59.86	83.80	119.72
8401020	Plain in situ concrete; grade C20								
840102010	Blinding; thickness:								
840102010A	not exceeding 150 mm	m³	1.25	14.94	–	53.60	68.54	95.96	137.09
840102010B	150 – 450 mm	m³	1.10	13.15	–	53.60	66.75	93.45	133.50
840102010C	450 – 600 mm	m³	0.95	11.36	–	53.60	64.96	90.94	129.91
840102010D	over 600 mm	m³	0.90	10.76	–	53.60	64.36	90.10	128.72
840102020	Base, footings, pile caps and ground slabs; thickness:								
840102020A	not exceeding 150 mm	m³	1.25	14.94	–	53.60	68.54	95.96	137.09
840102020B	150 – 450 mm	m³	1.10	13.15	–	53.60	66.75	93.45	133.50
840102020C	450 – 600 mm	m³	0.95	11.36	–	53.60	64.96	90.94	129.91
840102020D	over 600 mm	m³	0.90	10.76	–	53.60	64.36	90.10	128.72

Bridges

Rail 2003 Ref		Unit	Labour Hours	Labour Cost £	Plant Cost £	Materials Cost £	Unit Rate Base Cost £	Unit Rate Green Zone £	Unit Rate Red Zone £
840	**CONCRETE WORK**								
8401030	**Reinforced in situ concrete; grade C30; 20 mm aggregate**								
840103020	Base, footings, pile caps and ground slabs; thickness:								
840103020A	not exceeding 150 mm	m³	1.90	22.97	1.96	58.66	83.59	117.02	167.17
840103020B	150 – 450 mm	m³	1.70	20.55	1.68	58.66	80.89	113.24	161.78
840103020C	450 – 600 mm	m³	1.50	18.14	1.54	58.66	78.34	109.66	156.66
840103020D	over 600 mm	m³	1.40	16.93	1.40	58.66	76.99	107.78	153.96
840103030	Suspended slabs; thickness:								
840103030A	not exceeding 150 mm	m³	2.10	25.39	2.10	58.66	86.15	120.60	172.29
840103030B	150 – 450 mm	m³	1.80	21.76	1.82	58.66	82.24	115.14	164.47
840103030C	450 – 600 mm	m³	1.45	17.53	1.47	58.66	77.66	108.72	155.31
840103030D	over 600 mm	m³	1.35	16.32	1.40	58.66	76.38	106.93	152.75
840103040	Walls; thickness:								
840103040A	not exceeding 150 mm	m³	2.30	27.81	4.62	58.66	91.09	127.51	182.15
840103040B	150 – 450 mm	m³	2.05	24.79	4.06	58.66	87.51	122.50	174.99
840103040C	450 – 600 mm	m³	1.55	18.74	3.08	58.66	80.48	112.67	160.95
840103040D	over 600 mm	m³	1.40	16.93	2.80	58.66	78.39	109.74	156.76
840103050	Columns and piers; cross sectional area:								
840103050A	not exceeding 0.03 m²	m³	4.00	48.36	8.04	58.66	115.06	161.08	230.12
840103050B	0.03 – 0.1 m²	m³	2.95	35.67	5.88	58.66	100.21	140.28	200.39
840103050C	0.1 – 0.25 m²	m³	2.70	32.64	5.39	58.66	96.69	135.36	193.37
840103050D	0.25 – 1 m²	m³	2.40	29.02	4.83	58.66	92.51	129.50	184.99
840103050E	over 1 m²	m³	2.20	26.60	4.34	58.66	89.60	125.43	179.18
840103060	Beams; cross-sectional area:								
840103060A	not exceeding 0.03 m²	m³	3.60	43.52	7.28	58.66	109.46	153.24	218.91
840103060B	0.03 – 0.1 m²	m³	2.65	32.04	5.32	58.66	96.02	134.41	192.02
840103060C	0.1 – 0.25 m²	m³	2.45	29.62	4.83	58.66	93.11	130.35	186.20
840103060D	0.25 – 1 m²	m³	2.15	25.99	4.34	58.66	88.99	124.58	177.97
840103060E	over 1 m²	m³	2.00	24.18	3.92	58.66	86.76	121.45	173.50

Bridges

Rail 2003 Ref		Unit	Labour Hours	Labour Cost £	Plant Cost £	Materials Cost £	Unit Rate Base Cost £	Unit Rate Green Zone £	Unit Rate Red Zone £
840	**CONCRETE WORK**								
8401030	Reinforced in situ concrete; grade C30; 20 mm aggregate								
840103070	Casing to metal sections; cross-sectional area:								
840103070A	not exceeding 0.03 m^2	m^3	4.15	50.17	8.39	58.66	117.22	164.11	234.45
840103070B	0.03 – 0.1 m^2	m^3	3.05	36.88	6.09	58.66	101.63	142.27	203.23
840103070C	0.1 – 0.25 m^2	m^3	2.80	33.85	5.53	58.66	98.04	137.25	196.06
840103070D	0.25 – 1 m^2	m^3	2.45	29.62	4.97	58.66	93.25	130.54	186.48
840103070E	over 1 m^2	m^3	2.30	27.81	4.48	58.66	90.95	127.32	181.87
8401040	Reinforced in situ concrete; grade C40; 20 mm aggregate								
840104020	Base, footings, pile caps and ground slabs; thickness:								
840104020A	not exceeding 150 mm	m^3	1.90	22.97	1.96	65.17	90.10	126.14	180.21
840104020B	150 – 450 mm	m^3	1.70	20.55	1.68	65.17	87.40	122.36	174.82
840104020C	450 – 600 mm	m^3	1.50	18.14	1.54	65.17	84.85	118.78	169.70
840104020D	over 600 mm	m^3	1.40	16.93	1.40	65.17	83.50	116.90	167.00
840104030	Suspended slabs; thickness:								
840104030A	not exceeding 150 mm	m^3	2.10	25.39	2.10	65.17	92.66	129.72	185.33
840104030B	150 – 450 mm	m^3	1.80	21.76	1.82	65.17	88.75	124.26	177.51
840104030C	450 – 600 mm	m^3	1.45	17.53	1.47	65.17	84.17	117.84	168.35
840104030D	over 600 mm	m^3	1.25	15.11	1.40	65.17	81.68	114.36	163.38
840104040	Walls; thickness:								
840104040A	not exceeding 150 mm	m^3	2.30	27.81	4.62	65.17	97.60	136.63	195.19
840104040B	150 – 450 mm	m^3	2.05	24.79	4.06	65.17	94.02	131.62	188.03
840104040C	450 – 600 mm	m^3	1.55	18.74	3.08	65.17	86.99	121.79	173.99
840104040D	over 600 mm	m^3	1.40	16.93	2.80	65.17	84.90	118.86	169.80
840104050	Columns and piers; cross-sectional area:								
840104050A	not exceeding 0.03 m^2	m^3	4.00	48.36	8.04	65.17	121.57	170.20	243.16
840104050B	0.03 – 0.1 m^2	m^3	2.95	35.67	5.88	65.17	106.72	149.40	213.43
840104050C	0.1 – 0.25 m^2	m^3	2.70	32.64	5.39	65.17	103.20	144.48	206.41
840104050D	0.25 – 1 m^2	m^3	2.40	29.02	4.83	65.17	99.02	138.62	198.03
840104050E	over 1 m^2	m^3	2.20	26.60	4.34	65.17	96.11	134.55	192.22

Bridges

Rail 2003 Ref		Unit	Labour Hours	Labour Cost £	Plant Cost £	Materials Cost £	Unit Rate Base Cost £	Unit Rate Green Zone £	Unit Rate Red Zone £
840	**CONCRETE WORK**								
8401040	**Reinforced in situ concrete; grade C40; 20 mm aggregate**								
840104060	Beams; cross sectional area:								
840104060A	not exceeding 0.03 m²	m³	3.60	43.52	7.28	65.17	115.97	162.36	231.95
840104060B	0.03 – 0.1 m²	m³	3.10	37.48	7.00	65.17	109.65	153.50	219.30
840104060C	0.1 – 0.25 m²	m³	2.45	29.62	4.83	65.17	99.62	139.47	199.24
840104060D	0.25 – 1 m²	m³	2.25	27.20	5.25	65.17	97.62	136.66	195.25
840104060E	over 1 m²	m³	2.00	24.18	3.92	65.17	93.27	130.57	186.54
840104070	Casings to metal sections; cross-sectional area:								
840104070A	not exceeding 0.03 m²	m³	4.15	50.17	8.39	65.17	123.73	173.23	247.49
840104070B	0.03 – 0.1 m²	m³	3.05	36.88	6.09	65.17	108.14	151.39	216.27
840104070C	0.1 – 0.25 m²	m³	2.80	33.85	5.53	65.17	104.55	146.37	209.10
840104070D	0.25 – 1 m²	m³	2.45	29.62	4.97	65.17	99.76	139.66	199.52
840104070E	over 1 m²	m³	2.20	26.60	4.69	65.17	96.46	135.04	192.92

Bridges

Rail 2003 Ref		Unit	Labour Hours	Labour Cost £	Plant Cost £	Materials Cost £	Unit Rate Base Cost £	Unit Rate Green Zone £	Unit Rate Red Zone £

845 CONCRETE ANCILLARIES

84510 FORMWORK

8451010 Formwork to fair finish

845101010 Sides of foundation; width:

845101010A	not exceeding 250 mm	m	0.76	10.43	–	2.23	12.66	17.72	25.32
845101010B	250 – 500 mm	m	1.15	15.78	–	3.71	19.49	27.30	38.99
845101010C	500 mm – 1 m	m	2.04	28.00	–	6.66	34.66	48.53	69.32
845101010D	over 1 m	m²	1.86	25.53	–	7.01	32.54	45.55	65.06

845101015 Sides of ground beams and edges of beds; width:

845101015A	not exceeding 250 mm	m	0.76	10.43	–	2.23	12.66	17.72	25.32
845101015B	250 – 500 mm	m	1.15	15.78	–	3.71	19.49	27.30	38.99
845101015C	500 mm – 1 m	m	2.04	28.00	–	6.66	34.66	48.53	69.32
845101015D	over 1 m	m²	1.86	25.53	–	28.25	53.78	75.29	107.55

845101020 Edges of suspended slabs; width:

845101020A	not exceeding 250 mm	m	0.71	9.74	–	1.98	11.72	16.41	23.45
845101020B	250 – 500 mm	m	0.90	12.35	–	3.24	15.59	21.82	31.18
845101020C	500 mm – 1 m	m	1.60	21.96	2.89	5.61	30.46	42.64	60.93
845101020D	over 1 m	m²	1.46	20.04	5.78	6.45	32.27	45.17	64.53

845101025 Sides of upstands; width:

845101025A	not exceeding 250 mm	m	0.71	9.74	–	1.98	11.72	16.41	23.45
845101025B	250 – 500 mm	m	0.90	12.35	–	3.24	15.59	21.82	31.18
845101025C	500 mm – 1 m	m	1.60	21.96	2.89	8.26	33.11	46.36	66.23
845101025D	over 1 m	m²	1.46	20.04	5.78	6.45	32.27	45.17	64.53

845101030 Steps in top surfaces; width:

845101030A	not exceeding 250 mm	m	0.71	9.74	–	1.98	11.72	16.41	23.45
845101030B	250 – 500 mm	m	0.90	12.35	–	3.24	15.59	21.82	31.18
845101030C	500 mm – 1 m	m	1.59	21.82	2.89	5.61	30.32	42.45	60.65
845101030D	over 1 m	m²	1.46	20.04	5.78	6.45	32.27	45.17	64.53

Bridges

Rail 2003 Ref		Unit	Labour Hours	Labour Cost £	Plant Cost £	Materials Cost £	Unit Rate Base Cost £	Unit Rate Green Zone £	Unit Rate Red Zone £
845	**CONCRETE ANCILLARIES**								
8451010	Formwork to fair finish								
845101040	Machine bases and plinths; width:								
845101040A	not exceeding 250 mm	m	0.76	10.43	–	5.41	15.84	22.17	31.67
845101040B	250 – 500 mm	m	1.15	15.78	–	10.12	25.90	36.26	51.80
845101040C	500 mm – 1 m	m	2.04	28.00	–	19.42	47.42	66.39	94.83
845101040D	over 1 m	m^2	1.86	25.53	–	18.61	44.14	61.79	88.26
845101045	Soffits; not exceeding 1.5 m above floor level; thickness:								
845101045A	not exceeding 200 mm	m^2	1.46	20.04	2.41	6.86	29.31	41.02	58.60
845101045B	200 – 300 mm	m^2	2.04	28.00	2.41	6.86	37.27	52.17	74.52
845101048	Soffits; 1.5 – 3.0 m above floor level; thickness:								
845101048A	not exceeding 200 mm	m^2	1.79	24.57	2.41	6.86	33.84	47.36	67.66
845101048B	200 – 300 mm	m^2	2.50	34.31	2.41	18.61	55.33	77.45	110.65
845101050	Sloping soffits; not exceeding 1.5 m above floor level; thickness:								
845101050A	not exceeding 200 mm	m^2	1.46	20.04	2.41	6.86	29.31	41.02	58.60
845101050B	200 – 300 mm	m^2	2.04	28.00	2.41	6.86	37.27	52.17	74.52
845101052	Sloping soffits; 1.5 – 3.0 m above floor level; thickness:								
845101052A	not exceeding 200 mm	m^2	1.79	24.57	2.41	6.86	33.84	47.36	67.66
845101052B	200 – 300 mm	m^2	2.50	34.31	2.41	18.61	55.33	77.45	110.65
845101055	Top formwork; over 15 degrees from horizontal:								
845101055A	generally	m^2	1.86	25.53	–	5.02	30.55	42.77	61.09
845101060	Walls:								
845101060A	vertical surfaces	m^2	1.78	24.43	0.94	7.51	32.88	46.04	65.77
845101060B	curved surfaces	m^2	2.00	27.45	0.94	6.57	34.96	48.95	69.93
845101060C	conical surfaces	m^2	2.66	36.51	1.06	7.83	45.40	63.57	90.80
845101060D	battering surfaces	m^2	2.13	29.23	0.94	6.76	36.93	51.71	73.88

Bridges

Rail 2003 Ref		Unit	Labour Hours	Labour Cost £	Plant Cost £	Materials Cost £	Unit Rate Base Cost £	Unit Rate Green Zone £	Unit Rate Red Zone £
845	**CONCRETE ANCILLARIES**								
8451010	Formwork to fair finish								
845101065	Sides and soffits of attached beams; width:								
845101065A	not exceeding 250 mm	m	1.07	14.69	0.57	2.12	17.38	24.32	34.74
845101065B	250 – 500 mm	m	1.25	17.16	0.57	4.05	21.78	30.48	43.54
845101065C	500 mm – 1 m	m	1.54	21.14	1.13	6.51	28.78	40.30	57.57
845101065D	over 1 m	m^2	2.03	27.86	0.94	6.86	35.66	49.92	71.32
845101070	Sides and soffits of isolated beams; width:								
845101070A	not exceeding 250 mm	m	1.21	16.61	3.46	2.09	22.16	31.02	44.31
845101070B	250 – 500 mm	m	1.41	19.35	3.46	11.43	34.24	47.95	68.49
845101070C	500 mm – 1 m	m	1.74	23.88	6.91	5.81	36.60	51.24	73.20
845101070D	over 1 m	m^2	2.20	30.19	7.36	6.86	44.41	62.18	88.83
845101075	Sides of isolated rectangular columns; width:								
845101075A	not exceeding 250 mm	m	0.29	3.98	1.47	1.92	7.37	10.32	14.74
845101075B	250 – 500 mm	m	0.58	7.96	1.47	3.32	12.75	17.85	25.50
845101075C	500 mm – 1 m	m	0.73	10.02	1.78	5.34	17.14	23.99	34.26
845101075D	over 1 m	m^2	1.01	13.86	2.06	5.90	21.82	30.55	43.63
845101080	Sides of isolated circular columns; diameter:								
845101080A	not exceeding 300 mm	m^2	0.71	9.74	4.91	10.39	25.04	35.05	50.08
845101080B	300 – 600 mm	m^2	0.40	5.49	2.46	10.39	18.34	25.67	36.67
845101080C	600 – 900 mm	m^2	0.25	3.43	1.64	10.39	15.46	21.63	30.92
845101080D	over 900 mm	m^2	0.18	2.47	1.23	10.39	14.09	19.72	28.18

Bridges

Rail 2003 Ref		Unit	Labour Hours	Labour Cost £	Plant Cost £	Materials Cost £	Unit Rate Base Cost £	Unit Rate Green Zone £	Unit Rate Red Zone £
845	**CONCRETE ANCILLARIES**								
84520	**REINFORCEMENT**								
8452010	**Bar Reinforcement**								
845201010	Mild steel bars; BS 4449; nominal size:								
845201010A	6 mm	Tonne	72.00	898.56	–	447.61	1346.17	1884.64	2692.34
845201010B	8 mm	Tonne	54.00	673.92	–	435.72	1109.64	1553.50	2219.29
845201010C	10 mm	Tonne	44.00	549.12	–	405.00	954.12	1335.78	1908.25
845201010D	12 mm	Tonne	38.00	474.24	–	387.84	862.08	1206.92	1724.16
845201010E	16 mm	Tonne	30.00	374.40	–	363.04	737.44	1032.41	1474.88
845201010F	20 mm	Tonne	26.00	324.48	–	353.81	678.29	949.60	1356.58
845201010G	25 mm	Tonne	23.00	287.04	–	342.77	629.81	881.74	1259.62
845201010H	32 mm	Tonne	20.00	249.60	–	336.49	586.09	820.52	1172.17
845201020	High yield bars; BS 4449; nominal size:								
845201020A	6 mm	Tonne	72.00	898.56	–	464.35	1362.91	1908.06	2725.81
845201020B	8 mm	Tonne	54.00	673.92	–	452.08	1126.00	1576.40	2252.00
845201020C	10 mm	Tonne	44.00	549.12	–	420.39	969.51	1357.32	1939.03
845201020D	12 mm	Tonne	38.00	474.24	–	402.59	876.83	1227.56	1753.65
845201020E	16 mm	Tonne	30.00	374.40	–	376.83	751.23	1051.72	1502.45
845201020F	20 mm	Tonne	26.00	324.48	–	367.44	691.92	968.68	1383.83
845201020G	25 mm	Tonne	23.00	287.04	–	356.07	643.11	900.36	1286.23
845201020H	32 mm	Tonne	20.00	249.60	–	349.63	599.23	838.92	1198.46
845201030	Stainless steel bars; type 316 S66; nominal size:								
845201030A	6 mm	Tonne	44.00	549.12	–	6833.95	7383.07	10336.29	14766.13
845201030B	8 mm	Tonne	38.00	474.24	–	6709.51	7183.75	10057.25	14367.49
845201030C	10 mm	Tonne	30.00	374.40	–	6462.14	6836.54	9571.15	13673.07
845201030D	12 mm	Tonne	26.00	324.48	–	6317.69	6642.17	9299.03	13284.34
845201030E	16 mm	Tonne	23.00	287.04	–	6191.36	6478.40	9069.76	12956.79
845201030F	20 mm	Tonne	20.00	249.60	–	6065.78	6315.38	8841.53	12630.76
845201040	Steel fabric reinforcement; BS 4483; laid horizontally:								
845201040A	A142	m^2	0.03	0.37	–	1.82	2.19	3.07	4.39
845201040B	A193	m^2	0.03	0.37	–	2.42	2.79	3.91	5.60
845201040C	A252	m^2	0.03	0.37	–	3.13	3.50	4.90	7.00
845201040D	A393	m^2	0.05	0.62	–	4.91	5.53	7.75	11.08
845201040E	B283	m^2	0.03	0.37	–	3.13	3.50	4.90	7.00
845201040F	B385	m^2	0.04	0.50	–	3.74	4.24	5.94	8.49
845201040G	B503	m^2	0.05	0.62	–	4.76	5.38	7.54	10.78
845201040H	B785	m^2	0.05	0.62	–	6.05	6.67	9.33	13.34
845201040I	C385	m^2	0.03	0.37	–	2.88	3.25	4.55	6.51
845201040J	C503	m^2	0.04	0.50	–	3.55	4.05	5.67	8.10

Bridges

Rail 2003 Ref		Unit	Labour Hours	Labour Cost £	Plant Cost £	Materials Cost £	Unit Rate Base Cost £	Unit Rate Green Zone £	Unit Rate Red Zone £
845	**CONCRETE ANCILLARIES**								
84530	**JOINTS IN CONCRETE PAVINGS**								
8453010	Designed joints in in situ concrete								
845301010	Plain joints; horizontal; including formwork:								
845301010A	not exceeding 150 mm high	m	0.72	9.88	–	1.66	11.54	16.16	23.08
845301010B	150 – 300 mm high	m	0.82	11.25	–	4.14	15.39	21.55	30.78
845301020	Keyed joints; horizontal; including formwork:								
845301020A	not exceeding 150 mm high	m	0.77	10.57	–	1.83	12.40	17.35	24.79
845301020B	150 mm – 300 mm high	m	0.82	11.25	–	4.46	15.71	22.00	31.43
845301030	Movement joints; horizontal; including formwork; 25 mm mild steel dowel bars 600 mm long debonded for half length; dowel caps with compressible filler:								
845301030A	not exceeding 150 mm high	m	1.09	14.96	–	8.61	23.57	32.99	47.13
845301030B	150 – 300 mm high	m	1.19	16.33	–	11.08	27.41	38.37	54.82
845301040	Schlegal crack inducer; horizontal; placing in position:								
845301040A	generally	m	0.25	3.43	–	1.04	4.47	6.26	8.94
845301050	Serviseal PVC waterstops; flat dumbell; width:								
845301050A	100 mm	m	0.12	1.65	–	2.65	4.30	6.02	8.59
845301050B	170 mm	m	0.13	1.78	–	3.71	5.49	7.69	10.98
845301050C	210 mm	m	0.16	2.20	–	4.77	6.97	9.74	13.92
845301050D	250 mm	m	0.18	2.47	–	5.56	8.03	11.24	16.06

Bridges

Rail 2003 Ref		Unit	Labour Hours	Labour Cost £	Plant Cost £	Materials Cost £	Unit Rate Base Cost £	Unit Rate Green Zone £	Unit Rate Red Zone £
845	**CONCRETE ANCILLARIES**								
8453010	Designed joints in in situ concrete								
845301055	Serviseal PVC waterstops; centre bulb; width:								
845301055A	160 mm	m	0.12	1.65	–	3.71	5.36	7.50	10.70
845301055B	210 mm	m	0.15	2.06	–	4.77	6.83	9.55	13.65
845301055C	260 mm	m	0.18	2.47	–	5.56	8.03	11.24	16.06
845301060	Servi-tite flat dumbell CJ; width:								
845301060A	150 mm	m	0.12	1.65	–	7.28	8.93	12.50	17.85
845301060B	230 mm	m	0.15	2.06	–	9.93	11.99	16.78	23.98
845301060C	305 mm	m	0.18	2.47	–	18.54	21.01	29.41	42.01
845301065	Servi-tite flat centre bulb XJ; width:								
845301065A	150 mm	m	0.12	1.65	–	7.94	9.59	13.43	19.18
845301065B	230 mm	m	0.15	2.06	–	10.93	12.99	18.18	25.98
845301065C	305 mm	m	0.18	2.47	–	18.54	21.01	29.41	42.01
845301070	Expansion materials; Flexcell fibreboard compressible joint filler; 10 mm thick; width:								
845301070A	not exceeding 150 mm	m	0.10	1.37	–	1.21	2.58	3.61	5.16
845301070B	150 – 300 mm	m	0.18	2.47	–	2.35	4.82	6.74	9.63
845301071	Expansion materials; Flexcell fibreboard compressible joint filler; 13 mm thick; width:								
845301071A	not exceeding 150 mm	m	0.11	1.51	–	1.61	3.12	4.37	6.25
845301071B	150 – 300 mm	m	0.18	2.47	–	3.13	5.60	7.84	11.19
845301072	Expansion materials; Flexcell fibreboard compressible joint filler; 19 mm thick; width:								
845301072A	not exceeding 150 mm	m	0.20	2.75	–	2.02	4.77	6.66	9.52
845301072B	150 – 300 mm	m	0.23	3.16	–	3.91	7.07	9.89	14.13

Bridges

Rail 2003 Ref		Unit	Labour Hours	Labour Cost	Plant Cost	Materials Cost	Unit Rate	Unit Rate	Unit Rate
				£	£	£	Base Cost £	Green Zone £	Red Zone £

845 CONCRETE ANCILLARIES

8453010 Designed joints in in situ concrete

845301073 Expansion materials; Flexcell fibreboard compressible joint filler; 25 mm thick; width:

845301073A	not exceeding 150 mm	m	0.20	2.75	–	2.42	5.17	7.23	10.33
845301073B	150 – 300 mm	m	0.24	3.29	–	4.69	7.98	11.18	15.97

845301075 Joint sealants; hot poured bituminous rubber compound:

845301075A	10 × 25 mm	m	0.03	0.41	0.14	1.16	1.71	2.39	3.42
845301075B	13 × 25 mm	m	0.03	0.41	0.14	1.50	2.05	2.87	4.10
845301075C	19 × 25 mm	m	0.04	0.55	0.17	2.19	2.91	4.08	5.84
845301075D	25 × 25 mm	m	0.06	0.82	0.21	2.93	3.96	5.54	7.92

845301077 Joint sealants; cold poured polysulphide epoxy based compound:

845301077A	10 × 25 mm	m	0.03	0.41	–	0.40	0.81	1.14	1.63
845301077B	13 × 25 mm	m	0.03	0.41	–	0.56	0.97	1.36	1.93
845301077C	19 × 25 mm	m	0.04	0.55	–	0.76	1.31	1.83	2.61
845301077D	25 × 25 mm	m	0.06	0.82	–	1.26	2.08	2.92	4.17

845301079 Joint sealants; gun grade polysulphide rubber compound:

845301079A	10 × 25 mm	m	0.04	0.55	–	0.34	0.89	1.25	1.79
845301079B	13 × 25 mm	m	0.06	0.82	–	0.58	1.40	1.97	2.82
845301079C	19 × 25 mm	m	0.13	1.78	–	1.16	2.94	4.12	5.89
845301079D	25 × 25 mm	m	0.23	3.16	–	1.93	5.09	7.12	10.16

Bridges

Rail 2003 Ref		Unit	Labour Hours	Labour Cost £	Plant Cost £	Materials Cost £	Unit Rate Base Cost £	Unit Rate Green Zone £	Unit Rate Red Zone £
845	**CONCRETE ANCILLARIES**								
84580	**CONCRETE ACCESSORIES**								
8458010	**Surface finishes to concrete**								
845801010	Top surfaces; finishing over 300 mm wide with:								
845801010A	wood float	m^2	0.18	2.15	–	–	2.15	3.01	4.30
845801010B	steel trowel	m^2	0.16	1.91	–	–	1.91	2.68	3.83
845801010C	power float	m^2	0.15	1.79	0.21	–	2.00	2.80	4.01
8458020	**Grouting**								
845802010	Grouting under plates with proprietary compound; area:								
845802010A	not exceeding 0.1 m^2	Nr	0.20	2.39	–	0.34	2.73	3.82	5.45
845802010B	0.1 – 0.5 m^2	Nr	0.35	4.18	–	0.67	4.85	6.80	9.71
845802010C	0.5 – 1 m^2	Nr	0.50	5.98	–	0.67	6.65	9.31	13.29
845802010D	over 1 m^2	Nr	0.70	8.37	–	2.01	10.38	14.54	20.77

Bridges

Rail 2003 Ref		Unit	Composite Cost £	Unit Rate Base Cost £	Unit Rate Green Zone £	Unit Rate Red Zone £
849	**PRECAST CONCRETE**					
84910	**BEAMS AND COLUMNS**					
8491010	**Beams and columns**					
849101010	Length; not exceeding 5 m; mass:					
849101010A	not exceeding 250 kg	Nr	–	139.15	194.81	278.30
849101010B	250 – 500 kg	Nr	–	255.11	357.15	510.22
849101010C	500 kg – 1 t	Nr	–	463.83	649.36	927.66
849101010D	1 – 2 t	Nr	–	889.00	1244.60	1778.00
849101010E	2 – 5 t	Nr	–	1662.06	2326.88	3324.12
849101010F	5 – 10 t	Nr	–	3092.20	4329.08	6184.40
849101010G	over 10 t	Nr	–	5797.88	8117.03	11595.76
849101020	Length; 5 – 7 m; mass:					
849101020A	250 – 500 kg	Nr	–	239.65	335.51	479.30
849101020B	500 kg – 1 t	Nr	–	440.64	616.90	881.28
849101020C	1 – 2 t	Nr	–	850.36	1190.50	1700.72
849101020D	2 – 5 t	Nr	–	1584.75	2218.65	3169.50
849101020E	5 – 10 t	Nr	–	3030.36	4242.50	6060.72
849101020F	over 10 t	Nr	–	5542.77	7759.88	11085.54
849101030	Length; 7 – 10 m; mass:					
849101030A	250 – 500 kg	Nr	–	262.84	367.98	525.68
849101030B	500 kg – 1 t	Nr	–	494.75	692.65	989.50
849101030C	1 – 2 t	Nr	–	966.31	1352.83	1932.62
849101030D	2 – 5 t	Nr	–	1816.67	2543.34	3633.34
849101030E	5 – 10 t	Nr	–	3285.46	4599.64	6570.92
849101030F	over 10 t	Nr	–	6107.10	8549.94	12214.20
849101040	Length; 10 – 15 m; mass:					
849101040A	500 kg – 1 t	Nr	–	525.67	735.94	1051.34
849101040B	1 – 2 t	Nr	–	1028.16	1439.42	2056.32
849101040C	2 – 5 t	Nr	–	1893.97	2651.56	3787.94
849101040D	5 – 10 t	Nr	–	3535.84	4950.18	7071.68
849101040E	over 10 t	Nr	–	6570.93	9199.30	13141.86

Bridges

Rail 2003 Ref		Unit	Composite Cost £	Unit Rate Base Cost £	Unit Rate Green Zone £	Unit Rate Red Zone £
849	**PRECAST CONCRETE**					
8491010	**Beams and columns**					
849101050	Length; 15–20 m; mass:					
849101050A	500 kg – 1 t	Nr	– – –	525.67	525.67	735.94	1051.34
849101050B	1 – 2 t	Nr	– – –	1028.16	1028.16	1439.42	2056.32
849101050C	2 – 5 t	Nr	– – –	1893.97	1893.97	2651.56	3787.94
849101050D	5 – 10 t	Nr	– – –	3532.84	3532.84	4945.98	7065.68
849101050E	over 10 t	Nr	– – –	6570.93	6570.93	9199.30	13141.86
849101060	Length; over 20 m; mass:					
849101060A	1 – 2 t	Nr	– – –	1066.81	1066.81	1493.53	2133.62
849101060B	2 – 5 t	Nr	– – –	1983.74	1983.74	2777.24	3967.48
849101060C	5 – 10 t	Nr	– – –	3710.64	3710.64	5194.90	7421.28
849101060D	over 10 t	Nr	– – –	6957.45	6957.45	9740.43	13914.90
84950	**SLABS**					
8495010	**Precast slab units; hoisting and fixing on prepared bearings; grouting up joints**					
849501010	Slab thickness; not exceeding 200 mm; maximum loading 5 kN/m^2; size:					
849501010A	1.2 × 1.2 m	m^2	– – –	280.23	280.23	392.32	560.46
849501010B	2.4 × 1.2 m	m^2	– – –	252.01	252.01	352.81	504.02
849501010C	4.8 × 1.2 m	m^2	– – –	228.05	228.05	319.27	456.10
849501010D	2.4 × 2.4 m	m^2	– – –	193.26	193.26	270.56	386.52
849501010E	4.8 × 2.4 m	m^2	– – –	173.94	173.94	243.52	347.88
849501020	Slab thickness; 200–500 mm; maximum loading 10 kN/m^2; size:					
849501020A	1.2 × 1.2 m	m^2	– – –	610.71	610.71	854.99	1221.42
849501020B	2.4 × 1.2 m	m^2	– – –	548.87	548.87	768.42	1097.74
849501020C	4.8 × 1.2 m	m^2	– – –	487.02	487.02	681.83	974.04
849501020D	2.4 × 2.4 m	m^2	– – –	432.91	432.91	606.07	865.82
849501020E	4.8 × 2.4 m	m^2	– – –	386.53	386.53	541.14	773.06

Bridges

Rail 2003 Ref		Unit	Composite Cost £	Unit Rate Base Cost £	Unit Rate Green Zone £	Unit Rate Red Zone £
849	**PRECAST CONCRETE**					
8495010	Precast slab units; hoisting and fixing on prepared bearings; grouting up joints					
849501030	Slab thickness; 600 mm; maximum loading 10 kN/m^2; size:					
849501030A	1.2 × 1.2 m.	Nr	–	–	–	146.88	146.88	205.63	293.76
849501030B	2.4 × 1.2 m.	Nr	–	–	–	502.48	502.48	703.47	1004.96
849501030C	4.8 × 1.2 m.	Nr	–	–	–	533.40	533.40	746.76	1066.80
849501030D	2.4 × 2.4 m.	Nr	–	–	–	533.40	533.40	746.76	1066.80
849501030E	4.8 × 2.4 m.	Nr	–	–	–	927.66	927.66	1298.72	1855.32
84960	**CULVERTS, DUCTS AND COVERS**					
8496010	Precast units; bedding and jointing					
849601010	Culverts; 250 mm wall thickness; internal size:					
849601010A	1200 × 600 mm	m	–	–	–	386.53	386.53	541.14	773.06
849601010B	1500 × 900 mm	m	–	–	–	479.29	479.29	671.01	958.58
849601010C	1800 × 1200 mm	m	–	–	–	533.40	533.40	746.76	1066.80
849601010D	2400 × 1800 mm	m	–	–	–	788.51	788.51	1103.91	1577.02
849601020	Ducts and covers; nominal internal size:					
849601020A	200 × 100 mm	m	–	–	–	65.71	65.71	91.99	131.42
849601020B	300 × 200 mm	m	–	–	–	102.04	102.04	142.86	204.08
849601020C	450 × 300 mm	m	–	–	–	150.74	150.74	211.04	301.48
849601020D	600 × 450 mm	m	–	–	–	185.53	185.53	259.74	371.06

Bridges

Rail 2003 Ref		Unit	Composite Cost £	Unit Rate Base Cost £	Unit Rate Green Zone £	Unit Rate Red Zone £
849	**PRECAST CONCRETE**					
84970	**COPINGS AND SILLS**					
8497010	**Precast units; bedding and jointing**					
849701010	Copings; weathered; in 500 mm lengths; overall size:					
849701010A	250 mm wide × 100 mm high	m	– – –	14.69	20.57	29.38
849701010B	500 mm wide × 200 mm high	m	– – –	38.27	53.58	76.54
849701010C	750 mm wide × 200 mm high	m	– – –	54.50	76.30	109.00
849701010D	1000 mm wide × 300 mm high	m	– – –	103.20	144.48	206.40
849701020	Sills; weathered; in 500 mm lengths; overall size:					
849701020A	250 mm wide × 100 mm high	m	– – –	21.26	29.76	42.52
849701020B	500 mm wide × 200 mm high	m	– – –	54.18	75.85	108.36
849701020C	750 mm wide × 200 mm high	m	– – –	78.85	110.39	157.70
849701020D	1000 mm wide × 300 mm high	m	– – –	149.59	209.43	299.18
849701030	Parapet units; in 500 mm lengths; overall size:					
849701030A	250 mm wdie × 150 mm high	m	– – –	22.03	30.84	44.06
849701030B	500 mm wide × 225 mm high	m	– – –	42.90	60.06	85.80
849701030C	750 mm wide × 275 mm high	m	– – –	74.99	104.99	149.98
849701030D	1000 mm wide × 350 mm high	m	– – –	108.23	151.52	216.46

Note: Composite Cost column values for all rows shown as "–" (three dashes). Base Cost equals the values shown in the Composite Cost position: 14.69, 38.27, 54.50, 103.20, 21.26, 54.18, 78.85, 149.59, 22.03, 42.90, 74.99, 108.23.

Bridges

Rail 2003 Ref	Unit	Composite Cost £	Unit Rate Base Cost £	Unit Rate Green Zone £	Unit Rate Red Zone £

860 STRUCTURAL METALWORK

86010 FABRICATION OF MAIN MEMBERS

8601010 Framing fabrication of main members; bolted connections; hot rolled steel; B4360 Grade 43A; including shop and site black bolts; rolled sections

860101010 Straight on plan:

Ref	Item	Unit				Base	Green	Red
860101010A	beams	Tonne	–	–	–	1816.67	1816.67	2543.34	3633.34
860101010B	columns	Tonne	–	–	–	1662.06	1662.06	2326.88	3324.12
860101010C	channels	Tonne	–	–	–	1932.63	1932.63	2705.68	3865.26
860101010D	plates and flats	Tonne	–	–	–	2033.12	2033.12	2846.37	4066.24
860101010E	box and hollow sections	Tonne	–	–	–	1893.97	1893.97	2651.56	3787.94

860101020 Curved on plan:

860101020A	beams	Tonne	–	–	–	2543.33	2543.33	3560.66	5086.66
860101020B	columns	Tonne	–	–	–	2326.88	2326.88	3257.63	4653.76
860101020C	channels	Tonne	–	–	–	2705.68	2705.68	3787.95	5411.36
860101020D	plates and flats	Tonne	–	–	–	2844.82	2844.82	3982.75	5689.64
860101020E	box and hollow sections	Tonne	–	–	–	2651.56	2651.56	3712.18	5303.12

860101030 Straight on plan and cambered:

860101030A	beams	Tonne	–	–	–	2543.33	2543.33	3560.66	5086.66
860101030B	columns	Tonne	–	–	–	2326.88	2326.88	3257.63	4653.76
860101030C	channels	Tonne	–	–	–	2705.68	2705.68	3787.95	5411.36
860101030D	plates and flats	Tonne	–	–	–	2844.82	2844.82	3982.75	5689.64
860101030E	box and hollow sections	Tonne	–	–	–	2651.56	2651.56	3712.18	5303.12

860101040 Curved on plan and cambered:

860101040A	beams	Tonne	–	–	–	3385.96	3385.96	4740.34	6771.92
860101040B	columns	Tonne	–	–	–	3092.20	3092.20	4329.08	6184.40
860101040C	channels	Tonne	–	–	–	3594.68	3594.68	5032.55	7189.36
860101040D	plates and flats	Tonne	–	–	–	3787.95	3787.95	5303.13	7575.90
860101040E	box and hollow sections	Tonne	–	–	–	3525.11	3525.11	4935.15	7050.22

Bridges

Rail 2003 Ref		Unit	Composite Cost £	Unit Rate Base Cost £	Unit Rate Green Zone £	Unit Rate Red Zone £
860	**STRUCTURAL METALWORK**					
86050	**ERECTION OF MEMBERS FOR BRIDGES**					
8605010	Trial erection at fabrication works; subsequent dismantling					
860501010	Fabricated member sections:					
860501010A	generally	Tonne	– – –	301.49	301.49	422.09	602.98
8605020	Permanent erection; including supply of suitable lifting arrangements; setting baseplates on holding down bolt assemblies					
860502010	Fabricated member sections:					
860502010A	generally	Tonne	– – –	378.79	378.79	530.31	757.58
86060	**ERECTION OF FRAMES AND OTHER MEMBERS**					
8606010	Trial erection at fabrication works; subsequent dismantling					
860601010	Fabricated member sections:					
860601010A	generally	Tonne	– – –	270.57	270.57	378.80	541.14
8606020	Permanent erection; including supply of suitable lifting arrangments; setting baseplates on holding down bolt assemblies					
860602010	Fabricated member sections:					
860602010A	generally	Tonne	– – –	340.14	340.14	476.20	680.28

Bridges

Rail 2003 Ref		Unit	Labour Hours	Labour Cost	Plant Cost	Materials Cost	Unit Rate	Unit Rate	Unit Rate
				£	£	£	Base Cost £	Green Zone £	Red Zone £
860	**STRUCTURAL METALWORK**								
8606030	**Bolts; HSFG general grade; BS 4395; complete with nuts and washers**								
860603010	**M16; length:**								
860603010A	100 mm	Nr	0.18	3.19	–	1.07	4.26	5.96	8.51
860603010B	150 mm	Nr	0.19	3.36	–	1.78	5.14	7.20	10.28
860603010C	200 mm	Nr	0.20	3.54	–	2.66	6.20	8.69	12.41
860603020	**M20; length:**								
860603020A	100 mm	Nr	0.20	3.54	–	2.14	5.68	7.95	11.35
860603020B	150 mm	Nr	0.21	3.72	–	2.89	6.61	9.25	13.22
860603020C	200 mm	Nr	0.22	3.89	–	3.95	7.84	10.98	15.69
860603020D	300 mm	Nr	0.25	4.42	–	4.89	9.31	13.04	18.63
860603030	**M24; length:**								
860603030A	100 mm	Nr	0.22	3.89	–	3.80	7.69	10.77	15.39
860603030B	150 mm	Nr	0.23	4.07	–	4.75	8.82	12.35	17.64
860603030C	200 mm	Nr	0.25	4.42	–	5.05	9.47	13.27	18.95
860603030D	240 mm	Nr	0.27	4.78	–	5.75	10.53	14.74	21.06
860603030E	300 mm	Nr	0.30	5.31	–	6.65	11.96	16.74	23.92
8606040	**Holding down assemblies; mild steel rag or indented bolts; complete with nuts and washers**								
860604010	**M12; length:**								
860604010A	100 mm	Nr	–	–	–	3.40	3.40	4.76	6.80
860604010B	150 mm	Nr	–	–	–	3.78	3.78	5.29	7.56
860604020	**M16; length:**								
860604020A	100 mm	Nr	–	–	–	4.11	4.11	5.75	8.22
860604020B	150 mm	Nr	–	–	–	4.79	4.79	6.71	9.58
860604020C	200 mm	Nr	–	–	–	5.32	5.32	7.45	10.64
860604030	**M20; length:**								
860604030A	100 mm	Nr	–	–	–	4.84	4.84	6.78	9.68
860604030B	150 mm	Nr	–	–	–	5.60	5.60	7.84	11.20
860604030C	200 mm	Nr	–	–	–	6.37	6.37	8.92	12.74
860604030D	300 mm	Nr	–	–	–	7.65	7.65	10.71	15.30

Bridges

Rail 2003 Ref		Unit	Labour Hours	Labour Cost	Plant Cost	Materials Cost	Unit Rate	Unit Rate	Unit Rate
				£	£	£	Base Cost £	Green Zone £	Red Zone £

860 STRUCTURAL METALWORK

8606040 Holding down assemblies; mild steel rag or indented bolts; complete with nuts and washers

860604040 M24; length:

860604040A	100 mm	Nr	–	–	–	6.65	6.65	9.31	13.30
860604040B	150 mm	Nr	–	–	–	7.48	7.48	10.47	14.96
860604040C	200 mm	Nr	–	–	–	8.16	8.16	11.42	16.32
860604040D	240 mm	Nr	–	–	–	9.12	9.12	12.77	18.24
860604040E	300 mm	Nr	–	–	–	10.45	10.45	14.63	20.90

860604050 M30; length:

860604050A	100 mm	Nr	–	–	–	7.89	7.89	11.05	15.78
860604050B	150 mm	Nr	–	–	–	8.98	8.98	12.57	17.96
860604050C	200 mm	Nr	–	–	–	9.90	9.90	13.86	19.80
860604050D	240 mm	Nr	–	–	–	11.01	11.01	15.41	22.02
860604050E	300 mm	Nr	–	–	–	12.23	12.23	17.12	24.46

Rail 2003 Ref		Unit				Specialist Cost	Unit Rate	Unit Rate	Unit Rate
						£	Base Cost £	Green Zone £	Red Zone £

86070 OFF SITE SURFACE TREATMENT

8607010 Blast cleaning

860701010 Sand blasting:

860701010A	generally	Tonne	–	–	–	127.48	127.48	178.47	254.96

8607070 Painting

860707010 Etch primer; in one coat:

860707010A	generally	Tonne	–	–	–	148.82	148.82	208.35	297.64

Rail 2003 Ref		Unit				Composite Cost £	Unit Rate Base Cost £	Unit Rate Green Zone £	Unit Rate Red Zone £
865	**MISCELLANEOUS METALWORK**								
86510	**STAIR UNITS AND WALKWAYS**								
8651010	**Stairways; mild steel fabrication**								
865101010	**Straight flight staircase and balustrade; 10 × 185 mm flat stringers; 6 × 250 mm raised pattern treads set on angles; 915 mm high balustrade with D handrail and square balusters; welded amd bolted connections; fixing to masonry with bolts:**								
865101010A	750 mm wide × 2950 mm going × 2600 mm rise	Nr	–	–	–	3501.92	3501.92	4902.69	7003.84
865101010B	900 mm wide × 2950 mm going × 2600 mm rise	Nr	–	–	–	3671.99	3671.99	5140.79	7343.98
865101020	**Quarter landing staircase and balustrade; 10 × 185 mm flat stringers; 6 × 250 mm raised pattern treads set on angles; 915 mm high balustrade with D handrail and square balusters; welded and bolted connections; fixing to masonry with bolts:**								
865101020A	750 mm wide × 1475 mm going each flight × 2600 mm wide	Nr	–	–	–	5751.49	5751.49	8052.09	11502.98
865101020B	900 mm wide × 1475 mm going each flight × 2600 mm rise	Nr	–	–	–	6037.52	6037.52	8452.53	12075.04
865101030	**Half landing staircase and balustrade; 10 × 185 mm flat stringers; 6 × 250 mm raised pattern treads set on angles; 915 mm high balustrade with D handrail and square balusters; welded and bolted connections; fixing to masonry with bolts:**								
865101030A	750 mm wide × 1475 mm going each flight × 2600 mm rise	Nr	–	–	–	6478.16	6478.16	9069.42	12956.32
865101030B	900 mm wide × 1475 mm going each flight × 2600 mm rise	Nr	–	–	–	7305.32	7305.32	10227.45	14610.64

Bridges

Rail 2003 Ref	Unit	Composite Cost £	Unit Rate Base Cost £	Unit Rate Green Zone £	Unit Rate Red Zone £
865	**MISCELLANEOUS METALWORK**				
8651020	Walkways; mild steel fabrication				
865102010	Walkway; 8 mm chequer plate; 203 × 89 mm channel support bearers; bolted connections; 915 mm high balustrade one side of 25 mm square balusters and D handrail; fixing to masonry with bolts:				
865102010A	1000 mm wide .... m – – –	564.33	564.33	790.06	1128.66

Tunnels

Rail 2003 Ref		Unit	Composite Cost £	Unit Rate Base Cost £	Unit Rate Green Zone £	Unit Rate Red Zone £
85	**TUNNELS**					
8505	**GENERALLY**					
850505	**GENERAL ITEMS**					
85050510	**Tunnelling Equipment**					
8505051010	Tunnel boring equipment; delivery, setting up and removal on completion; size:					
8505051010A	8.60m min diameter cut............	Nr	–	–	– 9375000.00	9375000.00 13125000.00 18750000.00
8505051015	Tunnel boring equipment; moving to new location; dismantling, transporting and rebuilding; size:					
8505051015A	8.60m min diameter...............	Nr	–	–	– 625000.00	625000.00 875000.00 1250000.00
8505051020	Tunnel boring equipment; moving to new location; by sliding; size:					
8505051020A	8.60m min diameter...............	Nr	–	–	– 25000.00	25000.00 35000.00 50000.00

Tunnels

Rail 2003 Ref		Unit	Composite Cost £	Unit Rate Base Cost £	Unit Rate Green Zone £	Unit Rate Red Zone £
8515	**EXCAVATION**					
851510	**TUNNEL EXCAVATION**					
85151010	Excavation by tunnel boring machine					
8515101020	Excavation for tunnels by tunnel boring machine in natural material other than rock; tunnel diameter:					
8515101020A	exceeding 8.0 m	m^3	– – –	187.50	187.50	262.50	375.00
8515101025	Excavation for tunnels by tunnel boring machine in rock; tunnel diameter:					
8515101025A	exceeding 8.0 m	m^3	– – –	187.50	187.50	262.50	375.00
8515101030	Excavation for cross passages in natural material other than rock; passage diameter:					
8515101030A	2.0 – 3.0 m	m^3	– – –	250.00	250.00	350.00	500.00
8515101040	Excavation for cross passages in rock; passage diameter:					
8515101040A	2.0 – 3.0 m	m^3	– – –	312.50	312.50	437.50	625.00
8515101050	Excavation for vertical shafts in natural material other than rock; shaft diameter:					
8515101050A	not exceeding 2.0 m	m^3	– – –	312.50	312.50	437.50	625.00
8515101060	Excavation for vertical shafts in rock; shaft diameter:					
8515101060A	not exceeding 2.0 m	m^3	– – –	312.50	312.50	437.50	625.00

Tunnels

Rail 2003 Ref		Unit	Composite Cost £	Unit Rate Base Cost £	Unit Rate Green Zone £	Unit Rate Red Zone £
8520	**TUNNEL LINING**					
852010	**LINING TO CAVITIES**					
85201010	**Preformed segmental linings to tunnels**					
8520101010	Reinforced precast concrete bolted rings complete; size:					
8520101010A	8.0 m internal diameter × 750 mm long × 250 mm thick...............	Nr	– – –	5500.00	5500.00 7700.00	11000.00
8520101010B	8.0 m internal diameter × 750 mm long × 250 mm thick; semi-circular ..	Nr	– – –	2800.00	2800.00 3920.00	5600.00
8520101020	Precast concrete bolted rings; tunnel lining size:					
8520101020A	4.0 m long × 1.0 m wide × 300 mm thick............................	Nr	– – –	687.50	687.50 962.50	1375.00
8520101030	Precast concrete bolted rings; cross passage lining; size:					
8520101030A	1.8 m long × 0.5 wide × 300 mm thick	Nr	– – –	437.50	437.50 612.50	875.00
8520101040	Precast concrete bolted rings; vertical shaft lining; size:					
8520101040A	1.05 m long × 0.5 m wide × 300 mm thick............................	Nr	– – –	375.00	375.00 525.00	750.00

Tunnels

Rail 2003 Ref		Unit	Specialist Cost £	Unit Rate Base Cost £	Unit Rate Green Zone £	Unit Rate Red Zone £
8570	**SUPPORT AND STABILISATION**					
857030	**PRESSURE GROUTING**					
85703060	**Grouting behind lining; cement Bentonite injection grout mix**					
8570306010	Grout behind preformed segmental lining; diameter:					
8570306010A	8.0 m	Tonne	– – –	125.00	125.00	175.00	250.00
8570306010B	3.6 m	Tonne	– – –	125.00	125.00	175.00	250.00
8570306010C	2.0 m	Tonne	– – –	125.00	125.00	175.00	250.00

Tunnels

Rail 2003 Ref		Unit	Labour Hours	Labour Cost £	Plant Cost £	Materials Cost £	Unit Rate Base Cost £	Unit Rate Green Zone £	Unit Rate Red Zone £
8580	**REMEDIAL WORKS**								
858001	**BRICKWORK**								
85800101	Engineering brickwork, Class B, in cement mortar 1:3, vertical wall								
8580010101	Generally:								
8580010101A	half brick thick	m²	0.71	13.98	–	21.11	35.09	49.13	70.19
8580010101B	one brick thick	m²	1.41	27.76	–	42.90	70.66	98.93	141.31
8580010111	Projection of corbels, plinths, bands, oversailing courses and the like:								
8580010111A	215 × 102.5 mm	m	0.24	4.73	–	3.63	8.36	11.70	16.71
8580010111B	215 × 215 mm	m	0.37	7.29	–	3.97	11.26	15.75	22.50
8580010111C	327.5 × 102.5 mm	m	0.36	7.09	–	3.97	11.06	15.47	22.11
8580010111D	327.5 × 215 mm	m	0.71	13.98	–	7.26	21.24	29.74	42.48
8580010121	Extra over for fair face and flush pointing one side as the work proceeds:								
8580010121A	stretcher bond	m²	0.06	1.18	–	0.06	1.24	1.74	2.49
8580010121B	flemish bond	m²	0.06	1.18	–	0.06	1.24	1.74	2.49
8580010121C	margins	m²	0.06	1.08	–	0.06	1.14	1.61	2.30
85800111	Common brickwork, in gauged mortar 1:1:6, vertical walls								
8580011101	Generally:								
8580011101A	half brick thick	m²	0.68	13.39	–	11.45	24.84	34.77	49.68
8580011101B	one brick thick	m²	1.35	26.58	–	24.07	50.65	70.91	101.30
8580011121	Projection of corbels, plinths, bands, oversailing courses and the like:								
8580011121A	215 × 102.5 mm	m	0.24	4.73	–	2.04	6.77	9.47	13.52
8580011121B	215 mm × 215 mm	m	0.37	7.29	–	2.36	9.65	13.50	19.29
8580011121C	327.5 mm × 102.5 mm	m	0.36	7.09	–	2.36	9.45	13.22	18.90
8580011121D	327.5 mm × 215 mm	m	0.71	13.98	–	4.07	18.05	25.27	36.10

Tunnels

Rail 2003 Ref		Unit	Labour Hours	Labour Cost £	Plant Cost £	Materials Cost £	Unit Rate Base Cost £	Unit Rate Green Zone £	Unit Rate Red Zone £
8580	**REMEDIAL WORKS**								
85800121	**Sundries; vertical walls**								
8580012101	Wedge and pin up brickwork to underside of existing construction with slates in cement mortar (1:3):								
8580012101A	half brick thick	m	0.90	17.72	–	3.54	21.26	29.76	42.51
8580012101B	one brick thick	m	1.65	32.49	–	6.82	39.31	55.03	78.61
8580012101C	one and a half brick thick	m	2.30	45.29	–	9.67	54.96	76.94	109.91
8580012106	Cut chase in brickwork for:								
8580012106A	small pipe or conduit; not exceeding 55 mm diameter	m	0.35	6.89	–	–	6.89	9.65	13.78
8580012111	Pointing in gun-grade polysulphide based mastic sealant:								
8580012111A	one side	m	0.14	2.76	–	15.90	18.66	26.12	37.31
8580012111B	both sides	m	0.28	5.51	–	30.51	36.02	50.43	72.04
8580012116	Cut hole for small pipe and make good; hole not exceeding 55 mm diameter:								
8580012116A	half brick thick	Nr	0.40	7.88	–	–	7.88	11.03	15.75
8580012116B	one brick thick	Nr	0.55	10.83	–	–	10.83	15.16	21.66
8580012116C	one and a half brick thick	Nr	0.80	15.75	–	–	15.75	22.05	31.50
8580012121	Cut hole for large pipe and make good; hole 55 – 110 mm diameter:								
8580012121A	half brick thick	Nr	0.55	10.83	–	–	10.83	15.16	21.66
8580012121B	one brick thick	Nr	0.80	15.75	–	–	15.75	22.05	31.50
8580012121C	one and a half brick thick	Nr	1.05	20.67	–	–	20.67	28.94	41.35
8580012126	Cut hole for large pipe and make good; hole exceeding 110 mm diameter:								
8580012126A	half brick thick	Nr	0.67	13.19	–	–	13.19	18.47	26.38
8580012126B	one brick thick	Nr	0.98	19.30	–	–	19.30	27.01	38.59
8580012126C	one and a half brick thick	Nr	1.20	23.63	–	–	23.63	33.08	47.25

Tunnels

Rail 2003 Ref		Unit	Labour Hours	Labour Cost	Plant Cost	Materials Cost	Unit Rate	Unit Rate	Unit Rate
				£	£	£	Base Cost £	Green Zone £	Red Zone £
8580	**REMEDIAL WORKS**								
85800121	**Sundries; vertical walls**								
8580012131	Cut and pin in brickwork:								
8580012131A	bracket	Nr	0.28	5.51	–	–	5.51	7.72	11.03
8580012131B	end of steel joist; not exceeding 250 mm high	Nr	0.35	6.89	–	–	6.89	9.65	13.78
8580012131C	end of steel joist; 250 – 500 mm high	Nr	0.55	10.83	–	–	10.83	15.16	21.66

Rail 2003 Ref		Unit	Composite Cost	Unit Rate	Unit Rate	Unit Rate
			£	Base Cost £	Green Zone £	Red Zone £
85800131	**Repointing defective and open joints and resetting loose brickwork**					
8580013101	To side walls; rake out joints to average depth:					
8580013101A	12 mm	m²	17.86	17.86	25.00	35.72
8580013101B	50 mm	m²	25.90	25.90	36.26	51.80
8580013101C	70 mm	m²	49.01	49.01	68.61	98.02
8580013111	To haunches; rake out joints to average depth:					
8580013111A	12 mm	m²	34.63	34.63	48.48	69.26
8580013111B	50 mm	m²	50.25	50.25	70.35	100.50
8580013111C	70 mm	m²	95.00	95.00	133.00	190.00
8580013121	To crown; rake out joints to average depth:					
8580013121A	12 mm	m²	41.28	41.28	57.79	82.56
8580013121B	50 mm	m²	59.91	59.91	83.87	119.82
8580013121C	70 mm	m²	113.25	113.25	158.55	226.50
8580013131	Extra over to repoint brickwork in extreme wet conditions, including material additives and provisions for diverting running water; rake out joints average depth:					
8580013131A	12 mm	m²	11.44	11.44	16.02	22.88
8580013131B	50 mm	m²	16.62	16.62	23.27	33.24
8580013131C	70 mm	m²	31.39	31.39	43.95	62.78

Tunnels

Rail 2003 Ref		Unit	Composite Cost £	Unit Rate Base Cost £	Unit Rate Green Zone £	Unit Rate Red Zone £
8580	**REMEDIAL WORKS**					
85800141	Cut out defective brickwork, build in new brickwork and point up; engineering bricks in stretcher bond in 1:3 mix; single skin 112.5 mm thick; clear debris off site					
8580014101	Area of defective brickwork; location:					
8580014101A	side walls, not exceeding 3.0 m above track level	m^2	–	–	–	584.89	584.89	818.85	1169.78
8580014101B	haunches, 3.0 m to 4.5 m above track level	m^2	–	–	–	813.02	813.02	1138.23	1626.04
8580014101C	crown, 4.5 m to 6.0 m above track level.	m^2	–	–	–	943.04	943.04	1320.26	1886.08
8580014111	Single brick; location:					
8580014111A	side walls, not exceeding 3.0 m above track level	Nr	–	–	–	42.21	42.21	59.09	84.42
8580014111B	haunches, 3.0 m to 4.5 m above track level	Nr	–	–	–	58.67	58.67	82.14	117.34
8580014111C	crown, 4.5 m to 6.0 m above track level.	Nr	–	–	–	68.03	68.03	95.24	136.06
8580014121	Fracture in brickwork; stitch up and rebuild average 300 mm width; location:					
8580014121A	side walls, not exceeding 3.0 m above track level	m	–	–	–	357.54	357.54	500.56	715.08
8580014121B	haunches, 3.0 m to 4.5 m above track level	m	–	–	–	496.81	496.81	695.53	993.62
8580014121C	crown, 4.5 m to 6.0 m above track level.	m	–	–	–	576.31	576.31	806.83	1152.62
8580014191	Extra over to repair area of brickwork in extreme wet conditions; including material additives and provisions for diverting running water; location:					
8580014191A	side walls, not exceeding 3.0 m above track level	m^2	–	–	–	193.03	193.03	270.24	386.06
8580014191B	haunches, 3.0 to 4.5 m above track level	m^2	–	–	–	268.33	268.33	375.66	536.66
8580014191C	crown, 4.5 m to 6.0 m above track level.	m^2	–	–	–	311.23	311.23	435.72	622.46

Tunnels

Rail 2003 Ref		Unit	Composite Cost £	Unit Rate Base Cost £	Unit Rate Green Zone £	Unit Rate Red Zone £
8580	**REMEDIAL WORKS**					
85800151	Pin and grout hollow brickwork, drill holes and insert stainless steel bars in cementitious grout and repoint with mortar to match existing					
8580015101	Area in 500 × 500 mm grid; location:					
8580015101A	side walls; not exceeding 3.0 m above track level	m^2	– – –	314.79	440.71	629.58
8580015101B	haunches; 3.0 m to 4.5 m above track level	m^2	– – –	421.78	590.49	843.56
8580015101C	crown; 4.5 m to 6.0 m above track level.	m^2	– – –	444.66	622.52	889.32
8580015111	Extra over to pin and grout hollow brickwork in extreme wet conditions, including material additives and provisions for diverting running water; location:					
8580015111A	side walls; not exceeding 3.0 m above track level	m^2	– – –	103.98	145.57	207.96
8580015111B	haunches; 3.0 m to 4.5 m above track level	m^2	– – –	139.15	194.81	278.30
8580015111C	crown; 4.5 to 6.0 m above track level..	m^2	– – –	146.80	205.52	293.60

General Civils

Rail 2003 Ref		Unit	Labour Hours	Labour Cost £	Plant Cost £	Materials Cost £	Unit Rate Base Cost £	Unit Rate Green Zone £	Unit Rate Red Zone £
9	**GENERAL CIVILS**								
920	**GROUND INVESTIGATION**								
92010	**TRIAL HOLES**								
9201010	Excavation by machine; by number								
920101010	Excavate trial hole in material other than rock; backfill in excavated material; plan area 1 × 2 m; maximum depth:								
920101010A	not exceeding 1 m	Nr	0.60	7.17	13.42	–	20.59	28.82	41.17
920101010B	1 – 2 m	Nr	1.20	14.34	28.23	–	42.57	59.61	85.15
920101010C	2 – 3 m	Nr	2.10	25.10	46.80	–	71.90	100.65	143.79
920101010D	3 – 5 m	Nr	4.50	53.78	89.08	–	142.86	200.00	285.71
920101020	Excavate trial hole in material including rock; backfill in excavated material; plan area 1 × 2 m; maximum depth:								
920101020A	not exceeding 1 m	Nr	0.60	7.17	19.72	–	26.89	37.64	53.77
920101020B	1 – 2 m	Nr	1.20	14.34	42.06	–	56.40	78.96	112.80
920101020C	2 – 3 m	Nr	2.10	25.10	68.15	–	93.25	130.54	186.49
920101020D	3 – 5 m	Nr	4.50	53.78	129.51	–	183.29	256.60	366.56
9201020	Excavation by machine; by depth								
920102010	Excavate trial hole in material other than rock; backfill in excavated material; plan area 1 × 2 m; maximum depth:								
920102010A	not exceeding 1 m	m	0.60	7.17	13.42	–	20.59	28.82	41.17
920102010B	1 – 2 m	m	0.60	7.17	14.12	–	21.29	29.80	42.57
920102010C	2 – 3 m	m	0.70	8.37	15.60	–	23.97	33.55	47.93
920102010D	3 – 5 m	m	0.90	10.76	17.82	–	28.58	40.00	57.14
920102020	Excavate trial hole in material including rock; backfill in excavated material; plan area 1 × 2 m; maximum depth:								
920102020A	not exceeding 1 m	m	0.60	7.17	19.72	–	26.89	37.64	53.77
920102020B	1 – 2 m	m	0.60	7.17	21.12	–	28.29	39.60	56.57
920102020C	2 – 3 m	m	0.70	8.37	22.70	–	31.07	43.49	62.13
920102020D	3 – 5 m	m	0.90	10.76	25.76	–	36.52	51.12	73.03

General Civils

Rail 2003 Ref		Unit	Labour Hours	Labour Cost £	Plant Cost £	Materials Cost £	Unit Rate Base Cost £	Unit Rate Green Zone £	Unit Rate Red Zone £

920 GROUND INVESTIGATION

9201020 Excavation by machine; by depth

920102030 Excavate trial hole in material including rock; support to full depth; backfill in excavated material; plan area 1 × 2 m; maximum depth:

920102030A	not exceeding 1 m	m	4.20	50.20	37.64	3.76	91.60	128.26	183.22
920102030B	1 – 2 m	m	4.20	50.20	42.21	3.76	96.17	134.65	192.35
920102030C	2 – 3 m	m	5.50	65.74	45.46	3.76	114.96	160.95	229.92
920102030D	3 – 5 m	m	6.90	82.48	45.07	3.76	131.31	183.84	262.62

920102040 Excavate trial hole in material including rock; support to full depth; backfill in excavated material; plan area 1 × 2 m; maximum depth:

920102040A	not exceeding 1 m	m	4.20	50.20	43.94	3.76	97.90	137.08	195.82
920102040B	1 – 2 m	m	4.20	50.20	49.21	3.76	103.17	144.45	206.35
920102040C	2 – 3 m	m	5.50	65.74	52.56	3.76	122.06	170.89	244.12
920102040D	3 – 5 m	m	6.90	82.48	53.01	3.76	139.25	194.95	278.50

920102050 Excavate trial hole in material other than rock; backfill in imported hardcore; plan area 1 × 2 m; maximum depth:

920102050A	not exceeding 1 m	m	0.60	7.17	16.92	39.83	63.92	89.48	127.82
920102050B	1 – 2 m	m	0.60	7.17	18.55	39.83	65.55	91.77	131.09
920102050C	2 – 3 m	m	0.70	8.37	20.03	39.83	68.23	95.51	136.43
920102050D	3 – 5 m	m	0.90	10.76	21.32	39.83	71.91	100.66	143.79

920102060 Excavate trial hole in material including rock; backfill in imported hardcore; plan area 1 × 2 m; maximum depth:

920102060A	not exceeding 1 m	m	0.60	7.17	23.22	39.83	70.22	98.30	140.42
920102060B	1 – 2 m	m	0.60	7.17	24.62	39.83	71.62	100.26	143.22
920102060C	2 – 3 m	m	0.70	8.37	26.20	39.83	74.40	104.15	148.78
920102060D	3 – 5 m	m	0.90	10.76	29.26	39.83	79.85	111.78	159.68

General Civils

Rail 2003 Ref		Unit	Labour Hours	Labour Cost £	Plant Cost £	Materials Cost £	Unit Rate Base Cost £	Unit Rate Green Zone £	Unit Rate Red Zone £

920 GROUND INVESTIGATION

9201050 Excavation by hand; by number

920105010 Excavate trial hole in material other than rock; backfill in excavated material; plan area 1 × 2 m; maximum depth:

920105010A	not exceeding 1 m	Nr	6.90	82.46	0.47	–	82.93	116.09	165.84
920105010B	1–2 m	Nr	15.20	181.64	1.20	–	182.84	255.98	365.68
920105010C	2–3 m	Nr	26.40	315.48	2.00	–	317.48	444.47	634.96
920105010D	3–5 m	Nr	55.00	657.25	3.75	–	661.00	925.40	1322.00

920105020 Excavate trial hole in material including rock; backfill in excavated material; plan area 1 × 2 m; maximum depth:

920105020A	not exceeding 1 m	Nr	11.20	133.84	11.64	–	145.48	203.68	290.96
920105020B	1–2 m	Nr	38.72	462.70	25.78	–	488.48	683.88	976.97
920105020C	2–3 m	Nr	63.36	757.15	42.22	–	799.37	1119.12	1598.75
920105020D	3–5 m	Nr	122.40	1462.68	76.38	–	1539.06	2154.68	3078.11

9201060 Excavation by hand; by depth

920106010 Excavate trial hole in material other than rock; backfill in excavated material; plan area 1 × 2 m; maximum depth:

920106010A	not exceeding 1 m	m	6.90	82.46	0.47	–	82.93	116.09	165.84
920106010B	1–2 m	m	7.60	90.82	0.50	–	91.32	127.85	182.64
920106010C	2–3 m	m	8.90	106.35	0.53	–	106.88	149.65	213.78
920106010D	3–5 m	m	11.00	131.45	0.53	–	131.98	184.78	263.97

920106020 Excavate trial hole in material including rock; backfill in excavated material; plan area 1 × 2 m; maximum depth:

920106020A	not exceeding 1 m	m	14.10	168.50	11.64	–	180.14	252.19	360.27
920106020B	1–2 m	m	14.80	176.86	12.79	–	189.65	265.51	379.30
920106020C	2–3 m	m	16.10	192.39	13.94	–	206.33	288.87	412.67
920106020D	3–5 m	m	18.10	216.29	15.06	–	231.35	323.89	462.71

General Civils

Rail 2003 Ref		Unit	Labour Hours	Labour Cost	Plant Cost	Materials Cost	Unit Rate	Unit Rate	Unit Rate
				£	£	£	Base Cost £	Green Zone £	Red Zone £
920	**GROUND INVESTIGATION**								
9201060	Excavation by hand; by depth								
920106030	Excavate trial hole in material other than rock; support to full depth; backfill in excavated material; plan area 1 × 2 m; maximum depth:								
920106030A	not exceeding 1 m	m	10.50	125.49	24.69	3.76	153.94	215.52	307.89
920106030B	1–2 m	m	11.20	133.85	28.93	3.76	166.54	233.17	333.08
920106030C	2–3 m	m	13.70	163.73	31.06	3.76	198.55	277.98	397.11
920106030D	3–5 m	m	17.00	203.17	31.06	3.76	237.99	333.19	475.99
920106040	Excavate trial hole in material including rock; support to full depth; backfill in excavated material; plan area 1 × 2 m; maximum depth:								
920106040A	not exceeding 1 m	m	17.70	211.53	35.87	3.76	251.16	351.62	502.31
920106040B	1–2 m	m	18.40	219.89	41.22	3.76	264.87	370.83	529.75
920106040C	2–3 m	m	20.90	249.77	44.47	3.76	298.00	417.21	596.01
920106040D	3–5 m	m	24.20	289.21	45.59	3.76	338.56	473.99	677.12
9201090	Excavation ancillaries								
920109010	Extra over excavation for breaking out hard strata or surface obstructions:								
920109010A	hard rock	m^3	3.60	43.02	31.50	–	74.52	104.33	149.04
920109010B	concrete	m^3	3.00	35.85	26.25	–	62.10	86.94	124.20
920109010C	reinforced concrete	m^3	4.20	50.19	36.75	–	86.94	121.72	173.88

Rail 2003 Ref		Unit				Specialist Cost	Unit Rate	Unit Rate	Unit Rate
						£	Base Cost £	Green Zone £	Red Zone £
920109020	Standing time for excavation plant, equipment and crew for machine dug trial pit or trench (excluding PICOWs and Lookouts):								
920109020A	generally	hr	–	–	–	48.00	48.00	67.20	96.00
920109030	Provision of excavation plant, equipment and crew for machine dug trial pits and trenches as directed by the Employer (excluding PICOWs/Lookouts); maximum depth:								
920109030A	not exceeding 4.5 m	item	–	–	–	352.00	352.00	492.80	704.00
920109030B	4.5–6.0 m	item	–	–	–	428.00	428.00	599.20	856.00

General Civils

Rail 2003 Ref		Unit	Specialist Cost £	Unit Rate Base Cost £	Unit Rate Green Zone £	Unit Rate Red Zone £
920	**GROUND INVESTIGATION**					
9201090	**Excavation ancillaries**					
920109040	Bring pump to position of each exploratory pit or trench:					
920109040A	generally	Nr	– – –	10.00	14.00	20.00
920109050	Pump water from pit or trench:					
920109050A	generally	hr	– – –	2.00	2.80	4.00
920109060	Leave open trial pit or trench:					
920109060A	per day	m²	– – –	6.00	8.40	12.00
92030	**BOREHOLES**					
9203010	**Boreholes generally**					
920301010	Move boring plant and equipment to site of each hole:					
920301010A	and set up	Nr	– – –	30.00	42.00	60.00
920301010B	E.O. for set up on slope > 20 deg. (excluding temporary works or scaffolding)	Nr	– – –	50.00	70.00	100.00
920301020	Breaking out obstruction at exploratory:					
920301020A	generally	hr	– – –	60.00	84.00	120.00
920301020B	where present (alternative)	Nr	– – –	25.00	35.00	50.00
920301030	Advance borehole; depth:					
920301030A	not exceeding 10 m	m	– – –	11.50	16.10	23.00
920301030B	over 10 m and not exceeding 20 m	m	– – –	13.00	18.20	26.00
920301030C	over 20 m	m	– – –	15.00	21.00	30.00
920301040	Advance borehole through hard stratum or obstruction:					
920301040A	generally	hr	– – –	26.00	36.40	52.00

General Civils

Rail 2003 Ref		Unit	Specialist Cost £	Unit Rate Base Cost £	Unit Rate Green Zone £	Unit Rate Red Zone £
920	**GROUND INVESTIGATION**					
9203010	**Boreholes generally**					
920301050	Backfill borehole with grout:					
920301050A	cement/bentonite	m	–	–	–	30.00	30.00	42.00	60.00
920301060	Standing time for borehole plant, equipment and crew (excluding PICOWs/Lookouts):					
920301060A	generally	hr	–	–	–	48.00	48.00	67.20	96.00
920301070	Move equipment to site of each trial pit or trench:					
920301070A	and set up	Nr	–	–	–	315.00	315.00	441.00	630.00
92040	**SAMPLES**					
9204010	**Sampling**					
920401010	Small disturbed:					
920401010A	sample	Nr	–	–	–	0.50	0.50	0.70	1.00
920401020	Open tube:					
920401020A	sample	Nr	–	–	–	8.00	8.00	11.20	16.00
92050	**TESTING**					
9205010	**Site testing**					
920501010	Insitu testing:					
920501010A	standard penetration test in borehole	Nr	–	–	–	8.00	8.00	11.20	16.00
920501010B	hand penetrometer	Nr	–	–	–	1.00	1.00	1.40	2.00
920501010C	hand vane test	Nr	–	–	–	2.00	2.00	2.80	4.00
920501010D	bring hand-operated dynamic probing equipment (DPM) to site of each probe hole (Mackintosh Probe)	Nr	–	–	–	25.00	25.00	35.00	50.00
920501010E	machine operated	Nr	–	–	–	45.00	45.00	63.00	90.00
920501010F	carry out hand probing between existing ground level and 10 m depth	m	–	–	–	14.00	14.00	19.60	28.00

General Civils

Rail 2003 Ref		Unit	Specialist Cost £	Unit Rate Base Cost £	Unit Rate Green Zone £	Unit Rate Red Zone £
920	**GROUND INVESTIGATION**					
9205050	**Laboratory testing**					
920505010	**Classification:**					
920505010A	moisture content	Nr	– – –	1.80	2.52	3.60
920505010B	liquid limit, plastic limit and plasticity index	Nr	– – –	11.00	15.40	22.00
920505010C	particle size distribution by wet sieving	Nr	– – –	18.00	25.20	36.00
920505010D	particle size distribution by dry sieving	Nr	– – –	18.00	25.20	36.00
920505010E	sedimentation by pipette	Nr	– – –	21.00	29.40	42.00
920505010F	sedimentation by hydrometer	Nr	– – –	21.00	29.40	42.00
920505020	**Chemical and electrochemical:**					
920505020A	sulphate content of acid extract from soil	Nr	– – –	9.00	12.60	18.00
920505020B	sulphate content of water extract from soil	Nr	– – –	9.00	12.60	18.00
920505020C	sulphate content of groundwater	Nr	– – –	9.00	12.60	18.00
920505020D	sulphide content	Nr	– – –	9.00	12.60	18.00
920505020E	total dissolved solids	Nr	– – –	4.00	5.60	8.00
920505020F	pH value	Nr	– – –	2.00	2.80	4.00
920505020G	resistivity	Nr	– – –	5.00	7.00	10.00
920505020H	Redox potential	Nr	– – –	4.00	5.60	8.00
920505030	**Compunction related:**					
920505030A	maximum and minimum dry density for granular soils	Nr	– – –	48.00	67.20	96.00
920505040	**Compressibility, permeability, durability:**					
920505040A	one-dimensional consolidation properties test period 5 days	Nr	– – –	56.00	78.40	112.00
920505040B	extra over for test period in excess of 5 days	day	– – –	12.00	16.80	24.00
920505040C	measurement of swelling pressure, test period 2 days	Nr	– – –	22.00	30.80	44.00
920505040D	measurement of swelling, test period 2 days	Nr	– – –	22.00	30.80	44.00
920505040E	measurement of settlement on saturation, test period 1 day	Nr	– – –	22.00	30.80	44.00
920505040F	extra over last three items for test period in excess of 2 or 1 day(s)	day	– – –	22.00	30.80	44.00

General Civils

Rail 2003 Ref		Unit	Specialist Cost £	Unit Rate Base Cost £	Unit Rate Green Zone £	Unit Rate Red Zone £
920	**GROUND INVESTIGATION**					
9205050	**Laboratory testing**					
920505050	**Consolidation and permeability in Hydraulic cells:**					
920505050A	isotropic consolidation properties in triaxial cell, test period 4 days	Nr	–	–	–	360.00	360.00	504.00	720.00
920505050B	extra over for test periods in excess of 4 days	day	–	–	–	60.00	60.00	84.00	120.00
920505050C	permeability in triaxial cell, test period 4 days	Nr	–	–	–	68.00	68.00	95.20	136.00
920505050D	extra over for test periods in excess of 4 days	day	–	–	–	20.00	20.00	28.00	40.00
920505060	**Shear strength (total stress):**					
920505060A	shear strength by the Laboratory vane method	Nr	–	–	–	21.00	21.00	29.40	42.00
920505060B	shear strength by hand vane	Nr	–	–	–	2.00	2.00	2.80	4.00
920505060C	shear strength by hand penetrometer	Nr	–	–	–	2.00	2.00	2.80	4.00
920505060D	undrained shear strength of single 100 mm diameter specimen in triaxial compression with multi-stage loading without measurement of pore pressure	Nr	–	–	–	27.00	27.00	37.80	54.00
9205060	**Chemical testing for contaminated ground**					
920506010	**Soil samples – primary contaminants:**					
920506010A	Arsenic	Nr	–	–	–	4.00	4.00	5.60	8.00
920506010B	Cadmium	Nr	–	–	–	4.00	4.00	5.60	8.00
920506010C	Chromium	Nr	–	–	–	4.00	4.00	5.60	8.00
920506010D	Lead	Nr	–	–	–	4.00	4.00	5.60	8.00
920506010E	Mercury	Nr	–	–	–	4.00	4.00	5.60	8.00
920506010F	Borum – water soluble	Nr	–	–	–	4.00	4.00	5.60	8.00
920506010G	Copper	Nr	–	–	–	4.00	4.00	5.60	8.00
920506010H	Nickel	Nr	–	–	–	4.00	4.00	5.60	8.00
920506010I	Zinc	Nr	–	–	–	4.00	4.00	5.60	8.00
920506010J	Cyanide	Nr	–	–	–	6.00	6.00	8.40	12.00
920506010K	Thiocyanate	Nr	–	–	–	6.00	6.00	8.40	12.00
920506010L	Phenols	Nr	–	–	–	8.00	8.00	11.20	16.00
920506010M	Sulphide	Nr	–	–	–	9.00	9.00	12.60	18.00
920506010N	pH value	Nr	–	–	–	2.00	2.00	2.80	4.00
920506010O	Coal tar/polyaromatic hydrocarbon	Nr	–	–	–	28.00	28.00	39.20	56.00
920506010P	Asbestos	Nr	–	–	–	16.00	16.00	22.40	32.00

General Civils

Rail 2003 Ref		Unit	Specialist Cost £	Unit Rate Base Cost £	Unit Rate Green Zone £	Unit Rate Red Zone £
920	**GROUND INVESTIGATION**					
9205060	**Chemical testing for contaminated ground**					
920506020	Soil samples – secondary contaminants:					
920506020A	Antimony	Nr	– – –	4.00	5.60	8.00
920506020B	Barium	Nr	– – –	4.00	5.60	8.00
920506020C	Berylium	Nr	– – –	8.00	11.20	16.00
920506020D	Vanadium	Nr	– – –	8.00	11.20	16.00
920506020E	Freon extractable matter	Nr	– – –	8.00	11.20	16.00
920506020F	Mineral oils	Nr	– – –	8.00	11.20	16.00
920506020G	Chloride	Nr	– – –	8.00	11.20	16.00
920506030	Water samples:					
920506030A	Arsenic	Nr	– – –	4.00	5.60	8.00
920506030B	Cadmium	Nr	– – –	4.00	5.60	8.00
920506030C	Chromium	Nr	– – –	4.00	5.60	8.00
920506030D	Lead	Nr	– – –	4.00	5.60	8.00
920506030E	Mercury	Nr	– – –	4.00	5.60	8.00
920506030F	Boron – water soluble	Nr	– – –	4.00	5.60	8.00
920506030G	Copper	Nr	– – –	4.00	5.60	8.00
920506030H	Nickel	Nr	– – –	4.00	5.60	8.00
920506030I	Zinc	Nr	– – –	4.00	5.60	8.00
920506030J	Cyanide	Nr	– – –	6.00	8.40	12.00
920506030K	Thiocyanate	Nr	– – –	6.00	8.40	12.00
920506030L	Phenols	Nr	– – –	8.00	11.20	16.00
920506030M	Sulphur	Nr	– – –	9.00	12.60	18.00
920506030N	pH value	Nr	– – –	2.00	2.80	4.00
920506030O	Polyaromatic hydrocarbons	Nr	– – –	28.00	39.20	56.00
920506030P	Antimony	Nr	– – –	4.00	5.60	8.00
920506030Q	Barium	Nr	– – –	4.00	5.60	8.00
920506030R	Beryllium	Nr	– – –	8.00	11.20	16.00
920506030S	Chloride	Nr	– – –	8.00	11.20	16.00
920506030T	Iron	Nr	– – –	4.00	5.60	8.00
920506030U	Manganese	Nr	– – –	4.00	5.60	8.00
920506030V	Calcium	Nr	– – –	4.00	5.60	8.00
920506030W	Sodium	Nr	– – –	4.00	5.60	8.00
920506030X	Magnesium	Nr	– – –	4.00	5.60	8.00
920506030Y	Potassium	Nr	– – –	4.00	5.60	8.00

General Civils

Rail 2003 Ref		Unit	Specialist Cost £	Unit Rate Base Cost £	Unit Rate Green Zone £	Unit Rate Red Zone £
920	**GROUND INVESTIGATION**					
9205060	**Chemical testing for contaminated ground**					
920506040	**Gas samples:**					
920506040A	Carbon dioxide	Nr	–	–	–	6.00	6.00	8.40	12.00
920506040B	Hydrogen	Nr	–	–	–	6.00	6.00	8.40	12.00
920506040C	Hydrogen sulphide	Nr	–	–	–	14.00	14.00	19.60	28.00
920506040D	Methane	Nr	–	–	–	6.00	6.00	8.40	12.00
920506040E	Nitrogen	Nr	–	–	–	6.00	6.00	8.40	12.00
920506040F	Oxygen	Nr	–	–	–	6.00	6.00	8.40	12.00
920506040G	Ethane	Nr	–	–	–	6.00	6.00	8.40	12.00
920506040H	Propane	Nr	–	–	–	6.00	6.00	8.40	12.00
920506040I	Carbon monoxide	Nr	–	–	–	6.00	6.00	8.40	12.00

General Civils

Rail 2003 Ref		Unit	Specialist Cost £	Unit Rate Base Cost £	Unit Rate Green Zone £	Unit Rate Red Zone £
923	**GEOTECHNICAL AND OTHER SPECIALIST PROCESSES**					
92310	**DRILLING FOR GROUT HOLES**					
9231010	Drilling through material other than rock					
923101010	Vertically downwards; 100 mm diameter; in holes of depth:					
923101010A	not exceeding 5 m	m	–	–	–	7.25	7.25	10.15	14.50
923101010B	5 – 10 m	m	–	–	–	10.00	10.00	14.00	20.00
923101010C	10 – 20 m	m	–	–	–	15.00	15.00	21.00	30.00
923101010D	20 – 30 m	m	–	–	–	21.50	21.50	30.10	43.00
923101010E	over 30 m	m	–	–	–	30.75	30.75	43.05	61.50
923101020	Downwards at an angle 0 – 45 degrees to the vertical; 100 mm diameter; in holes of depth:					
923101020A	not exceeding 5 m	m	–	–	–	7.75	7.75	10.85	15.50
923101020B	5 – 10 m	m	–	–	–	10.75	10.75	15.05	21.50
923101020C	10 – 20 m	m	–	–	–	16.00	16.00	22.40	32.00
923101020D	20 – 30 m	m	–	–	–	23.00	23.00	32.20	46.00
923101020E	over 30 m	m	–	–	–	33.00	33.00	46.20	66.00
923101030	Vertically upwards; 100 mm diameter; in holes of depth:					
923101030A	not exceeding 5 m	m	–	–	–	11.60	11.60	16.24	23.20
923101030B	5 – 10 m	m	–	–	–	16.00	16.00	22.40	32.00
923101030C	10 – 20 m	m	–	–	–	24.50	24.50	34.30	49.00
923101030D	20 – 30 m	m	–	–	–	34.50	34.50	48.30	69.00
923101030E	over 30 m	m	–	–	–	49.50	49.50	69.30	99.00

General Civils

Rail 2003 Ref		Unit	Specialist Cost £	Unit Rate Base Cost £	Unit Rate Green Zone £	Unit Rate Red Zone £
923	**GEOTECHNICAL AND OTHER SPECIALIST PROCESSES**					
92340	**GROUT HOLES**					
9234010	General grout holes					
923401010	Number of holes; diameter:					
923401010A	75 mm	Nr	–	–	–	125.00	125.00	175.00	250.00
923401010B	100 mm	Nr	–	–	–	225.00	225.00	315.00	450.00
923401010C	150 mm	Nr	–	–	–	425.00	425.00	595.00	850.00
92350	**GROUT MATERIALS AND INJECTION**					
9235010	Materials generally					
923501010	Materials:					
923501010A	cement grout	Tonne	–	–	–	46.96	46.96	65.74	93.92
923501010B	cement/PFA	Tonne	–	–	–	27.94	27.94	39.11	55.88
923501010C	PFA	Tonne	–	–	–	17.19	17.19	24.07	34.39
923501010D	sand	Tonne	–	–	–	5.74	5.74	8.03	11.47
923501010E	pea gravel	Tonne	–	–	–	6.05	6.05	8.47	12.10
923501010F	Bentonite	Tonne	–	–	–	94.56	94.56	132.38	189.12
9235020	**Injection**					
923502010	Injection:					
923502010A	number	Nr	–	–	–	47.50	47.50	66.50	95.00
923502010B	neat cement grout	Tonne	–	–	–	27.00	27.00	37.80	54.00
923502010C	chemical grout	Tonne	–	–	–	27.00	27.00	37.80	54.00

General Civils

Rail 2003 Ref		Unit	Labour Hours	Labour Cost	Plant Cost	Materials Cost	Unit Rate	Unit Rate	Unit Rate
				£	£	£	Base Cost £	Green Zone £	Red Zone £
925	**DEMOLITION AND SITE CLEARANCE**								
92510	**DEMOLITION OF STRUCTURES**								
9251010	**Masonry structural elements**								
925101010	**Brick walls and partitions:**								
925101010A	half brick thick	m²	0.80	9.56	0.62	–	10.18	14.25	20.36
925101010B	one brick thick	m²	1.40	16.74	1.08	–	17.82	24.94	35.63
925101010C	one and a half brick thick	m²	2.10	25.10	1.62	–	26.72	37.41	53.45
925101010D	two brick thick	m²	2.70	32.28	2.08	–	34.36	48.10	68.71
925101015	**Block walls and partitions:**								
925101015A	100 mm thick	m²	0.30	3.59	0.93	–	4.52	6.32	9.02
925101015B	125 mm thick	m²	0.33	3.95	1.00	–	4.95	6.92	9.89
925101015C	150 mm thick	m²	0.35	4.18	1.08	–	5.26	7.37	10.53
925101015D	200 mm thick	m²	0.40	4.78	1.23	–	6.01	8.42	12.03
925101015E	225 mm thick	m²	0.55	6.58	1.70	–	8.28	11.58	16.54
925101018	**Attached piers and buttresses:**								
925101018A	half brick	m²	0.45	5.38	1.39	–	6.77	9.47	13.53
925101018B	one brick	m²	0.80	9.56	2.47	–	12.03	16.84	24.06
925101018C	one and a half brick	m²	1.15	13.75	3.54	–	17.29	24.21	34.58
925101018D	two brick	m²	1.40	16.74	4.32	–	21.06	29.47	42.10
9251020	**Concrete structural elements**								
925102020	**Reinforced walls and attached columns:**								
925102020A	not exceeding 150 mm thick	m³	12.00	143.45	83.80	–	227.25	318.15	454.50
925102020B	150 – 300 mm thick	m³	14.00	167.36	97.77	–	265.13	371.18	530.25
925102020C	over 300 mm thick	m³	18.00	215.17	125.70	–	340.87	477.22	681.75
925102022	**Reinforced concrete ground slabs:**								
925102022A	not exceeding 150 mm thick	m³	6.00	71.72	41.90	–	113.62	159.07	227.25
925102022B	150 – 300 mm thick	m³	8.00	95.63	55.87	–	151.50	212.10	303.00
925102022C	over 300 mm thick	m³	10.00	119.54	69.84	–	189.38	265.13	378.75

General Civils

Rail 2003 Ref		Unit	Labour Hours	Labour Cost £	Plant Cost £	Materials Cost £	Unit Rate Base Cost £	Unit Rate Green Zone £	Unit Rate Red Zone £
925	**DEMOLITION AND SITE CLEARANCE**								
9251020	**Concrete structural elements**								
925102025	Reinforced concrete suspended slabs and attached beams:								
925102025A	not exceeding 150 mm thick	m^3	10.00	119.54	69.84	–	189.38	265.13	378.75
925102025B	150 – 300 mm thick	m^3	16.00	191.26	111.74	–	303.00	424.20	606.00
925102025C	over 300 mm thick	m^3	19.00	227.13	132.69	–	359.82	503.74	719.62
925102027	Reinforced isolated columns; sectional area:								
925102027A	not exceeding 0.1 m^2	m^3	10.00	119.54	69.84	–	189.38	265.13	378.75
925102027B	0.10 – 0.25 m^2	m^3	13.00	155.40	90.79	–	246.19	344.66	492.37
925102027C	over 0.25 m^2	m^3	18.00	215.17	125.70	–	340.87	477.22	681.75
925102029	Reinforced isolated beams; sectional area:								
925102029A	not exceeding 0.1 m^2	m^3	12.00	143.45	83.80	–	227.25	318.15	454.50
925102029B	0.10 – 0.25 m^2	m^3	14.00	167.36	97.77	–	265.13	371.18	530.25
925102029C	over 0.25 m^2	m^3	19.00	227.13	132.69	–	359.82	503.74	719.62
9251030	**Steelwork structural elements**								
925103010	Steel beams, joists and lintels:								
925103010A	not exceeding 10 kg/m	Tonne	50.00	597.70	22.99	–	620.69	868.97	1241.38
925103010B	10 – 20 kg/m	Tonne	46.00	549.88	20.90	–	570.78	799.10	1141.57
925103010C	20 – 50 kg/m	Tonne	40.00	478.16	18.81	–	496.97	695.75	993.94
925103015	Steel columns and stanchions:								
925103015A	not exceeding 10 kg/m	Tonne	60.00	717.24	22.99	–	740.23	1036.33	1480.46
925103015B	10 – 20 kg/m	Tonne	56.00	669.42	20.90	–	690.32	966.45	1380.65
925103015C	20 – 50 kg/m	Tonne	50.00	597.70	18.81	–	616.51	863.11	1233.02
925103018	Steel rails and purlins:								
925103018A	not exceeding 10 kg/m	Tonne	50.00	597.70	20.90	–	618.60	866.04	1237.20
925103018B	10 – 20 kg/m	Tonne	46.00	549.88	18.81	–	568.69	796.17	1137.39

General Civils

Rail 2003 Ref		Unit	Labour Hours	Labour Cost £	Plant Cost £	Materials Cost £	Unit Rate Base Cost £	Unit Rate Green Zone £	Unit Rate Red Zone £
925	**DEMOLITION AND SITE CLEARANCE**								
92520	**REMOVAL OF OLD WORK**								
9252010	General metalwork								
925201010	Stairs, steps, balustrades and handrails; including disposal:								
925201010A	stair or step unit; straight flight	Nr	3.50	41.84	18.78	–	60.62	84.86	121.24
925201010B	stair or step unit; two or more flights	Nr	5.00	59.77	23.48	–	83.25	116.55	166.49
925201010C	balustrading	m	0.80	9.56	7.04	–	16.60	23.25	33.22
925201010D	handrail	m	0.30	3.59	2.35	–	5.94	8.31	11.87
925201020	Signage; including disposal:								
925201020A	road signs complete with posts	Nr	0.70	8.37	9.39	–	17.76	24.87	35.52
925201020B	general signage complete with fixing brackets	Nr	0.55	6.58	9.39	–	15.97	22.36	31.93
925201025	Signage; setting aside for re-use:								
925201025A	road signs complete with posts	Nr	0.75	8.97	–	–	8.97	12.55	17.93
925201025B	general signage complete with fixing brackets	Nr	1.00	11.95	–	–	11.95	16.74	23.91
9252020	General elements								
925202010	Timber decking; setting aside for re-use:								
925202010A	not exceeding 50 m^3	m^2	0.75	8.97	–	–	8.97	12.55	17.93
925202010B	over 50 m^3	m^2	0.68	8.13	–	–	8.13	11.38	16.26
925202015	Timber decking; including disposal:								
925202015A	not exceeding 50 m^3	m^2	0.33	3.95	2.35	–	6.30	8.81	12.59
925202015B	over 50 m^3	m^2	0.30	3.59	2.35	–	5.94	8.31	11.87

General Civils

Rail 2003 Ref		Unit	Labour Hours	Labour Cost £	Plant Cost £	Materials Cost £	Unit Rate Base Cost £	Unit Rate Green Zone £	Unit Rate Red Zone £
925	**DEMOLITION AND SITE CLEARANCE**								
9252030	**Fencing**								
925203010	Open post and rail; including disposal:								
925203010A	not exceeding 1.2 m high	m	0.28	3.35	1.88	–	5.23	7.32	10.45
925203010B	1.2 – 1.8 m high	m	0.31	3.71	2.82	–	6.53	9.13	13.04
925203010C	over 1.8 m high	m	0.34	4.06	3.76	–	7.82	10.95	15.64
925203020	Post and wire; including disposal:								
925203020A	not exceeding 1.2 m high	m	0.22	2.63	1.41	–	4.04	5.65	8.08
925203020B	1.2 – 1.8 m high	m	0.24	2.87	2.07	–	4.94	6.91	9.87
925203020C	over 1.8 m high	m	0.26	3.11	3.29	–	6.40	8.95	12.79
925203030	Post and chain link; including disposal:								
925203030A	not exceeding 1.2 m high	m	0.30	3.59	1.69	–	5.28	7.39	10.55
925203030B	1.2 – 1.8 m high	m	0.33	3.95	2.44	–	6.39	8.94	12.77
925203030C	over 1.8 m high	m	0.36	4.30	3.57	–	7.87	11.02	15.75
925203040	Post and palisade; including disposal:								
925203040A	not exceeding 1.2 m high	m	0.60	7.17	3.10	–	10.27	14.38	20.54
925203040B	1.2 – 1.8 m high	m	0.66	7.89	4.51	–	12.40	17.36	24.79
925203040C	over 1.8 m high	m	0.73	8.73	5.54	–	14.27	19.98	28.53
925203050	Post and weldmesh; including disposal:								
925203050A	not exceeding 1.2 m high	m	0.45	5.38	3.10	–	8.48	11.87	16.96
925203050B	1.2 – 1.8 m high	m	0.50	5.98	4.51	–	10.49	14.68	20.96
925203050C	over 1.8 m high	m	0.55	6.58	5.54	–	12.12	16.97	24.23
9252040	**Removal of hard access constructions**								
925204010	Break out and remove hard surfacings; dispose of off-site:								
925204010A	asphalt wearing and base courses	m^2	0.40	4.78	1.40	–	6.18	8.65	12.35
925204010B	macadam wearing and base courses	m^2	0.35	4.18	1.40	–	5.58	7.82	11.16
925204010C	reinforced concrete pavings	m^2	1.30	15.54	7.28	–	22.82	31.94	45.63
925204010D	plain concrete pavings	m^2	1.30	15.54	6.98	–	22.52	31.53	45.04

General Civils

Rail 2003 Ref		Unit	Labour Hours	Labour Cost £	Plant Cost £	Materials Cost £	Unit Rate Base Cost £	Unit Rate Green Zone £	Unit Rate Red Zone £
925	**DEMOLITION AND SITE CLEARANCE**								
9252040	Removal of hard access constructions								
925204020	Take up kerbs and edgings complete with foundations; dispose of off-site; size:								
925204020A	125 × 255 mm	m	0.56	6.69	0.94	–	7.63	10.68	15.26
925204020B	150 × 305 mm	m	0.80	9.56	1.32	–	10.88	15.22	21.75
925204020C	50 × 150 mm	m	0.20	2.39	0.38	–	2.77	3.88	5.53
925204020D	50 × 250 mm	m	0.30	3.59	0.56	–	4.15	5.81	8.30
9252050	Pipelines and ducts								
925205010	General drainage and duct pipework; disposal; size:								
925205010A	not exceeding 300 mm diameter	m	0.30	3.59	0.56	–	4.15	5.81	8.30
925205010B	300 – 500 mm diameter	m	0.30	3.59	0.56	–	4.15	5.81	8.30
925205010C	over 500 mm diameter	m	0.30	3.59	0.56	–	4.15	5.81	8.30
92530	**GENERAL SITE CLEARANCE**								
9253010	Trees, stumps and general vegetation								
925301010	Trees; cutting off at ground level; including disposal; girth:								
925301010A	600 mm – 1.50 m	Nr	2.40	28.68	25.35	–	54.03	75.65	108.07
925301010B	1.50 – 3.00 m	Nr	9.60	114.72	101.42	–	216.14	302.59	432.27
925301010C	3.00 – 4.50 m	Nr	14.35	171.48	150.20	–	321.68	450.35	643.36
925301020	Stumps; including disposal; backfilling with imported materials; diameter:								
925301020A	not exceeding 600 mm	Nr	5.50	65.73	125.06	–	190.79	267.10	381.57
925301020B	600 mm – 1.0 m	Nr	9.00	107.55	196.34	–	303.89	425.45	607.78
925301020C	over 1.0 m	Nr	14.00	167.30	294.51	–	461.81	646.53	923.62
925301030	Vegetation, undergrowth, bushes and the like; remove from site:								
925301030A	generally	m^2	0.02	0.24	0.81	–	1.05	1.46	2.09

General Civils

Rail 2003 Ref		Unit	Labour Hours	Labour Cost £	Plant Cost £	Materials Cost £	Unit Rate Base Cost £	Unit Rate Green Zone £	Unit Rate Red Zone £
925	**DEMOLITION AND SITE CLEARANCE**								
9253010	**Trees, stumps and general vegetation**								
925301040	Cut down hedges; grub up roots and remove from site; height:								
925301040A	not exceeding 2.0 m	m	1.00	11.95	31.59	–	43.54	60.95	87.07
925301040B	2.0 – 3.0 m	m	2.50	29.88	76.62	–	106.50	149.09	212.98
925301040C	3.0 – 4.0 m	m	4.00	47.80	116.95	–	164.75	230.65	329.50
925301040D	4.0 – 5.0 m	m	5.00	59.75	157.93	–	217.68	304.75	435.35
925301040E	over 5.0 m	m	7.00	83.65	207.01	–	290.66	406.92	581.32

General Civils

Rail 2003 Ref		Unit	Labour Hours	Labour Cost £	Plant Cost £	Materials Cost £	Unit Rate Base Cost £	Unit Rate Green Zone £	Unit Rate Red Zone £
928	**EARTHWORKS**								
92810	**EXCAVATION**								
9281010	**Machine excavation**								
928101010	Oversite excavation to remove topsoil; average depth:								
928101010A	150 mm	m³	–	–	0.35	–	0.35	0.49	0.70
928101010B	300 mm	m³	–	–	0.53	–	0.53	0.74	1.05
928101020	Excavation for cuttings in topsoil; average depth:								
928101020A	not exceeding 0.25 m	m³	0.10	1.20	2.00	–	3.20	4.47	6.39
928101020B	0.25 – 0.50 m	m³	0.08	0.96	1.60	–	2.56	3.58	5.11
928101020C	0.50 – 1.0 m	m³	0.10	1.20	2.00	–	3.20	4.47	6.39
928101020D	1.0 – 2.0 m	m³	0.12	1.43	2.40	–	3.83	5.37	7.67
928101020E	2.0 – 5.0 m	m³	0.14	1.67	2.80	–	4.47	6.26	8.95
928101020F	5.0 – 10.0 m	m³	0.15	1.79	3.00	–	4.79	6.71	9.59
928101020G	over 10.0 m	m³	0.16	1.91	3.20	–	5.11	7.16	10.23
928101023	Excavation for cuttings in material other than topsoil, rock or other artificial hard material; average depth:								
928101023A	not exceeding 0.25 m	m³	0.45	5.38	9.00	–	14.38	20.13	28.76
928101023B	0.25 – 0.50 m	m³	0.43	5.14	8.60	–	13.74	19.24	27.48
928101023C	0.50 – 1.0 m	m³	0.45	5.38	9.00	–	14.38	20.13	28.76
928101023D	1.0 – 2.0 m	m³	0.47	5.62	9.40	–	15.02	21.03	30.04
928101023E	2.0 – 5.0 m	m³	0.49	5.86	9.80	–	15.66	21.92	31.31
928101023F	5.0 – 10.0 m	m³	0.50	5.98	10.00	–	15.98	22.37	31.95
928101023G	over 10.0 m	m³	0.53	6.34	10.60	–	16.94	23.71	33.87
928101026	Excavation for cuttings in rock; average depth:								
928101026A	not exceeding 0.25 m	m³	0.65	7.77	11.37	–	19.14	26.81	38.29
928101026B	0.25 – 0.50 m	m³	0.63	7.53	11.03	–	18.56	25.98	37.11
928101026C	0.50 – 1.0 m	m³	0.63	7.53	11.03	–	18.56	25.98	37.11
928101026D	1.0 – 2.0 m	m³	0.65	7.77	11.37	–	19.14	26.81	38.29
928101026E	2.0 – 5.0 m	m³	0.67	8.01	11.73	–	19.74	27.63	39.47
928101026F	5.0 – 10.0 m	m³	0.69	8.25	12.08	–	20.33	28.46	40.65
928101026G	over 10.0 m	m³	0.73	8.73	12.78	–	21.51	30.11	43.00

General Civils

Rail 2003 Ref		Unit	Labour Hours	Labour Cost £	Plant Cost £	Materials Cost £	Unit Rate Base Cost £	Unit Rate Green Zone £	Unit Rate Red Zone £
928	**EARTHWORKS**								
9281010	**Machine excavation**								
928101028	Excavation for tunnels in material other than topsoil, rock or other artificial hard material; average diameter of tunnel:								
928101028B	over 10.0 m	m³	2.09	24.92	39.85	–	64.77	90.67	129.53
928101029	Excavation for tunnels in material in rock; average diameter of tunnel:								
928101029B	over 10.0 m	m³	2.78	33.23	53.23	–	86.46	121.05	172.93
928101030	Excavation for foundations in topsoil; average depth:								
928101030A	not exceeding 0.25 m	m³	0.30	3.59	5.25	–	8.84	12.37	17.67
928101030B	0.25 – 0.50 m	m³	0.28	3.35	4.90	–	8.25	11.55	16.49
928101030C	0.50 – 1.0 m	m³	0.28	3.35	4.90	–	8.25	11.55	16.49
928101030D	1.0 – 2.0 m	m³	0.30	3.59	5.25	–	8.84	12.37	17.67
928101030E	2.0 – 5.0 m	m³	0.31	3.71	5.43	–	9.14	12.79	18.26
928101030F	5.0 – 10.0 m	m³	0.33	3.95	5.78	–	9.73	13.61	19.44
928101030G	over 10.0 m	m³	0.35	4.18	6.13	–	10.31	14.44	20.62
928101033	Excavation for foundations in material other than topsoil, rock or other artificial material; average depth:								
928101033A	not exceeding 0.25 m	m³	0.65	7.77	11.37	–	19.14	26.81	38.29
928101033B	0.25 – 0.50 m	m³	0.63	7.53	11.03	–	18.56	25.98	37.11
928101033C	0.50 – 1.0 m	m³	0.65	7.77	11.37	–	19.14	26.81	38.29
928101033D	1.0 – 2.0 m	m³	0.65	7.77	11.37	–	19.14	26.81	38.29
928101033E	2.0 – 5.0 m	m³	0.66	7.89	11.55	–	19.44	27.22	38.88
928101033F	5.0 – 10.0 m	m³	0.68	8.13	11.90	–	20.03	28.04	40.06
928101033G	over 10.0 m	m³	0.70	8.37	12.25	–	20.62	28.87	41.24
928101035	Excavation for foundation in rock; average depth:								
928101035A	not exceeding 0.25 m	m³	0.85	10.16	14.88	–	25.04	35.06	50.07
928101035B	0.25 – 0.50 m	m³	0.83	9.92	14.53	–	24.45	34.23	48.89
928101035C	0.50 – 1.0 m	m³	0.85	10.16	14.88	–	25.04	35.06	50.07
928101035D	1.0 – 2.0 m	m³	0.85	10.16	14.88	–	25.04	35.06	50.07
928101035E	2.0 – 5.0 m	m³	0.86	10.28	15.05	–	25.33	35.46	50.66
928101035F	5.0 – 10.0 m	m³	0.88	10.52	15.40	–	25.92	36.29	51.84

General Civils

Rail 2003 Ref		Unit	Labour Hours	Labour Cost £	Plant Cost £	Materials Cost £	Unit Rate Base Cost £	Unit Rate Green Zone £	Unit Rate Red Zone £
928	**EARTHWORKS**								
9281010	Machine excavation								
928101035	Excavation for foundation in rock; average depth:								
928101035G	over 10.0 m....................	m^3	0.90	10.76	15.75	–	26.51	37.11	53.02
92850	**EXCAVATION ANCILLARIES**								
9285010	Trimming excavated surfaces								
928501010	Topsoil:								
928501010A	horizontal.....................	m^2	0.01	0.12	0.18	–	0.30	0.42	0.59
928501010B	sloping, not exceeding 45 degrees to horizontal.....................	m^2	0.01	0.12	0.18	–	0.30	0.42	0.59
928501020	Material other than topsoil, rock or artificial hard material:								
928501020A	horizontal.....................	m^2	0.02	0.24	0.35	–	0.59	0.82	1.18
928501020B	sloping, not exceeding 45 degrees to horizontal.....................	m^2	0.02	0.24	0.35	–	0.59	0.82	1.18
928501020C	sloping, over 45 degrees to horizontal.	m^2	0.03	0.36	0.53	–	0.89	1.24	1.77
928501030	Rock:								
928501030A	horizontal.....................	m^2	0.22	2.63	1.54	–	4.17	5.83	8.33
928501030B	sloping, not exceeding 45 degrees to horizontal.....................	m^2	0.24	2.87	1.68	–	4.55	6.37	9.09
928501030C	sloping, over 45 degrees to horizontal.	m^2	0.32	3.82	2.24	–	6.06	8.48	12.12
928501030D	vertical........................	m^2	0.42	5.02	2.93	–	7.95	11.14	15.91
9285020	Preparation of excavated surfaces								
928502010	Topsoil:								
928502010A	generally	m^2	0.02	0.24	0.35	–	0.59	0.82	1.18

General Civils

Rail 2003 Ref		Unit	Labour Hours	Labour Cost £	Plant Cost £	Materials Cost £	Unit Rate Base Cost £	Unit Rate Green Zone £	Unit Rate Red Zone £
928	**EARTHWORKS**								
9285020	**Preparation of excavated surfaces**								
928502020	Material other than topsoil, rock or artificial hard material:								
928502020A	generally	m^2	0.02	0.24	0.35	–	0.59	0.82	1.18
928502020B	sloping, not exceeding 45 degrees to horizontal	m^2	0.02	0.24	0.35	–	0.59	0.82	1.18
928502020C	sloping, over 45 degrees to horizontal	m^2	0.02	0.24	0.35	–	0.59	0.82	1.18
928502030	Rock:								
928502030A	generally	m^2	0.22	2.63	1.54	–	4.17	5.83	8.33
928502030B	sloping, not exceeding 45 degrees to horizontal	m^2	0.24	2.87	1.68	–	4.55	6.37	9.09
928502030C	sloping, over 45 degrees to horizontal	m^2	0.32	3.82	2.24	–	6.06	8.48	12.12
928502030D	vertical	m^2	0.42	5.02	2.93	–	7.95	11.14	15.91
9285030	**Disposal of excavated material**								
928503010	Topsoil; remove from site to tip:								
928503010A	loading by hand	m^3	1.37	16.37	7.05	–	23.42	32.79	46.84
928503010B	loading by machine	m^3	–	–	8.28	–	8.28	11.59	16.55
928503020	Material other than topsoil, rock or artificial hard material; remove from site to tip:								
928503020A	loading by machine	m^3	–	–	8.63	–	8.63	12.08	17.25
928503030	Rock; remove from site to tip:								
928503030A	loading by machine	m^3	–	–	8.98	–	8.98	12.57	17.95
928503040	Contaminated excavated material; remove from site to approved tip:								
928503040A	loading by machine	m^3	–	–	24.90	–	24.90	34.86	49.80

General Civils

Rail 2003 Ref		Unit	Labour Hours	Labour Cost £	Plant Cost £	Materials Cost £	Unit Rate Base Cost £	Unit Rate Green Zone £	Unit Rate Red Zone £
928	**EARTHWORKS**								
92860	**FILLING**								
9286010	Filling to excavations with imported materials								
928601010	Filled into excavation; by machine compacting in 150 mm layers:								
928601010A	sand	m³	0.30	3.59	3.75	10.25	17.59	24.61	35.16
928601010B	hardcore	m³	0.33	3.94	6.40	14.18	24.52	34.34	49.06
928601010C	topsoil	m³	0.32	3.82	4.64	7.45	15.91	22.28	31.83
928601010D	type 1 material	m³	0.33	3.94	6.40	13.62	23.96	33.54	47.92
928601010E	type 2 material	m³	0.33	3.94	6.40	13.20	23.54	32.96	47.09
9286020	Filling to excavations with selected excavated materials								
928602010	Filled into excavation; by machine; compacting in 150 mm layers:								
928602010A	generally	m³	0.30	3.59	4.63	–	8.22	11.50	16.42
92870	**FILLING ANCILLARIES**								
9287010	Trimming filled surfaces								
928701010	Topsoil:								
928701010A	horizontal	m²	0.01	0.12	0.18	–	0.30	0.42	0.59
928701010B	sloping, not exceeding 45 degrees to horizontal	m²	0.01	0.12	0.18	–	0.30	0.42	0.59
928701020	Material other than topsoil, rock or artificial hard material:								
928701020A	horizontal	m²	0.02	0.24	0.35	–	0.59	0.82	1.18
928701020B	sloping, not exceeding 45 degrees to horizontal	m²	0.02	0.24	0.35	–	0.59	0.82	1.18
928701020C	sloping, over 45 degrees to horizontal	m²	0.03	0.36	0.53	–	0.89	1.24	1.77

General Civils

Rail 2003 Ref		Unit	Labour Hours	Labour Cost £	Plant Cost £	Materials Cost £	Unit Rate Base Cost £	Unit Rate Green Zone £	Unit Rate Red Zone £

928 EARTHWORKS

9287030 Geotextiles

928703010 Terram 4000 geotextile stabilising matting; 150 mm laps; laid:

928703010A	horizontally	m²	0.02	0.24	–	2.23	2.47	3.45	4.94
928703010B	sloping	m²	0.03	0.36	–	2.23	2.59	3.62	5.18

930 IN SITU CONCRETE

93010 PROVISION OF CONCRETE

9301010 Ordinary prescribed mix

930101010 OPC; 14 mm aggregate; grade:

930101010A	C7P	m³	–	–	–	45.05	45.05	63.06	90.09
930101010B	C10P	m³	–	–	–	45.75	45.75	64.05	91.50
930101010C	C15P	m³	–	–	–	50.32	50.32	70.44	100.63
930101010D	C20P	m³	–	–	–	57.48	57.48	80.48	114.97
930101010E	C25P	m³	–	–	–	61.07	61.07	85.50	122.14
930101010F	C30P	m³	–	–	–	64.65	64.65	90.51	129.31
930101010G	C35P	m³	–	–	–	68.24	68.24	95.53	136.47
930101010H	C40P	m³	–	–	–	71.89	71.89	100.64	143.77

930101020 OPC; 20 mm aggregate; grade:

930101020A	C7P	m³	–	–	–	41.25	41.25	57.75	82.50
930101020B	C10P	m³	–	–	–	42.32	42.32	59.25	84.64
930101020C	C15P	m³	–	–	–	47.76	47.76	66.87	95.53
930101020D	C20P	m³	–	–	–	52.14	52.14	72.99	104.28
930101020E	C25P	m³	–	–	–	55.40	55.40	77.57	110.81
930101020F	C30P	m³	–	–	–	58.66	58.66	82.12	117.31
930101020G	C35P	m³	–	–	–	61.92	61.92	86.69	123.84
930101020H	C40P	m³	–	–	–	65.17	65.17	91.24	130.35

General Civils

Rail 2003 Ref		Unit	Labour Hours	Labour Cost £	Plant Cost £	Materials Cost £	Unit Rate Base Cost £	Unit Rate Green Zone £	Unit Rate Red Zone £

930 IN SITU CONCRETE

93040 PLACING OF CONCRETE

9304010 Mass concrete

930401010 Blinding; in layers; thickness:

930401010A	not exceeding 150 mm	m³	1.25	14.94	–	–	14.94	20.92	29.89
930401010B	150 – 450 mm	m³	1.10	13.15	–	–	13.15	18.41	26.30
930401010C	450 – 600 mm	m³	0.95	11.36	–	–	11.36	15.90	22.71
930401010D	over 600 mm	m³	0.90	10.76	–	–	10.76	15.06	21.52

930401020 Bases, footings, pile caps and ground slabs; thickness:

930401020A	not exceeding 150 mm	m³	1.25	14.94	–	–	14.94	20.92	29.89
930401020B	150 – 450 mm	m³	1.10	13.15	–	–	13.15	18.41	26.30
930401020C	450 – 600 mm	m³	0.95	11.36	–	–	11.36	15.90	22.71
930401020D	over 600 mm	m³	0.90	10.76	–	–	10.76	15.06	21.52

9304020 Reinforced concrete

930402010 Bases, footings, pile caps and ground slabs; thickness:

930402010A	not exceeding 150 mm	m³	1.90	22.97	1.96	–	24.93	34.90	49.86
930402010B	150 – 450 mm	m³	1.70	20.55	1.68	–	22.23	31.12	44.47
930402010C	450 – 600 mm	m³	1.50	18.14	1.54	–	19.68	27.54	39.35
930402010D	over 600 mm	m³	1.40	16.93	1.40	–	18.33	25.66	36.65

930402020 Suspended slabs; thickness:

930402020A	not exceeding 150 mm	m³	2.10	25.39	2.10	–	27.49	38.48	54.98
930402020B	150 – 450 mm	m³	1.80	21.76	1.82	–	23.58	33.02	47.16
930402020C	450 – 600 mm	m³	1.45	17.53	1.47	–	19.00	26.60	38.00
930402020D	over 600 mm	m³	1.35	16.32	1.40	–	17.72	24.81	35.44

930402030 Walls; thickness:

930402030A	not exceeding 150 mm	m³	2.30	27.81	4.62	–	32.43	45.39	64.84
930402030B	150 – 450 mm	m³	2.05	24.79	4.06	–	28.85	40.38	57.68
930402030C	450 – 600 mm	m³	1.55	18.74	3.08	–	21.82	30.55	43.64
930402030D	over 600 mm	m³	1.40	16.93	2.80	–	19.73	27.62	39.45

General Civils

Rail 2003 Ref		Unit	Labour Hours	Labour Cost	Plant Cost	Materials Cost	Unit Rate	Unit Rate	Unit Rate
				£	£	£	Base Cost £	Green Zone £	Red Zone £
930	**IN SITU CONCRETE**								
9304020	**Reinforced concrete**								
930402040	Columns and piers; cross sectional area:								
930402040A	not exceeding 0.03 m²	m³	4.00	48.36	8.04	–	56.40	78.96	112.81
930402040B	0.03 – 0.1 m²	m³	2.95	35.67	5.88	–	41.55	58.16	83.08
930402040C	0.1 – 0.25 m²	m³	2.70	32.64	5.39	–	38.03	53.24	76.06
930402040D	0.25 – 1 m²	m³	2.40	29.02	4.83	–	33.85	47.38	67.68
930402040E	over 1 m²	m³	2.20	26.60	4.34	–	30.94	43.31	61.87
930402050	Beams; cross sectional area:								
930402050A	not exceeding 0.03 m²	m³	3.60	43.52	7.28	–	50.80	71.12	101.60
930402050B	0.03 – 0.1 m²	m³	2.65	32.04	5.32	–	37.36	52.29	74.71
930402050C	0.1 – 0.25 m²	m³	2.45	29.62	4.83	–	34.45	48.23	68.89
930402050D	0.25 – 1 m²	m³	2.15	25.99	4.34	–	30.33	42.46	60.66
930402050E	over 1 m²	m³	2.00	24.18	3.92	–	28.10	39.33	56.19
930402060	Casings to metal sections; cross sectional area:								
930402060A	not exceeding 0.03 m²	m³	4.15	50.17	8.39	–	58.56	81.99	117.14
930402060B	0.03 – 0.1 m²	m³	3.05	36.88	6.09	–	42.97	60.15	85.92
930402060C	0.1 – 0.25 m²	m³	2.80	33.85	5.53	–	39.38	55.13	78.75
930402060D	0.25 – 1 m²	m³	2.45	29.62	4.97	–	34.59	48.42	69.17
930402060E	over 1 m²	m³	2.30	27.81	4.48	–	32.29	45.20	64.56

General Civils

Rail 2003 Ref		Unit	Labour Hours	Labour Cost £	Plant Cost £	Materials Cost £	Unit Rate Base Cost £	Unit Rate Green Zone £	Unit Rate Red Zone £

935 CONCRETE ANCILLARIES

93520 FORMWORK

9352010 Formwork to fair finish

935201010 Sides of foundation; width:

935201010A	not exceeding 250 mm	m	0.76	10.43	–	2.23	12.66	17.72	25.32
935201010B	250 – 500 mm	m	1.15	15.78	–	3.71	19.49	27.30	38.99
935201010C	500 mm – 1 m	m	2.04	28.00	–	6.66	34.66	48.53	69.32
935201010D	over 1 m	m²	1.86	25.53	–	7.01	32.54	45.55	65.06

935201015 Sides of ground beams and edges of beds; width:

935201015A	not exceeding 250 mm	m	0.76	10.43	–	2.23	12.66	17.72	25.32
935201015B	250 – 500 mm	m	1.15	15.78	–	3.71	19.49	27.30	38.99
935201015C	500 mm – 1 m	m	2.04	28.00	–	6.66	34.66	48.53	69.32
935201015D	over 1 m	m²	1.86	25.53	–	28.25	53.78	75.29	107.55

935201020 Edges of suspended slabs; width:

935201020A	not exceeding 250 mm	m	0.71	9.74	–	1.98	11.72	16.41	23.45
935201020B	250 – 500 mm	m	0.90	12.35	–	3.24	15.59	21.82	31.18
935201020C	500 mm – 1 m	m	1.60	21.96	2.89	5.61	30.46	42.64	60.93
935201020D	over 1 m	m²	1.46	20.04	5.78	6.45	32.27	45.17	64.53

935201025 Sides of upstands; width:

935201025A	not exceeding 250 mm	m	0.71	9.74	–	1.98	11.72	16.41	23.45
935201025B	250 – 500 mm	m	0.90	12.35	–	3.24	15.59	21.82	31.18
935201025C	500 mm – 1 m	m	1.60	21.96	2.89	8.26	33.11	46.36	66.23
935201025D	over 1 m	m²	1.46	20.04	5.78	6.45	32.27	45.17	64.53

935201030 Steps in top surfaces; width:

935201030A	not exceeding 250 mm	m	0.71	9.74	–	1.98	11.72	16.41	23.45
935201030B	250 – 500 mm	m	0.90	12.35	–	3.24	15.59	21.82	31.18
935201030C	500 mm – 1 m	m	1.59	21.82	2.89	5.61	30.32	42.45	60.65
935201030D	over 1 m	m²	1.46	20.04	5.78	6.45	32.27	45.17	64.53

General Civils

Rail 2003 Ref		Unit	Labour Hours	Labour Cost £	Plant Cost £	Materials Cost £	Unit Rate Base Cost £	Unit Rate Green Zone £	Unit Rate Red Zone £
935	**CONCRETE ANCILLARIES**								
9352010	Formwork to fair finish								
935201040	Machine bases and plinths; width:								
935201040A	not exceeding 250 mm	m	0.76	10.43	–	5.41	15.84	22.17	31.67
935201040B	250 – 500 mm	m	1.15	15.78	–	10.12	25.90	36.26	51.80
935201040C	500 mm – 1 m	m	2.04	28.00	–	19.42	47.42	66.39	94.83
935201040D	over 1 m	m^2	1.86	25.53	–	18.61	44.14	61.79	88.26
935201045	Soffits; not exceeding 1.5 m above floor level; thickness:								
935201045A	not exceeding 200 mm	m^2	1.46	20.04	2.41	6.86	29.31	41.02	58.60
935201045B	200 – 300 mm	m^2	2.04	28.00	2.41	6.86	37.27	52.17	74.52
935201048	Soffits; 1.5 – 3.0 m above floor level; thickness:								
935201048A	not exceeding 200 mm	m^2	1.79	24.57	2.41	6.86	33.84	47.36	67.66
935201048B	200 – 300 mm	m^2	2.50	34.31	2.41	18.61	55.33	77.45	110.65
935201050	Sloping soffits; not exceeding 1.5 m above floor level; thickness:								
935201050A	not exceeding 200 mm	m^2	1.46	20.04	2.41	6.86	29.31	41.02	58.60
935201050B	200 – 300 mm	m^2	2.04	28.00	2.41	6.86	37.27	52.17	74.52
935201052	Sloping soffits; 1.5 – 3.0 m above floor level; thickness:								
935201052A	not exceeding 200 mm	m^2	1.79	24.57	2.41	6.86	33.84	47.36	67.66
935201052B	200 – 300 mm	m^2	2.50	34.31	2.41	18.61	55.33	77.45	110.65
935201055	Top formwork; over 15 degrees from horizontal:								
935201055A	generally	m^2	1.86	25.53	–	5.02	30.55	42.77	61.09
935201060	Walls:								
935201060A	vertical surfaces	m^2	1.78	24.43	0.94	7.51	32.88	46.04	65.77
935201060B	curved surfaces	m^2	2.00	27.45	0.94	6.57	34.96	48.95	69.93
935201060C	conical surfaces	m^2	2.66	36.51	1.06	7.83	45.40	63.57	90.80
935201060D	battering surfaces	m^2	2.13	29.23	0.94	6.76	36.93	51.71	73.88

General Civils

Rail 2003 Ref		Unit	Labour Hours	Labour Cost £	Plant Cost £	Materials Cost £	Unit Rate Base Cost £	Unit Rate Green Zone £	Unit Rate Red Zone £
935	**CONCRETE ANCILLARIES**								
9352010	**Formwork to fair finish**								
935201065	Sides and soffits of attached beams; width:								
935201065A	not exceeding 250 mm	m	1.07	14.69	0.57	2.12	17.38	24.32	34.74
935201065B	250 – 500 mm	m	1.25	17.16	0.57	4.05	21.78	30.48	43.54
935201065C	500 mm – 1 m	m	1.54	21.14	1.13	6.51	28.78	40.30	57.57
935201065D	over 1 m	m^2	2.03	27.86	0.94	6.86	35.66	49.92	71.32
935201070	Sides and soffits of isolated beams; width:								
935201070A	not exceeding 250 mm	m	1.21	16.61	3.46	2.09	22.16	31.02	44.31
935201070B	250 – 500 mm	m	1.41	19.35	3.46	11.43	34.24	47.95	68.49
935201070C	500 mm – 1 m	m	1.74	23.88	6.91	5.81	36.60	51.24	73.20
935201070D	over 1 m	m^2	2.20	30.19	7.36	6.86	44.41	62.18	88.83
935201075	Sides of isolated rectangular columns; width:								
935201075A	not exceeding 250 mm	m	0.29	3.98	1.47	1.92	7.37	10.32	14.74
935201075B	250 – 500 mm	m	0.58	7.96	1.47	3.32	12.75	17.85	25.50
935201075C	500 mm – 1 m	m	0.73	10.02	1.78	5.34	17.14	23.99	34.26
935201075D	over 1 m	m^2	1.01	13.86	2.06	5.90	21.82	30.55	43.63
935201080	Sides of isolated circular columns; diameter:								
935201080A	not exceeding 300 mm	m^2	0.71	9.74	4.91	10.39	25.04	35.05	50.08
935201080B	300 – 600 mm	m^2	0.40	5.49	2.46	10.39	18.34	25.67	36.67
935201080C	600 – 900 mm	m^2	0.25	3.43	1.64	10.39	15.46	21.63	30.92
935201080D	over 900 mm	m^2	0.18	2.47	1.23	10.39	14.09	19.72	28.18

General Civils

Rail 2003 Ref		Unit	Labour Hours	Labour Cost £	Plant Cost £	Materials Cost £	Unit Rate Base Cost £	Unit Rate Green Zone £	Unit Rate Red Zone £

935 CONCRETE ANCILLARIES

93550 REINFORCEMENT

9355010 Bar reinforcement

935501010 Mild steel bars; BS 4449; nominal size:

Ref	Size	Unit	Hrs	Lab	Plant	Mat	Base	Green	Red
935501010A	6 mm	Tonne	72.00	898.56	–	447.61	1346.17	1884.64	2692.34
935501010B	8 mm	Tonne	54.00	673.92	–	435.72	1109.64	1553.50	2219.29
935501010C	10 mm	Tonne	44.00	549.12	–	405.00	954.12	1335.78	1908.25
935501010D	12 mm	Tonne	38.00	474.24	–	387.84	862.08	1206.92	1724.16
935501010E	16 mm	Tonne	30.00	374.40	–	363.04	737.44	1032.41	1474.88
935501010F	20 mm	Tonne	26.00	324.48	–	356.39	680.87	953.22	1361.74
935501010G	25 mm	Tonne	23.00	287.04	–	345.35	632.39	885.35	1264.78
935501010H	32 mm	Tonne	20.00	249.60	–	339.07	588.67	824.14	1177.34

935501020 High yield bars; BS 4449; nominal size:

Ref	Size	Unit	Hrs	Lab	Plant	Mat	Base	Green	Red
935501020A	6 mm	Tonne	72.00	898.56	–	464.35	1362.91	1908.06	2725.81
935501020B	8 mm	Tonne	54.00	673.92	–	452.08	1126.00	1576.40	2252.00
935501020C	10 mm	Tonne	44.00	549.12	–	420.39	969.51	1357.32	1939.03
935501020D	12 mm	Tonne	38.00	474.24	–	402.59	876.83	1227.56	1753.65
935501020E	16 mm	Tonne	30.00	374.40	–	376.83	751.23	1051.72	1502.45
935501020F	20 mm	Tonne	26.00	324.48	–	370.02	694.50	972.29	1388.99
935501020G	25 mm	Tonne	23.00	287.04	–	358.66	645.70	903.98	1291.39
935501020H	32 mm	Tonne	20.00	249.60	–	352.21	601.81	842.54	1203.62

935501030 Stainless steel bars; type 316 S66; nominal size:

Ref	Size	Unit	Hrs	Lab	Plant	Mat	Base	Green	Red
935501030A	10 mm	Tonne	44.00	549.12	–	6833.95	7383.07	10336.29	14766.13
935501030B	12 mm	Tonne	38.00	474.24	–	6709.51	7183.75	10057.25	14367.49
935501030C	16 mm	Tonne	30.00	374.40	–	6462.14	6836.54	9571.15	13673.07
935501030D	20 mm	Tonne	26.00	324.48	–	6336.18	6660.66	9324.93	13321.33
935501030E	25 mm	Tonne	23.00	287.04	–	6209.85	6496.89	9095.65	12993.78
935501030F	32 mm	Tonne	20.00	249.60	–	6084.28	6333.88	8867.43	12667.75

9355020 Fabric reinforcement

935502010 Steel fabric reinforcement; BS 4483; laid horizontally:

Ref	Type	Unit	Hrs	Lab	Plant	Mat	Base	Green	Red
935502010A	A142	m^2	0.03	0.37	–	1.82	2.19	3.07	4.39
935502010B	A193	m^2	0.03	0.37	–	2.42	2.79	3.91	5.60
935502010C	A252	m^2	0.03	0.37	–	3.13	3.50	4.90	7.00
935502010D	A393	m^2	0.05	0.62	–	4.91	5.53	7.75	11.08
935502010E	B283	m^2	0.03	0.37	–	3.13	3.50	4.90	7.00
935502010F	B385	m^2	0.04	0.50	–	3.74	4.24	5.94	8.49
935502010G	B503	m^2	0.05	0.62	–	4.76	5.38	7.54	10.78

General Civils

Rail 2003 Ref		Unit	Labour Hours	Labour Cost	Plant Cost	Materials Cost	Unit Rate	Unit Rate	Unit Rate
				£	£	£	Base Cost £	Green Zone £	Red Zone £

935 CONCRETE ANCILLARIES

9355020 Fabric reinforcement

935502010 Steel fabric reinforcement; BS 4483; laid horizontally:

935502010H	B785	m²	0.05	0.62	–	6.05	6.67	9.33	13.34
935502010I	C385	m²	0.03	0.37	–	2.88	3.25	4.55	6.51
935502010J	C503	m²	0.04	0.50	–	3.55	4.05	5.67	8.10

93560 JOINTS IN CONCRETE PAVINGS

9356050 Designed joints in in situ concrete

935605010 Plain joints; horizontal; including formwork:

935605010A	not exceeding 150 mm high	m	0.72	9.88	–	1.66	11.54	16.16	23.08
935605010B	150 – 300 mm high	m	0.82	11.25	–	4.14	15.39	21.55	30.78

935605020 Keyed joints; horizontal; including formwork:

935605020A	not exceeding 150 mm high	m	0.77	10.57	–	1.83	12.40	17.35	24.79
935605020B	150 – 300 mm high	m	0.82	11.25	–	4.46	15.71	22.00	31.43

935605030 Movement joints; horizontal; including formwork; 25 mm mild steel dowel bars 600 mm long debonded for half length; dowel caps with compressible filler:

935605030A	not exceeding 150 mm high	m	1.09	14.96	–	8.61	23.57	32.99	47.13
935605030B	150 – 300 mm high	m	1.19	16.33	–	11.08	27.41	38.37	54.82

935605040 Schlegal crack inducer; horizontal; placing in position:

935605040A	generally	m	0.25	3.43	–	1.04	4.47	6.26	8.94

935605050 Serviseal PVC waterstops; flat dumbell; width:

935605050A	100 mm	m	0.12	1.65	–	2.65	4.30	6.02	8.59
935605050B	170 mm	m	0.13	1.78	–	3.71	5.49	7.69	10.98
935605050C	210 mm	m	0.16	2.20	–	4.77	6.97	9.74	13.92
935605050D	250 mm	m	0.18	2.47	–	5.56	8.03	11.24	16.06

General Civils

Rail 2003 Ref		Unit	Labour Hours	Labour Cost £	Plant Cost £	Materials Cost £	Unit Rate Base Cost £	Unit Rate Green Zone £	Unit Rate Red Zone £
935	**CONCRETE ANCILLARIES**								
9356050	**Designed joints in in situ concrete**								
935605055	Serviseal PVC waterstops; centre bulb; width:								
935605055A	160 mm	m	0.12	1.65	–	3.71	5.36	7.50	10.70
935605055B	210 mm	m	0.15	2.06	–	4.77	6.83	9.55	13.65
935605055C	260 mm	m	0.18	2.47	–	5.56	8.03	11.24	16.06
935605060	Servi-tite flat dumbell CJ; width:								
935605060A	150 mm	m	0.12	1.65	–	7.28	8.93	12.50	17.85
935605060B	230 mm	m	0.15	2.06	–	9.93	11.99	16.78	23.98
935605060C	305 mm	m	0.18	2.47	–	18.54	21.01	29.41	42.01
935605065	Servi-tite flat centre bulb XJ; width:								
935605065A	150 mm	m	0.12	1.65	–	7.94	9.59	13.43	19.18
935605065B	230 mm	m	0.15	2.06	–	10.93	12.99	18.18	25.98
935605065C	305 mm	m	0.18	2.47	–	18.54	21.01	29.41	42.01
935605070	Expansion materials; Flexcell fibreboard compressible joint filler; 10 mm thick; width:								
935605070A	not exceeding 150 mm	m	0.10	1.37	–	1.21	2.58	3.61	5.16
935605070B	150 – 300 mm	m	0.18	2.47	–	2.35	4.82	6.74	9.63
935605071	Expansion materials; Flexcell fibreboard compressible joint filler; 13 mm thick; width:								
935605071A	not exceeding 150 mm	m	0.11	1.51	–	1.61	3.12	4.37	6.25
935605071B	150 – 300 mm	m	0.18	2.47	–	3.13	5.60	7.84	11.19
935605072	Expansion materials; Flexcell fibreboard compressible joint filler; 19 mm thick; width:								
935605072A	not exceeding 150 mm	m	0.20	2.75	–	2.02	4.77	6.66	9.52
935605072B	150 – 300 mm	m	0.23	3.16	–	3.91	7.07	9.89	14.13

General Civils

Rail 2003 Ref		Unit	Labour Hours	Labour Cost £	Plant Cost £	Materials Cost £	Unit Rate Base Cost £	Unit Rate Green Zone £	Unit Rate Red Zone £
935	**CONCRETE ANCILLARIES**								
9356050	Designed joints in in situ concrete								
935605073	Expansion materials; Flexcell fibreboard compressible joint filler; 25 mm thick; width:								
935605073A	not exceeding 150 mm	m	0.20	2.75	–	2.42	5.17	7.23	10.33
935605073B	150 – 300 mm	m	0.24	3.29	–	4.69	7.98	11.18	15.97
935605075	Joint sealants; hot poured bituminous rubber compound:								
935605075A	10 × 25 mm	m	0.03	0.41	0.14	1.16	1.71	2.39	3.42
935605075B	13 × 25 mm	m	0.03	0.41	0.14	1.50	2.05	2.87	4.10
935605075C	19 × 25 mm	m	0.04	0.55	0.17	2.19	2.91	4.08	5.84
935605075D	25 × 25 mm	m	0.06	0.82	0.21	2.93	3.96	5.54	7.92
935605077	Joint sealants; cold poured polysulphide epoxy based compound:								
935605077A	10 × 25 mm	m	0.03	0.41	–	0.40	0.81	1.14	1.63
935605077B	13 × 25 mm	m	0.03	0.41	–	0.56	0.97	1.36	1.93
935605077C	19 × 25 mm	m	0.04	0.55	–	0.76	1.31	1.83	2.61
935605077D	25 × 25 mm	m	0.06	0.82	–	1.26	2.08	2.92	4.17
935605079	Joints sealant; gun grade polysulphide rubber compound:								
935605079A	10 × 25 mm	m	0.04	0.55	–	0.51	1.06	1.49	2.13
935605079B	13 × 25 mm	m	0.06	0.82	–	0.75	1.57	2.21	3.16
935605079C	19 × 25 mm	m	0.13	1.78	–	1.33	3.11	4.36	6.23
935605079D	25 × 25 mm	m	0.23	3.16	–	2.10	5.26	7.35	10.50

General Civils

Rail 2003 Ref		Unit	Labour Hours	Labour Cost £	Plant Cost £	Materials Cost £	Unit Rate Base Cost £	Unit Rate Green Zone £	Unit Rate Red Zone £
935	**CONCRETE ANCILLARIES**								
93580	**CONCRETE ACCESSORIES**								
9358010	Surface finishes to concrete								
935801010	Top surfaces; finishing over 300 mm wide with:								
935801010A	wood float	m^2	0.18	2.15	–	–	2.15	3.01	4.30
935801010B	steel trowel	m^2	0.16	1.91	–	–	1.91	2.68	3.83
935801010C	power float	m^2	0.15	1.79	0.21	–	2.00	2.80	4.01
9358020	Grouting								
935802010	Grouting under plates with proprietary compound; area:								
935802010A	not exceeding 0.1 m^2	Nr	0.20	2.39	–	0.34	2.73	3.82	5.45
935802010B	0.1 – 0.5 m^2	Nr	0.35	4.18	–	0.67	4.85	6.80	9.71
935802010C	0.5 – 1 m^2	Nr	0.50	5.98	–	0.67	6.65	9.31	13.29
935802010D	over 1 m^2	Nr	0.70	8.37	–	2.01	10.38	14.54	20.77

General Civils

Rail 2003 Ref		Unit	Composite Cost £	Unit Rate Base Cost £	Unit Rate Green Zone £	Unit Rate Red Zone £
938	**PRECAST CONCRETE**					
93810	**BEAMS AND COLUMNS**					
9381010	Precast beams and columns generally					
938101010	Length; not exceeding 5 m; mass:					
938101010A	not exceeding 250 kg	Nr	–	–	–	90.00	90.00	126.00	180.00
938101010B	250 – 500 kg	Nr	–	–	–	165.00	165.00	231.00	330.00
938101010C	500 kg – 1 t	Nr	–	–	–	300.00	300.00	420.00	600.00
938101010D	1 – 2 t	Nr	–	–	–	575.00	575.00	805.00	1150.00
938101010E	2 – 5 t	Nr	–	–	–	1075.00	1075.00	1505.00	2150.00
938101010F	5 – 10 t	Nr	–	–	–	2000.00	2000.00	2800.00	4000.00
938101010G	over 10 t	Nr	–	–	–	3750.00	3750.00	5250.00	7500.00
938101020	Length; 5 – 7 m; mass:					
938101020A	250 – 500 kg	Nr	–	–	–	155.00	155.00	217.00	310.00
938101020B	500 – 1 t	Nr	–	–	–	550.00	550.00	770.00	1100.00
938101020C	1 – 2 t	Nr	–	–	–	1025.00	1025.00	1435.00	2050.00
938101020D	2 – 5 t	Nr	–	–	–	1960.00	1960.00	2744.00	3920.00
938101020E	5 – 10 t	Nr	–	–	–	3585.00	3585.00	5019.00	7170.00
938101030	Length; 7 – 10 m; mass:					
938101030A	250 – 500 kg	Nr	–	–	–	170.00	170.00	238.00	340.00
938101030B	500 – 1 t	Nr	–	–	–	625.00	625.00	875.00	1250.00
938101030C	1 – 2 t	Nr	–	–	–	1175.00	1175.00	1645.00	2350.00
938101030D	2 – 5 t	Nr	–	–	–	2125.00	2125.00	2975.00	4250.00
938101030E	5 – 10 t	Nr	–	–	–	3950.00	3950.00	5530.00	7900.00
938101040	Length; 10 – 15 m; mass:					
938101040A	500 kg – 1 t	Nr	–	–	–	340.00	340.00	476.00	680.00
938101040B	1 – 2 t	Nr	–	–	–	665.00	665.00	931.00	1330.00
938101040C	2 – 5 t	Nr	–	–	–	4250.00	4250.00	5950.00	8500.00
938101050	Length; 15 – 20 m; mass:					
938101050A	500 – 1 t	Nr	–	–	–	340.00	340.00	476.00	680.00
938101050B	1 – 2 t	Nr	–	–	–	665.00	665.00	931.00	1330.00
938101050C	2 – 5 t	Nr	–	–	–	4250.00	4250.00	5950.00	8500.00

General Civils

Rail 2003 Ref	Unit	Composite Cost £	Unit Rate Base Cost £	Unit Rate Green Zone £	Unit Rate Red Zone £
938	**PRECAST CONCRETE**				
9381010	**Precast beams and columns generally**				
938101060	Length; over 20 m; mass:				
938101060A	1–2 t............................ Nr	– – – 690.00	690.00	966.00	1380.00
93850	**SLABS**				
9385010	**Precast slab units; hoisting and fixing on prepared bearings; grouting up joints**				
938501010	Slab thickness; not exceeding 200 mm; maximum loading 5 kN/m^2; size:				
938501010A	1.2 × 1.2 m....................... m^2	– – – 39.00	39.00	54.60	78.00
938501010B	2.4 × 1.2 m....................... m^2	– – – 72.00	72.00	100.80	144.00
938501010C	4.8 × 1.2 m....................... m^2	– – – 162.00	162.00	226.80	324.00
938501010D	2.4 × 2.4 m....................... m^2	– – – 162.00	162.00	226.80	324.00
938501010E	4.8 × 2.4 m....................... m^2	– – – 265.00	265.00	371.00	530.00
938501020	Slab thickness; 450 mm; maximum loading 10 kN/m^2; size:				
938501020A	1.2 × 1.2 m....................... m^2	– – – 72.00	72.00	100.80	144.00
938501020B	2.4 × 1.2 m....................... m^2	– – – 133.00	133.00	186.20	266.00
938501020C	4.8 × 1.2 m....................... m^2	– – – 258.00	258.00	361.20	516.00
938501020D	2.4 × 2.4 m....................... m^2	– – – 258.00	258.00	361.20	516.00
938501020E	4.8 × 2.4 m....................... m^2	– – – 422.00	422.00	590.80	844.00
938501030	Slab thickness; 600 mm; maximum loading 10 kN/m^2; size:				
938501030A	1.2 × 1.2 m....................... Nr	– – – 95.00	95.00	133.00	190.00
938501030B	2.4 × 1.2 m....................... Nr	– – – 325.00	325.00	455.00	650.00
938501030C	4.8 × 1.2 m....................... Nr	– – – 345.00	345.00	483.00	690.00
938501030D	2.4 × 2.4 m....................... Nr	– – – 345.00	345.00	483.00	690.00
938501030E	4.8 × 2.4 m....................... Nr	– – – 600.00	600.00	840.00	1200.00

General Civils

Rail 2003 Ref		Unit			Composite Cost £	Unit Rate Base Cost £	Unit Rate Green Zone £	Unit Rate Red Zone £

938 PRECAST CONCRETE

93860 CULVERTS, DUCTS AND COVERS

9386010 Precast units; bedding and jointing

938601010 Culverts; 250 mm wall thickness; internal size:

938601010A	1200 × 600 mm	m	–	–	–	250.00	250.00	350.00	500.00
938601010B	1800 × 1200 mm	m	–	–	–	310.00	310.00	434.00	620.00
938601010C	2400 × 1800 mm	m	–	–	–	345.00	345.00	483.00	690.00
938601010D	3000 × 2400 mm	m	–	–	–	510.00	510.00	714.00	1020.00

938601020 Ducts and covers; nominal internal size:

938601020A	200 × 100 mm	m	–	–	–	42.50	42.50	59.50	85.00
938601020B	300 × 200 mm	m	–	–	–	66.00	66.00	92.40	132.00
938601020C	450 × 300 mm	m	–	–	–	97.50	97.50	136.50	195.00
938601020D	600 × 450 mm	m	–	–	–	120.00	120.00	168.00	240.00

93870 COPINGS AND SILLS

9387010 Precast units; bedding and jointing

938701010 Copings; weathered; in 500 mm lengths; overall size:

938701010A	250 mm wide × 100 mm high	m	–	–	–	9.50	9.50	13.30	19.00
938701010B	500 mm wide × 200 mm high	m	–	–	–	24.75	24.75	34.65	49.50
938701010C	750 mm wide × 200 mm high	m	–	–	–	35.25	35.25	49.35	70.50
938701010D	1000 mm wide × 300 mm high	m	–	–	–	66.75	66.75	93.45	133.50

938701020 Sills; weathered; in 500 mm lengths; overall size:

938701020A	250 mm wide × 100 mm high	m	–	–	–	13.75	13.75	19.25	27.50
938701020B	500 mm wide × 200 mm high	m	–	–	–	35.75	35.75	50.05	71.50
938701020C	750 mm wide × 200 mm high	m	–	–	–	51.00	51.00	71.40	102.00
938701020D	1000 mm wide × 300 mm high	m	–	–	–	96.75	96.75	135.45	193.50

938701030 Parapet units; in 500 mm lengths; overall size:

938701030A	250 mm wide × 150 mm high	m	–	–	–	14.25	14.25	19.95	28.50
938701030B	500 mm wide × 225 mm high	m	–	–	–	27.75	27.75	38.85	55.50
938701030C	700 mm wide × 275 mm high	m	–	–	–	48.50	48.50	67.90	97.00
938701030D	1000 mm wide × 350 mm high	m	–	–	–	70.00	70.00	98.00	140.00

General Civils

Rail 2003 Ref		Unit	Labour Hours	Labour Cost £	Plant Cost £	Materials Cost £	Unit Rate Base Cost £	Unit Rate Green Zone £	Unit Rate Red Zone £
940	**PIPEWORK – PIPES**								
94010	**CLAY PIPES**								
9401010	HepSeal; vitrified clay pipes; BS EN295: 1991; spigot and socket flexible joints;								
940101010	Nominal bore 100 mm; in trenches; depth:								
940101010A	not exceeding 1.5 m	m	1.68	20.08	3.98	7.52	31.58	44.22	63.17
940101010B	1.5 – 2.0 m	m	1.91	22.83	3.98	7.68	34.49	48.29	68.99
940101010C	2.0 – 2.5 m	m	2.17	25.94	5.26	7.84	39.04	54.64	78.08
940101010D	2.5 – 3.0 m	m	2.44	29.17	8.88	8.00	46.05	64.46	92.08
940101020	Nominal bore 150 mm; in trenches; depth:								
940101020A	not exceeding 1.5 m	m	1.73	20.68	3.98	9.44	34.10	47.74	68.20
940101020B	1.5 – 2.0 m	m	1.96	23.43	5.26	9.59	38.28	53.59	76.57
940101020C	2.0 – 2.5 m	m	2.20	26.30	7.23	9.75	43.28	60.61	86.57
940101020D	2.5 – 3.0 m	m	2.49	29.76	8.88	9.91	48.55	67.98	97.11
940101030	Nominal bore 225 mm; in trenches; depth:								
940101030A	not exceeding 1.5 m	m	1.93	23.07	5.23	17.38	45.68	63.96	91.36
940101030B	1.5 – 2.0 m	m	2.18	26.06	6.86	17.54	50.46	70.63	100.90
940101030C	2.0 – 2.5 m	m	2.48	29.64	9.53	17.70	56.87	79.62	113.75
940101030D	2.5 – 3.0 m	m	2.75	32.87	11.18	17.86	61.91	86.67	123.80
940101040	Nominal bore 300 mm; in trenches; depth:								
940101040A	not exceeding 1.5 m	m	2.17	25.94	6.48	26.57	58.99	82.58	117.96
940101040B	1.5 – 2.0 m	m	2.44	29.17	8.45	26.73	64.35	90.08	128.69
940101040C	2.0 – 2.5 m	m	2.78	33.23	11.66	26.89	71.78	100.49	143.56
940101040D	2.5 – 3.0 m	m	3.08	36.81	13.48	27.05	77.34	108.28	154.68
940101050	Nominal bore 400 mm; in trenches; depth:								
940101050A	not exceeding 1.5 m	m	2.54	30.36	9.13	52.92	92.41	129.37	184.82
940101050B	1.5 – 2.0 m	m	2.86	34.19	11.98	53.08	99.25	138.95	198.49
940101050C	2.0 – 2.5 m	m	3.30	39.44	16.43	53.24	109.11	152.76	218.23
940101050D	2.5 – 3.0 m	m	3.64	43.51	19.65	53.40	116.56	163.17	233.11

General Civils

Rail 2003 Ref		Unit	Labour Hours	Labour Cost £	Plant Cost £	Materials Cost £	Unit Rate Base Cost £	Unit Rate Green Zone £	Unit Rate Red Zone £
940	**PIPEWORK – PIPES**								
9401010	HepSeal; vitrified clay pipes; BS EN295: 1991; spigot and socket flexible joints;								
940101060	Nominal bore 450 mm; in trenches; depth:								
940101060A	not exceeding 1.5 m	m	2.62	31.32	16.13	68.25	115.70	161.98	231.40
940101060B	1.5 – 2.0 m	m	2.94	35.14	18.98	68.41	122.53	171.55	245.06
940101060C	2.0 – 2.5 m	m	3.38	40.40	23.43	68.57	132.40	185.36	264.80
940101060D	2.5 – 3.0 m	m	3.72	44.46	26.65	68.73	139.84	195.78	279.68
940101070	Nominal bore 500 mm; in trenches; depth:								
940101070A	not exceeding 1.5 m	m	2.35	28.09	18.25	75.70	122.04	170.85	244.08
940101070B	1.5 – 2.0 m	m	2.70	32.27	23.03	75.86	131.16	183.63	262.33
940101070C	2.0 – 2.5 m	m	3.17	37.89	26.43	76.02	140.34	196.48	280.68
940101070D	2.5 – 3.0 m	m	3.55	42.43	30.01	76.18	148.62	208.07	297.24
940101080	Nominal bore 600 mm; in trenches; depth:								
940101080A	not exceeding 1.5 m	m	2.93	35.02	20.19	119.55	174.76	244.68	349.53
940101080B	1.5 – 2.0 m	m	3.33	39.80	23.76	119.71	183.27	256.58	366.55
940101080C	2.0 – 2.5 m	m	3.82	45.66	29.43	119.87	194.96	272.95	389.94
940101080D	2.5 – 3.0 m	m	4.24	50.68	33.18	120.03	203.89	285.46	407.80
9401020	HepLine; vitrified clay subsoil drainage pipes; perforated; BS 65: 1991:								
940102010	Nominal bore 225 mm; in trenches; depth:								
940102010A	not exceeding 1.5 m	m	1.93	23.07	5.23	14.73	43.03	60.25	86.07
940102010B	1.5 – 2.0 m	m	2.18	26.06	6.86	14.89	47.81	66.92	95.61
940102010C	2.0 – 2.5 m	m	2.48	29.64	9.53	15.05	54.22	75.92	108.46
940102010D	2.5 – 3.0 m	m	2.75	32.87	11.18	15.21	59.26	82.96	118.51
940102020	Nominal bore 300 mm; in trenches; depth:								
940102020A	not exceeding 1.5 m	m	2.17	25.94	6.48	26.98	59.40	83.15	118.78
940102020B	1.5 – 2.0 m	m	2.44	29.17	8.45	27.14	64.76	90.66	129.51
940102020C	2.0 – 2.5 m	m	2.78	33.23	11.66	27.30	72.19	101.06	144.38
940102020D	2.5 – 3.0 m	m	3.08	36.81	13.48	27.46	77.75	108.85	155.50

General Civils

Rail 2003 Ref		Unit	Labour Hours	Labour Cost £	Plant Cost £	Materials Cost £	Unit Rate Base Cost £	Unit Rate Green Zone £	Unit Rate Red Zone £
940	**PIPEWORK – PIPES**								
9401020	HepLine; vitrified clay subsoil drainage pipes; perforated; BS 65: 1991:								
940102030	Nominal bore 400 mm; in trenches; depth:								
940102030A	not exceeding 1.5 m	m	2.54	30.36	9.13	84.56	124.05	173.67	248.10
940102030B	1.5 – 2.0 m	m	2.86	34.19	11.98	84.72	130.89	183.24	261.77
940102030C	2.0 – 2.5 m	m	3.30	39.44	16.43	84.88	140.75	197.06	281.51
940102030D	2.5 – 3.0 m	m	3.64	43.51	19.65	85.04	148.20	207.47	296.39
940102040	Nominal bore 450 mm; in trenches; depth:								
940102040A	not exceeding 1.5 m	m	2.62	31.32	16.13	107.47	154.92	216.89	309.84
940102040B	1.5 – 2.0 m	m	2.94	35.14	18.98	107.63	161.75	226.46	323.51
940102040C	2.0 – 2.5 m	m	3.38	40.40	23.43	107.79	171.62	240.27	343.25
940102040D	2.5 – 3.0 m	m	3.72	44.46	26.65	107.95	179.06	250.69	358.13
94020	**CONCRETE PIPES**								
9402010	Concrete pipes; BS 5911; rebated flexible joints with mastic seal								
940201010	Nominal bore 300 mm; in trenches; depth:								
940201010A	not exceeding 1.5 m	m	2.17	25.94	6.48	8.78	41.20	57.68	82.39
940201010B	1.5 – 2.0 m	m	2.44	29.17	8.45	8.94	46.56	65.18	93.12
940201010C	2.0 – 2.5 m	m	2.78	33.23	11.66	9.10	53.99	75.58	107.99
940201010D	2.5 – 3.0 m	m	3.08	36.81	13.48	9.26	59.55	83.38	119.11
940201020	Nominal bore 375 mm; in trenches; depth:								
940201020A	not exceeding 1.5 m	m	2.54	30.36	9.13	11.37	50.86	71.21	101.73
940201020B	1.5 – 2.0 m	m	2.86	34.19	11.98	11.53	57.70	80.78	115.40
940201020C	2.0 – 2.5 m	m	3.30	39.44	16.43	11.69	67.56	94.60	135.14
940201020D	2.5 – 3.0 m	m	3.64	43.51	19.65	11.85	75.01	105.01	150.02
940201030	Nominal bore 450 mm; in trenches; depth:								
940201030A	not exceeding 1.5 m	m	2.62	31.32	16.13	13.68	61.13	85.59	122.27
940201030B	1.5 – 2.0 m	m	2.94	35.14	18.98	13.84	67.96	95.16	135.93
940201030C	2.0 – 2.5 m	m	3.38	40.40	23.43	14.00	77.83	108.97	155.67
940201030D	2.5 – 3.0 m	m	3.72	44.46	26.65	14.16	85.27	119.39	170.55

General Civils

Rail 2003 Ref		Unit	Labour Hours	Labour Cost £	Plant Cost £	Materials Cost £	Unit Rate Base Cost £	Unit Rate Green Zone £	Unit Rate Red Zone £
940	**PIPEWORK – PIPES**								
9402010	Concrete pipes; BS 5911; rebated flexible joints with mastic seal								
940201040	Nominal bore 600 mm; in trenches; depth:								
940201040A	1.5 – 2.0 m	m	2.93	35.02	20.19	25.20	80.41	112.58	160.82
940201040B	2.0 – 2.5 m	m	3.33	39.80	23.76	25.36	88.92	124.48	177.84
940201040C	2.5 – 3.0 m	m	3.82	45.66	29.43	25.52	100.61	140.86	201.23
940201040D	3.0 – 3.5 m	m	4.24	50.68	33.18	25.68	109.54	153.36	219.09
9402050	Concrete pipes; laid drainage system; ogee joints								
940205010	Nominal bore 225 mm; in trenches; depth:								
940205010A	not exceeding 1.5 m	m	1.93	23.07	5.23	5.23	33.53	46.96	67.08
940205010B	1.5 – 2.0 m	m	2.18	26.06	6.86	5.39	38.31	53.63	76.62
940205010C	2.0 – 2.5 m	m	2.48	29.64	9.53	5.55	44.72	62.62	89.47
940205010D	2.5 – 3.0 m	m	2.75	32.87	11.18	5.71	49.76	69.67	99.52
940205020	Nominal bore 300 mm; in trenches; depth:								
940205020A	not exceeding 1.5 m	m	2.17	25.94	6.48	7.97	40.39	56.53	80.75
940205020B	1.5 – 2.0 m	m	2.44	29.17	8.45	8.13	45.75	64.04	91.48
940205020C	2.0 – 2.5 m	m	2.78	33.23	11.66	8.29	53.18	74.44	106.35
940205020D	2.5 – 3.0 m	m	3.08	36.81	13.48	8.45	58.74	82.23	117.47
940205030	Nominal bore 375 mm; in trenches; depth:								
940205030A	not exceeding 1.5 m	m	2.54	30.36	9.13	10.28	49.77	69.68	99.54
940205030B	1.5 – 2.0 m	m	2.86	34.19	11.98	10.44	56.61	79.25	113.21
940205030C	2.0 – 2.5 m	m	3.30	39.44	16.43	10.60	66.47	93.06	132.95
940205030D	2.5 – 3.0 m	m	3.64	43.51	19.65	10.75	73.91	103.48	147.83
940205040	Nominal bore 450 mm; in trenches; depth:								
940205040A	1.5 – 2.0 m	m	2.62	31.32	16.13	12.32	59.77	83.68	119.55
940205040B	2.0 – 2.5 m	m	2.94	35.14	18.98	12.48	66.60	93.25	133.21
940205040C	2.5 – 3.0 m	m	3.38	40.40	23.43	12.64	76.47	107.07	152.95
940205040D	3.0 – 3.5 m	m	3.72	44.46	26.65	12.80	83.91	117.48	167.83

General Civils

Rail 2003 Ref		Unit	Labour Hours	Labour Cost £	Plant Cost £	Materials Cost £	Unit Rate Base Cost £	Unit Rate Green Zone £	Unit Rate Red Zone £
940	**PIPEWORK – PIPES**								
94060	**PLASTIC PIPES**								
9406010	**UPVC pipes; BS 5481; flexible joints**								
940601010	Nominal bore 200 mm; in trenches; depth:								
940601010A	not exceeding 1.5 m	m	1.73	20.68	3.98	31.39	56.05	78.47	112.10
940601010B	1.5 – 2.0 m	m	1.96	23.43	5.26	31.55	60.24	84.32	120.47
940601010C	2.0 – 2.5 m	m	2.20	26.30	7.23	31.71	65.24	91.34	130.47
940601010D	2.5 – 3.0 m	m	2.49	29.76	8.88	31.87	70.51	98.71	141.01
940601020	Nominal bore 250 mm; in trenches; depth:								
940601020A	not exceeding 1.5 m	m	1.93	23.07	5.23	47.33	75.63	105.89	151.27
940601020B	1.5 – 2.0 m	m	2.18	26.06	6.86	47.49	80.41	112.57	160.81
940601020C	2.0 – 2.5 m	m	2.48	29.64	9.53	47.65	86.82	121.56	173.66
940601020D	2.5 – 3.0 m	m	2.75	32.87	11.18	47.81	91.86	128.61	183.71
940601030	Nominal bore 315 mm; in trenches; depth:								
940601030A	not exceeding 1.5 m	m	2.17	25.94	6.48	70.85	103.27	144.57	206.52
940601030B	1.5 – 2.0 m	m	2.44	29.17	8.45	71.01	108.63	152.07	217.24
940601030C	2.0 – 2.5 m	m	2.78	33.23	11.66	71.17	116.06	162.47	232.11
940601030D	2.5 – 3.0 m	m	3.08	36.81	13.48	71.33	121.62	170.27	243.23
940601040	Nominal bore 400 mm; in trenches; depth:								
940601040A	not exceeding 1.5 m	m	2.54	30.36	9.13	111.94	151.43	212.00	302.86
940601040B	1.5 – 2.0 m	m	2.86	34.19	11.98	112.10	158.27	221.57	316.53
940601040C	2.0 – 2.5 m	m	3.30	39.44	16.43	112.25	168.12	235.39	336.27
940601040D	2.5 – 3.0 m	m	3.64	43.51	19.65	112.41	175.57	245.80	351.15
9406050	**UPVC pipes; slotted; BS 5481; flexible joints**								
940605010	Nominal bore 200 mm; in trenches; depth:								
940605010A	not exceeding 1.5 m	m	1.73	20.68	3.98	6.98	31.64	44.30	63.29
940605010B	1.5 – 2.0 m	m	1.96	23.43	5.26	7.14	35.83	50.15	71.66
940605010C	2.0 – 2.5 m	m	2.20	26.30	7.23	7.30	40.83	57.17	81.66
940605010D	2.5 – 3.0 m	m	2.49	29.76	8.88	7.46	46.10	64.54	92.20

General Civils

Rail 2003 Ref		Unit	Labour Hours	Labour Cost £	Plant Cost £	Materials Cost £	Unit Rate Base Cost £	Unit Rate Green Zone £	Unit Rate Red Zone £
940	**PIPEWORK – PIPES**								
9406050	**UPVC pipes; slotted; BS 5481; flexible joints**								
940605020	Nominal bore 250 mm; in trenches; depth:								
940605020A	not exceeding 1.5 m	m	1.73	20.68	3.98	11.28	35.94	50.32	71.88
940605020B	1.5 – 2.0 m	m	1.96	23.43	5.26	11.44	40.13	56.17	80.25
940605020C	2.0 – 2.5 m	m	2.20	26.30	7.23	11.60	45.13	63.18	90.25
940605020D	2.5 – 3.0 m	m	2.49	29.76	8.88	11.76	50.40	70.56	100.79
940605030	Nominal bore 300 mm; in trenches; depth:								
940605030A	not exceeding 1.5 m	m	1.94	23.15	4.90	21.93	49.98	69.97	99.95
940605030B	1.5 – 2.0 m	m	2.53	30.19	6.53	22.29	59.01	82.63	118.04
940605030C	2.0 – 2.5 m	m	3.17	37.84	8.17	22.66	68.67	96.14	137.34
940605030D	2.5 – 3.0 m	m	3.80	45.45	9.80	23.03	78.28	109.58	156.54

General Civils

Rail 2003 Ref		Unit	Labour Hours	Labour Cost £	Plant Cost £	Materials Cost £	Unit Rate Base Cost £	Unit Rate Green Zone £	Unit Rate Red Zone £
943	**PIPEWORK – FITTINGS AND VALVES**								
94330	**PIPE FITTINGS**								
9433010	HepSeal vitrified clay fittings, BS EN295: 1991								
943301010	Bends; size:								
943301010A	100 mm	Nr	0.20	2.39	–	9.21	11.60	16.24	23.19
943301010B	150 mm	Nr	0.20	2.39	–	15.19	17.58	24.61	35.16
943301010C	225 mm	Nr	0.22	2.63	–	31.15	33.78	47.28	67.55
943301010D	300 mm	Nr	0.35	4.18	–	61.43	65.61	91.86	131.22
943301010E	400 mm	Nr	0.45	5.38	–	138.45	143.83	201.36	287.66
943301010F	450 mm	Nr	0.55	6.57	3.50	182.32	192.39	269.35	384.79
943301010G	500 mm	Nr	0.60	7.17	3.50	282.93	293.60	411.04	587.20
943301010H	600 mm	Nr	0.70	8.37	3.50	596.26	608.13	851.37	1216.25
943301020	Junctions; size:								
943301020A	100 × 100 mm	Nr	0.25	2.99	–	12.79	15.78	22.09	31.57
943301020B	150 × 100 mm	Nr	0.25	2.99	–	18.92	21.91	30.67	43.82
943301020C	150 × 150 mm	Nr	0.25	2.99	–	19.83	22.82	31.95	45.65
943301020D	225 × 150 mm	Nr	0.27	3.23	–	47.47	50.70	70.97	101.38
9433020	Concrete fittings, BS 5911								
943302010	Bends; size:								
943302010A	300 mm	Nr	0.60	7.17	3.50	38.39	49.06	68.69	98.12
943302010B	375 mm	Nr	0.75	8.96	3.50	48.00	60.46	84.64	120.92
943302010C	450 mm	Nr	0.85	10.16	3.50	58.25	71.91	100.67	143.82
943302010D	600 mm	Nr	1.00	11.95	4.38	67.85	84.18	117.85	168.35
943302010E	750 mm	Nr	1.20	14.34	4.90	77.45	96.69	135.36	193.37
943302020	Junctions; size:								
943302020A	300 × 300 mm	Nr	0.70	8.37	3.50	84.73	96.60	135.23	193.18
943302020B	300 × 450 mm	Nr	0.80	9.56	3.50	95.32	108.38	151.72	216.75
943302020C	450 × 450 mm	Nr	0.90	10.76	3.50	63.54	77.80	108.92	155.60
943302020D	525 × 450 mm	Nr	1.10	13.15	4.38	73.15	90.68	126.94	181.34
943302020E	525 × 525 mm	Nr	1.30	15.54	4.90	82.74	103.18	144.45	206.35

General Civils

Rail 2003 Ref		Unit	Labour Hours	Labour Cost £	Plant Cost £	Materials Cost £	Unit Rate Base Cost £	Unit Rate Green Zone £	Unit Rate Red Zone £

943 PIPEWORK – FITTINGS AND VALVES

9433030 UPVC fittings; BS 5481; flexible joints

943303010 Bends; size:

943303010A	200 mm	Nr	0.17	2.03	–	6.51	8.54	11.96	17.08
943303010B	250 mm	Nr	0.20	2.39	–	9.30	11.69	16.37	23.39
943303010C	315 mm	Nr	0.20	2.39	–	14.51	16.90	23.67	33.81
943303010D	400 mm	Nr	0.27	3.23	–	36.06	39.29	55.00	78.56
943303010E	450 mm	Nr	0.31	3.71	–	58.54	62.25	87.14	124.49

943303020 Junctions; size:

943303020A	200 × 200 mm	Nr	0.20	2.39	–	9.16	11.55	16.17	23.10
943303020B	250 × 200 mm	Nr	0.25	2.99	–	19.27	22.26	31.16	44.52
943303020C	250 × 250 mm	Nr	0.30	3.59	–	22.64	26.23	36.71	52.45
943303020D	315 × 250 mm	Nr	0.25	2.99	–	19.68	22.67	31.73	45.33
943303020E	315 × 315 mm	Nr	0.35	4.18	–	25.12	29.30	41.02	58.60
943303020F	400 × 315 mm	Nr	0.35	4.18	–	50.31	54.49	76.29	108.98
943303020G	400 × 400 mm	Nr	0.35	4.18	–	51.86	56.04	78.46	112.09
943303020H	450 × 400 mm	Nr	0.37	4.42	–	51.86	56.28	78.79	112.56
943303020I	450 × 450 mm	Nr	0.37	4.42	–	64.60	69.02	96.63	138.05

General Civils

Rail 2003 Ref		Unit	Labour Hours	Labour Cost £	Plant Cost £	Materials Cost £	Unit Rate Base Cost £	Unit Rate Green Zone £	Unit Rate Red Zone £

945	PIPEWORK – MANHOLES AND PIPEWORK ANCILLARIES								
94510	MANHOLES								
9451040	Precast concrete manhole units; complete								
945104010	1050 mm diameter; complete with Grade A cover and frame; depth to invert:								
945104010A	not exceeding 1.5 m	Nr	18.32	226.96	65.93	538.81	831.70	1164.37	1663.38
945104010B	1.5 – 2.0 m	Nr	26.94	329.99	88.68	644.17	1062.84	1487.98	2125.68
945104010C	2.0 – 2.5 m	Nr	31.20	382.15	99.46	743.97	1225.58	1715.82	2451.18
945104020	1200 mm diameter; complete with Grade A cover and frame; depth to invert:								
945104020A	not exceeding 1.5 m	Nr	18.50	229.11	76.60	592.90	898.61	1258.06	1797.23
945104020B	1.5 – 2.0 m	Nr	29.87	365.64	102.13	701.99	1169.76	1637.66	2339.52
945104020C	2.0 – 2.5 m	Nr	34.60	422.80	115.76	804.00	1342.56	1879.57	2685.10
9451050	UPVC inspection chamber units; complete								
945105010	475 mm diameter; standard steel push fit access cover and frame; depth to invert:								
945105010A	not exceeding 1.5 m	Nr	1.80	21.51	–	100.25	121.76	170.45	243.51
94530	SUMP UNITS								
9453010	UPVC sump units; complete								
945301010	375 mm diameter; untrapped; standard steel push fit access cover and frame; depth:								
945301010A	not exceeding 1.5 m	Nr	1.80	21.51	–	30.86	52.37	73.32	104.75
945301020	510 mm diameter; untrapped; standard steel push fit access cover and frame; depth:								
945301020A	not exceeding 1.5 m	Nr	2.25	26.89	–	38.36	65.25	91.35	130.51

General Civils

Rail 2003 Ref		Unit	Labour Hours	Labour Cost £	Plant Cost £	Materials Cost £	Unit Rate Base Cost £	Unit Rate Green Zone £	Unit Rate Red Zone £

945 PIPEWORK – MANHOLES AND PIPEWORK ANCILLARIES

94540 FRENCH AND RUBBLE DRAINS

9454010 Filling to drainage trenches

945401010 Filling French and rubble drains imported materials; type:

Ref	Description	Unit	Hrs	Lab	Plant	Mat	Base	Green	Red
945401010A	stone aggregate; 40 mm nominal	m³	0.75	8.96	–	9.58	18.54	25.96	37.09
945401010B	clean brick rubble	m³	0.85	10.16	–	10.40	20.56	28.78	41.12

9454060 Excavation of drainage trenches

945406010 Excavate for unpiped drains; by machine; cross sectional area:

Ref	Description	Unit	Hrs	Lab	Plant	Mat	Base	Green	Red
945406010A	not exceeding 0.25 m²	m	0.05	0.58	2.00	–	2.58	3.62	5.17
945406010B	0.25 – 0.50 m²	m	0.10	1.16	4.01	–	5.17	7.24	10.35
945406010C	0.50 – 0.75 m²	m	0.15	1.75	6.01	–	7.76	10.86	15.52
945406010D	0.75 – 1.00 m²	m	0.19	2.33	8.02	–	10.35	14.49	20.70
945406010E	1.00 – 1.50 m²	m	0.29	3.49	12.03	–	15.52	21.73	31.03
945406010F	1.50 – 2.00 m²	m	0.28	3.30	16.04	–	19.34	27.07	38.67

945406020 Excavate for unpiped drainage trenches by hand; cross sectional area of trench:

Ref	Description	Unit	Hrs	Lab	Plant	Mat	Base	Green	Red
945406020A	not exceeding 0.25 m²	m	1.26	15.09	1.76	–	16.85	23.60	33.72
945406020B	0.25 – 0.50 m²	m	2.28	27.22	3.53	–	30.75	43.05	61.49
945406020C	0.50 – 0.75 m²	m	3.49	41.74	5.29	–	47.03	65.84	94.06
945406020D	0.75 – 1.00 m²	m	4.88	58.27	7.05	–	65.32	91.45	130.64
945406020E	1.00 – 1.50 m²	m	7.97	95.23	10.58	–	105.81	148.13	211.62
945406020F	1.50 – 2.00 m²	m	11.32	135.27	14.10	–	149.37	209.13	298.76

General Civils

Rail 2003 Ref		Unit	Labour Hours	Labour Cost £	Plant Cost £	Materials Cost £	Unit Rate Base Cost £	Unit Rate Green Zone £	Unit Rate Red Zone £
945	**PIPEWORK – MANHOLES AND PIPEWORK ANCILLARIES**								
94560	**CROSSINGS**								
9456010	HepSeal; vitrified clay pipes; BS EN295: 1991: spigot and socket flexible joints;								
945601010	Nominal bore 225 mm; crossing width:								
945601010A	not exceeding 3.0 m	Nr	5.79	69.21	15.70	52.13	137.04	191.85	274.07
945601010B	3.0 – 5.0 m	Nr	4.65	55.58	26.17	86.88	168.63	236.08	337.25
945601010C	5.0 – 10.0 m	Nr	9.30	111.15	52.33	173.77	337.25	472.15	674.50
945601020	Nominal bore 300 mm; crossing width:								
945601020A	not exceeding 3.0 m	Nr	6.51	77.81	19.43	79.71	176.95	247.74	353.90
945601020B	3.0 – 5.0 m	Nr	10.85	129.69	32.38	132.86	294.93	412.89	589.83
945601020C	5.0 – 10.0 m	Nr	21.70	259.37	64.75	265.72	589.84	825.77	1179.68
94570	**SOAKAWAYS**								
9457010	Precast concrete soakaway units; complete								
945701010	600 mm diameter; complete with Grade A cover and frame; depth to invert:								
945701010A	not exceeding 1.5 m	Nr	15.35	188.98	82.29	817.08	1088.35	1523.69	2176.71
945701010B	1.5 – 2.0 m	Nr	19.47	238.23	100.52	999.11	1337.86	1873.01	2675.73
945701010C	2.0 – 2.5 m	Nr	29.93	363.27	133.63	1213.07	1709.97	2393.95	3419.93
945701010D	2.5 – 3.0 m	Nr	35.22	426.50	152.84	1368.29	1947.63	2726.69	3895.28

General Civils

Rail 2003 Ref		Unit	Labour Hours	Labour Cost £	Plant Cost £	Materials Cost £	Unit Rate Base Cost £	Unit Rate Green Zone £	Unit Rate Red Zone £

948 PIPEWORK – SUPPORTS AND PROTECTION

94830 BEDS, HAUNCHES AND SURROUNDS

9483010 Pipework bedding

948301010 Sand; 100 mm bed; nominal pipe bore:

948301010A	not exceeding 200 mm	m	0.12	1.43	0.15	0.49	2.07	2.90	4.14
948301010B	200 – 300 mm	m	0.16	1.91	0.29	0.74	2.94	4.12	5.88
948301010C	300 – 600 mm	m	0.25	2.99	0.44	1.39	4.82	6.74	9.64

948301015 Sand; 150 mm bed; nominal pipe bore:

948301015A	not exceeding 200 mm	m	0.18	2.15	0.22	0.66	3.03	4.24	6.05
948301015B	200 – 300 mm	m	0.23	2.75	0.36	1.15	4.26	5.97	8.52
948301015C	300 – 600 mm	m	0.37	4.42	0.58	2.05	7.05	9.87	14.10

948301020 Selected excavated granular material; 100 mm bed; nominal pipe bore:

948301020A	not exceeding 200 mm	m	0.12	1.43	0.15	–	1.58	2.21	3.16
948301020B	200 – 300 mm	m	0.16	1.91	0.29	–	2.20	3.09	4.40
948301020C	300 – 600 mm	m	0.25	2.99	0.44	–	3.43	4.79	6.85

948301025 Selected excavated granular material; 150 mm bed; nominal pipe bore:

948301025A	not exceeding 200 mm	m	0.18	2.15	0.22	–	2.37	3.32	4.74
948301025B	200 – 300 mm	m	0.23	2.75	0.36	–	3.11	4.36	6.23
948301025C	300 – 600 mm	m	0.37	4.42	0.58	–	5.00	7.00	10.00

948301030 Imported granular material; 100 mm bed; nominal pipe bore:

948301030A	not exceeding 200 mm	m	0.12	1.43	0.15	0.60	2.18	3.06	4.37
948301030B	200 – 300 mm	m	0.16	1.91	0.29	0.91	3.11	4.36	6.22
948301030C	300 – 600 mm	m	0.25	2.99	0.44	1.72	5.15	7.19	10.28

948301035 Imported granular material; 150 mm bed; nominal pipe bore:

948301035A	not exceeding 200 mm	m	0.18	2.15	0.22	0.81	3.18	4.45	6.35
948301035B	200 – 300 mm	m	0.23	2.75	0.36	1.41	4.52	6.34	9.05
948301035C	300 – 600 mm	m	0.37	4.42	0.58	2.52	7.52	10.53	15.04

General Civils

Rail 2003 Ref		Unit	Labour Hours	Labour Cost £	Plant Cost £	Materials Cost £	Unit Rate Base Cost £	Unit Rate Green Zone £	Unit Rate Red Zone £
948	**PIPEWORK – SUPPORTS AND PROTECTION**								
9483010	**Pipework bedding**								
948301040	Mass concrete; grade C15P; 100 mm bed; nominal pipe bore:								
948301040A	not exceeding 200 mm	m	0.08	0.96	0.29	2.23	3.48	4.87	6.95
948301040B	200 – 300 mm	m	0.13	1.55	0.51	0.45	2.51	3.51	5.02
948301040C	300 – 600 mm	m	0.15	1.79	0.58	4.46	6.83	9.57	13.68
948301040D	600 – 900 mm								
948301045	Mass concrete; grade C15P; 150 mm bed; nominal pipe bore:								
948301045A	not exceeding 200 mm	m	0.12	1.44	0.51	3.13	5.08	7.10	10.14
948301045B	200 – 300 mm	m	0.15	1.79	0.44	5.36	7.59	10.62	15.17
948301045C	300 – 600 mm	m	0.23	2.75	0.87	6.25	9.87	13.82	19.74
948301045D	600 – 900 mm								
948301050	Reinforced concrete; grade C25P; 100 mm bed; nominal pipe bore:								
948301050A	not exceeding 200 mm	m	0.30	3.70	0.29	5.51	9.50	13.31	19.01
948301050B	200 – 300 mm	m	0.34	4.18	0.51	7.47	12.16	17.02	24.32
948301050C	300 – 600 mm	m	0.37	4.54	0.58	9.57	14.69	20.55	29.37
948301050D	600 – 900 mm								
948301055	Reinforced concrete; grade C25P; 150 mm bed; nominal; pipe bore:								
948301055A	not exceeding 200 mm	m	0.35	4.30	0.51	7.07	11.88	16.62	23.75
948301055B	200 – 300 mm	m	0.43	5.26	0.87	10.32	16.45	23.02	32.88
948301055C	300 – 600 mm	m	0.48	5.85	1.02	12.42	19.29	27.00	38.57
9483020	**Pipework bedding and haunching**								
948302010	Sand; 100 mm bed; nominal pipe bore:								
948302010A	not exceeding 200 mm	m	0.18	2.15	0.22	0.66	3.03	4.24	6.05
948302010B	200 – 300 mm	m	0.23	2.75	0.36	1.15	4.26	5.97	8.52
948302010C	300 – 600 mm	m	0.39	4.66	0.73	2.05	7.44	10.42	14.87

General Civils

Rail 2003 Ref		Unit	Labour Hours	Labour Cost £	Plant Cost £	Materials Cost £	Unit Rate Base Cost £	Unit Rate Green Zone £	Unit Rate Red Zone £
948	**PIPEWORK – SUPPORTS AND PROTECTION**								
9483020	**Pipework bedding and haunching**								
948302015	Sand; 150 mm bed; nominal pipe bore:								
948302015A	not exceeding 200 mm	m	0.27	3.23	0.29	1.07	4.59	6.42	9.16
948302015B	200 – 300 mm	m	0.34	4.06	0.51	1.72	6.29	8.81	12.59
948302015C	300 – 600 mm	m	0.56	6.69	0.94	3.11	10.74	15.05	21.51
948302020	Selected excavated granular material; 100 mm bed; nominal pipe bore:								
948302020A	not exceeding 200 mm	m	0.18	2.15	0.22	–	2.37	3.32	4.74
948302020B	200 – 300 mm	m	0.23	2.75	0.36	–	3.11	4.36	6.23
948302020C	300 – 600 mm	m	0.39	4.66	0.73	–	5.39	7.55	10.77
948302025	Selected excavated granular material; 150 mm bed; nominal pipe bore:								
948302025A	not exceeding 200 mm	m	0.27	3.23	0.29	–	3.52	4.93	7.03
948302025B	200 – 300 mm	m	0.34	4.06	0.51	–	4.57	6.40	9.15
948302025C	300 – 600 mm	m	0.56	6.69	0.94	–	7.63	10.69	15.28
948302030	Imported granular material; 100 mm bed; nominal pipe bore:								
948302030A	not exceeding 200 mm	m	0.18	2.15	0.22	0.81	3.18	4.45	6.35
948302030B	200 – 300 mm	m	0.23	2.75	0.36	1.41	4.52	6.34	9.05
948302030C	300 – 600 mm	m	0.38	4.54	0.65	2.52	7.71	10.80	15.43
948302035	Imported granular material; 150 mm bed; nominal pipe bore:								
948302035A	not exceeding 200 mm	m	0.27	3.23	0.29	1.31	4.83	6.77	9.65
948302035B	200 – 300 mm	m	0.34	4.06	0.51	2.12	6.69	9.37	13.39
948302035C	300 – 600 mm	m	0.56	6.69	0.94	3.83	11.46	16.06	22.95
948302040	Mass concrete; grade C15P; 100 mm bed; nominal pipe bore:								
948302040A	not exceeding 200 mm	m	0.11	1.32	0.44	3.13	4.89	6.83	9.75
948302040B	200 – 300 mm	m	0.19	2.27	0.73	0.89	3.89	5.45	7.78
948302040C	300 – 600 mm	m	0.23	2.75	0.87	6.25	9.87	13.82	19.74

General Civils

Rail 2003 Ref		Unit	Labour Hours	Labour Cost £	Plant Cost £	Materials Cost £	Unit Rate Base Cost £	Unit Rate Green Zone £	Unit Rate Red Zone £

948 PIPEWORK – SUPPORTS AND PROTECTION

9483020 Pipework bedding and haunching

948302045 Mass concrete; grade C15P; 150 mm bed; nominal pipe bore:

948302045A	not exceeding 200 mm	m	0.19	2.27	0.80	4.91	7.98	11.17	15.96
948302045B	200 – 300 mm	m	0.23	2.75	0.87	8.04	11.66	16.32	23.31
948302045C	300 – 600 mm	m	0.34	4.06	1.31	9.82	15.19	21.27	30.38

948302050 Reinforced concrete; grade C25P; 100 mm bed; nominal pipe bore:

948302050A	not exceeding 200 mm	m	0.41	5.02	0.73	9.14	14.89	20.83	29.76
948302050B	200 – 300 mm	m	0.53	6.45	1.31	13.94	21.70	30.38	43.39
948302050C	300 – 600 mm	m	0.60	7.29	1.53	17.07	25.89	36.24	51.78

948302055 Reinforced concrete; grade C25P; 150 mm bed; nominal pipe bore:

948302055A	not exceeding 200 mm	m	0.39	4.78	0.65	10.34	15.77	22.08	31.55
948302055B	200 – 300 mm	m	0.50	6.09	1.16	14.37	21.62	30.28	43.25
948302055C	300 – 600 mm	m	0.57	6.93	1.38	19.62	27.93	39.10	55.87

9483030 Pipework surrounds

948303010 Sand; 100 mm cover; nominal pipe bore:

948303010A	not exceeding 200 mm	m	0.32	3.82	0.29	1.31	5.42	7.60	10.85
948303010B	200 – 300 mm	m	0.44	5.26	0.73	2.13	8.12	11.36	16.23
948303010C	300 – 600 mm	m	0.71	8.49	1.31	3.85	13.65	19.10	27.28

948303015 Sand; 150 mm cover; nominal pipe bore:

948303015A	not exceeding 200 mm	m	0.45	5.38	0.51	1.72	7.61	10.65	15.22
948303015B	200 – 300 mm	m	0.57	6.81	0.87	2.87	10.55	14.78	21.10
948303015C	300 – 600 mm	m	0.92	11.00	1.53	5.16	17.69	24.76	35.37

948303020 Selected excavated granular material; 100 mm cover; nominal pipe bore:

948303020A	not exceeding 200 mm	m	0.34	4.06	0.44	–	4.50	6.30	9.00
948303020B	200 – 300 mm	m	0.44	5.26	0.73	–	5.99	8.38	11.97
948303020C	300 – 600 mm	m	0.71	8.49	1.31	–	9.80	13.71	19.58

General Civils

Rail 2003 Ref		Unit	Labour Hours	Labour Cost £	Plant Cost £	Materials Cost £	Unit Rate Base Cost £	Unit Rate Green Zone £	Unit Rate Red Zone £

948 PIPEWORK – SUPPORTS AND PROTECTION

9483030 Pipework surrounds

948303025 Selected excavated granular material; 150 mm cover; nominal pipe bore:

948303025A	not exceeding 200 mm	m	0.45	5.38	0.51	–	5.89	8.24	11.78
948303025B	200 – 300 mm	m	0.57	6.81	0.87	–	7.68	10.76	15.36
948303025C	300 – 600 mm	m	0.92	11.00	1.53	–	12.53	17.53	25.04

948303030 Imported granular material; 100 mm cover; nominal pipe bore:

948303030A	not exceeding 200 mm	m	0.34	4.06	0.44	1.61	6.11	8.56	12.23
948303030B	200 – 300 mm	m	0.44	5.26	0.73	2.62	8.61	12.05	17.21
948303030C	300 – 600 mm	m	0.71	8.49	1.31	4.74	14.54	20.35	29.06

948303035 Imported granular material; 150 mm cover; nominal pipe bore:

948303035A	not exceeding 200 mm	m	0.45	5.38	0.51	2.12	8.01	11.21	16.02
948303035B	200 – 300 mm	m	0.57	6.81	0.87	3.53	11.21	15.70	22.42
948303035C	300 – 600 mm	m	0.92	11.00	1.53	6.35	18.88	26.43	37.75

948303040 Mass concrete; grade C15P; 100 mm cover; nominal pipe bore:

948303040A	not exceeding 200 mm	m	0.34	4.06	0.94	10.71	15.71	22.01	31.45
948303040B	200 – 300 mm	m	0.52	6.22	1.82	13.84	21.88	30.61	43.74
948303040C	300 – 600 mm	m	1.08	12.91	4.50	25.44	42.85	59.99	85.71

948303045 Mass concrete; grade C15P; 150 mm cover; nominal pipe bore:

948303045A	not exceeding 200 mm	m	0.38	4.54	1.02	12.50	18.06	25.28	36.11
948303045B	200 – 300 mm	m	0.58	6.93	2.03	15.62	24.58	34.42	49.18
948303045C	300 – 600 mm	m	1.19	14.22	4.94	28.12	47.28	66.19	94.56

948303050 Reinforced concrete; grade C25P; 100 mm cover; nominal pipe bore:

948303050A	not exceeding 200 mm	m	0.56	6.81	0.94	19.19	26.94	37.71	53.89
948303050B	200 – 300 mm	m	0.74	8.96	1.82	23.02	33.80	47.32	67.59
948303050C	300 – 600 mm	m	1.30	15.65	4.50	37.58	57.73	80.83	115.47

General Civils

Rail 2003 Ref		Unit	Labour Hours	Labour Cost £	Plant Cost £	Materials Cost £	Unit Rate Base Cost £	Unit Rate Green Zone £	Unit Rate Red Zone £
948	**PIPEWORK – SUPPORTS AND PROTECTION**								
9483030	Pipework surrounds								
948303055	Reinforced concrete; grade C25P; 150 mm cover; nominal pipe bore:								
948303055A	not exceeding 200 mm	m	0.60	7.29	1.02	21.94	30.25	42.33	60.48
948303055B	200 – 300 mm	m	0.74	8.96	1.82	26.45	37.23	52.12	74.45
948303055C	300 – 600 mm	m	1.41	16.97	4.94	40.90	62.81	87.93	125.60

General Civils

Rail 2003 Ref	Unit	Composite Cost £	Unit Rate Base Cost £	Unit Rate Green Zone £	Unit Rate Red Zone £

950 STRUCTURAL METALWORK

95010 FABRICATION OF MAIN MEMBERS

9501010 Framing fabrication of main members; bolted connections; hot rolled steel; B4360 Grade 43A; including shop and site black bolts; rolled sections

950101005 Beams; straight on plan; weight:

Ref	Description	Unit				Base	Green	Red
950101005A	not exceeding 50 kg/m	Tonne	–	–	–	2043.75	2861.25	4087.50
950101005B	50 – 150 kg/m	Tonne	–	–	–	1816.67	2543.34	3633.34
950101005C	exceeding 150 kg/m	Tonne	–	–	–	1725.84	2416.18	3451.68

950101010 Beams; curved on plan; weight:

950101010A	not exceeding 50 kg/m	Tonne	–	–	–	2861.25	4005.75	5722.50
950101010B	50 – 150 kg/m	Tonne	–	–	–	2543.33	3560.66	5086.66
950101010C	exceeding 150 kg/m	Tonne	–	–	–	2416.16	3382.62	4832.32

950101015 Beams; straight on plan and cambered; weight:

950101015A	not exceeding 50 kg/m	Tonne	–	–	–	2861.25	4005.75	5722.50
950101015B	50 – 150 kg/m	Tonne	–	–	–	2543.33	3560.66	5086.66
950101015C	exceeding 150 kg/m	Tonne	–	–	–	2416.16	3382.62	4832.32

950101020 Beams; curved on plan and cambered; weight:

950101020A	not exceeding 50 kg/m	Tonne	–	–	–	3809.21	5332.89	7618.42
950101020B	50 – 150 kg/m	Tonne	–	–	–	3385.96	4740.34	6771.92
950101020C	exceeding 150 kg/m	Tonne	–	–	–	3216.66	4503.32	6433.32

950101025 Columns; straight on plan; weight:

950101025A	not exceeding 50 kg/m	Tonne	–	–	–	1869.82	2617.75	3739.64
950101025B	50 – 150 kg/m	Tonne	–	–	–	1662.06	2326.88	3324.12
950101025C	exceeding 150 kg/m	Tonne	–	–	–	1579.96	2211.94	3159.92

950101030 Columns; curved on plan; weight:

950101030A	not exceeding 50 kg/m	Tonne	–	–	–	2617.74	3664.84	5235.48
950101030B	50 – 150 kg/m	Tonne	–	–	–	2326.88	3257.63	4653.76
950101030C	exceeding 150 kg/m	Tonne	–	–	–	2210.54	3094.76	4421.08

General Civils

Rail 2003 Ref		Unit	Composite Cost £	Unit Rate Base Cost £	Unit Rate Green Zone £	Unit Rate Red Zone £
950	**STRUCTURAL METALWORK**					
9501010	Framing fabrication of main members; bolted connections; hot rolled steel; B4360 Grade 43A; including shop and site black bolts; rolled sections					
950101035	Columns; straight on plan and cambered; weight:					
950101035A	not exceeding 50 kg/m.............	Tonne	– – –	2617.74	2617.74	3664.84	5235.48
950101035B	50 – 150 kg/m....................	Tonne	– – –	2326.88	2326.88	3257.63	4653.76
950101035C	exceeding 150 kg/m...............	Tonne	– – –	2210.54	2210.54	3094.76	4421.08
950101040	Columns; curved on plan and cambered; weight:					
950101040A	not exceeding 50 kg/m.............	Tonne	– – –	3478.73	3478.73	4870.22	6957.46
950101040B	50 – 150 kg/m....................	Tonne	– – –	3092.20	3092.20	4329.08	6184.40
950101040C	exceeding 150 kg/m...............	Tonne	– – –	2937.59	2937.59	4112.63	5875.18
950101045	Box and hollow sections; straight on plan; weight:					
950101045A	not exceeding 50 kg/m.............	Tonne	– – –	2130.72	2130.72	2983.01	4261.44
950101045B	50 – 150 kg/m....................	Tonne	– – –	1893.97	1893.97	2651.56	3787.94
950101045C	exceeding 150 kg/m...............	Tonne	– – –	1799.27	1799.27	2518.98	3598.54
950101050	Box and hollow sections; curved on plan; weight:					
950101050A	not exceeding 50 kg/m.............	Tonne	– – –	3965.75	3965.75	5552.05	7931.50
950101050B	50 – 150 kg/m....................	Tonne	– – –	3525.11	3525.11	4935.15	7050.22
950101050C	exceeding 150 kg/m...............	Tonne	– – –	3348.85	3348.85	4688.39	6697.70
950101055	Box and hollow sections; straight on plan and cambered; weight:					
950101055A	not exceeding 50 kg/m.............	Tonne	– – –	2983.01	2983.01	4176.21	5966.02
950101055B	50 – 150 kg/m....................	Tonne	– – –	2651.56	2651.56	3712.18	5303.12
950101055C	exceeding 150 kg/m...............	Tonne	– – –	2518.98	2518.98	3526.57	5037.96

General Civils

Rail 2003 Ref		Unit	Composite Cost £	Unit Rate Base Cost £	Unit Rate Green Zone £	Unit Rate Red Zone £

950 STRUCTURAL METALWORK

9501010 Framing fabrication of main members; bolted connections; hot rolled steel; B4360 Grade 43A; including shop and site black bolts; rolled sections

950101060 Box and hollow sections; curved on plan and cambered; weight:

950101060A	not exceeding 50 kg/m.............	Tonne	– – –	3965.75	3965.75	5552.05	7931.50
950101060B	50 – 150 kg/m....................	Tonne	– – –	3525.11	3525.11	4935.15	7050.22
950101060C	exceeding 150 kg/m...............	Tonne	– – –	3348.85	3348.85	4688.39	6697.70

95060 FABRICATION OF SECONDARY MEMBERS

9506010 Framing fabrication of secondary members; bolted connections; hot rolled steel; BS4360 Grade 43A; including shop and site black bolts

950601005 Bracings; straight on plan; weight:

950601005A	not exceeding 50 kg/m.............	Tonne	– – –	2287.26	2287.26	3202.16	4574.52
950601005B	50 – 150 kg/m....................	Tonne	– – –	2033.12	2033.12	2846.37	4066.24
950601005C	exceeding 150 kg/m...............	Tonne	– – –	1931.46	1931.46	2704.04	3862.92

950601010 Bracings; curved on plan; weight:

950601010A	not exceeding 50 kg/m.............	Tonne	– – –	3200.42	3200.42	4480.59	6400.84
950601010B	50 – 150 kg/m....................	Tonne	– – –	2844.82	2844.82	3982.75	5689.64
950601010C	exceeding 150 kg/m...............	Tonne	– – –	2702.58	2702.58	3783.61	5405.16

950601015 Bracings; straight on plan and cambered; weight:

950601015A	not exceeding 50 kg/m.............	Tonne	– – –	3200.42	3200.42	4480.59	6400.84
950601015B	50 – 150 kg/m....................	Tonne	– – –	2844.82	2844.82	3982.75	5689.64
950601015C	exceeding 150 kg/m...............	Tonne	– – –	2702.58	2702.58	3783.61	5405.16

950601020 Bracings; curved on plan and cambered; weight:

950601020A	not exceeding 50 kg/m.............	Tonne	– – –	4261.44	4261.44	5966.02	8522.88
950601020B	50 – 150 kg/m....................	Tonne	– – –	3787.95	3787.95	5303.13	7575.90
950601020C	exceeding 150 kg/m...............	Tonne	– – –	3787.95	3787.95	5303.13	7575.90

Rail 2003 Ref		Unit	Composite Cost £	Unit Rate Base Cost £	Unit Rate Green Zone £	Unit Rate Red Zone £

950 STRUCTURAL METALWORK

9506010 Framing fabrication of secondary members; bolted connections; hot rolled steel; BS4360 Grade 43A; including shop and site black bolts

950601025 Purlins; straight on plan; weight:

950601025A	not exceeding 50 kg/m.............	Tonne	– – –	2287.26	3202.16	4574.52
950601025B	50 – 150 kg/m....................	Tonne	– – –	2033.12	2846.37	4066.24
950601025C	exceeding 150 kg/m...............	Tonne	– – –	1931.46	2704.04	3862.92

950601030 Purlins; curved on plan; weight:

950601030A	not exceeding 50 kg/m.............	Tonne	– – –	4261.44	5966.02	8522.88
950601030B	50 – 150 kg/m....................	Tonne	– – –	3787.95	5303.13	7575.90
950601030C	exceeding 150 kg/m...............	Tonne	– – –	3598.55	5037.97	7197.10

950601035 Purlins; straight on plan and cambered; weight:

950601035A	not exceeding 50 kg/m.............	Tonne	– – –	3200.42	4480.59	6400.84
950601035B	50 – 150 kg/m....................	Tonne	– – –	2844.82	3982.75	5689.64
950601035C	exceeding 150 kg/m...............	Tonne	– – –	2702.58	3783.61	5405.16

950601040 Purlins; curved on plan and cambered; weight:

950601040A	not exceeding 50 kg/m.............	Tonne	– – –	4261.44	5966.02	8522.88
950601040B	50 – 150 kg/m....................	Tonne	– – –	3787.95	5303.13	7575.90
950601040C	exceeding 150 kg/m...............	Tonne	– – –	3585.55	5019.77	7171.10

950601045 Channels; straight on plan; weight:

950601045A	not exceeding 50 kg/m.............	Tonne	– – –	2174.21	3043.89	4348.42
950601045B	50 – 150 kg/m....................	Tonne	– – –	1932.63	2705.68	3865.26
950601045C	exceeding 150 kg/m...............	Tonne	– – –	1835.00	2569.00	3670.00

950601050 Channels; curved on plan; weight:

950601050A	not exceeding 50 kg/m.............	Tonne	– – –	4044.02	5661.63	8088.04
950601050B	50 – 150 kg/m....................	Tonne	– – –	3594.68	5032.55	7189.36
950601050C	exceeding 150 kg/m...............	Tonne	– – –	3414.95	4780.93	6829.90

General Civils

Rail 2003 Ref		Unit	Composite Cost £	Unit Rate Base Cost £	Unit Rate Green Zone £	Unit Rate Red Zone £
950	**STRUCTURAL METALWORK**					
9506010	Framing fabrication of secondary members; bolted connections; hot rolled steel; BS4360 Grade 43A; including shop and site black bolts					
950601055	Channels; straight on plan and cambered; weight:					
950601055A	not exceeding 50 kg/m	Tonne	– – –	3043.89	4261.45	6087.78
950601055B	50 – 150 kg/m	Tonne	– – –	2705.68	3787.95	5411.36
950601055C	exceeding 150 kg/m	Tonne	– – –	2570.40	3598.56	5140.80
950601060	Channels; curved on plan and cambered; weight:					
950601060A	not exceeding 50 kg/m	Tonne	– – –	4044.02	5661.63	8088.04
950601060B	50 – 150 kg/m	Tonne	– – –	3594.68	5032.55	7189.36
950601060C	exceeding 150 kg/m	Tonne	– – –	3414.95	4780.93	6829.90
950601065	Flats; weight:					
950601065A	not exceeding 25 kg/m	Tonne	– – –	2287.26	3202.16	4574.52
950601065B	25 – 50 kg/m	Tonne	– – –	2033.12	2846.37	4066.24
950601065C	exceeding 50 kg/m	Tonne	– – –	1931.46	2704.04	3862.92
950601070	Plates; weight:					
950601070A	not exceeding 50 kg/m^2	Tonne	– – –	2287.26	3202.16	4574.52
950601070B	50 – 150 kg/m^2	Tonne	– – –	2033.12	2846.37	4066.24
950601070C	exceeding 150 kg/m^2	Tonne	– – –	1931.46	2704.04	3862.92

General Civils

Rail 2003 Ref	Unit	Composite Cost £	Unit Rate Base Cost £	Unit Rate Green Zone £	Unit Rate Red Zone £
950	**STRUCTURAL METALWORK**				
95070	**ERECTION OF FABRICATED MEMBERS**				
9507010	Trial erection at fabrication works; subsequent dismantling				
950701010	Main members:				
950701010A	not exceeding 50 kg/m............ Tonne	–	–	–	339.09	339.09	474.73	678.18
950701010B	50 – 150 kg/m.................... Tonne	–	–	–	301.41	301.41	421.97	602.82
950701010C	exceeding 150 kg/m............... Tonne	–	–	–	286.34	286.34	400.88	572.68
950701060	Secondary members:				
950701060A	not exceeding 50 kg/m............ Tonne	–	–	–	304.39	304.39	426.15	608.78
950701060B	50 – 150 kg/m.................... Tonne	–	–	–	270.57	270.57	378.80	541.14
950701060C	exceeding 150 kg/m............... Tonne	–	–	–	257.04	257.04	359.86	514.08
9507030	Permanent erection; including supply of suitable lifting arrangements; setting baseplates on holding down bolt assemblies				
950703010	Main members:				
950703010A	not exceeding 50 kg/m............ Tonne	–	–	–	426.40	426.40	596.96	852.80
950703010B	50 – 150 kg/m.................... Tonne	–	–	–	378.79	378.79	530.31	757.58
950703010C	exceeding 150 kg/m............... Tonne	–	–	–	359.85	359.85	503.79	719.70
950703060	Secondary members:				
950703060A	not exceeding 50 kg/m............ Tonne	–	–	–	382.76	382.76	535.86	765.52
950703060B	50 – 150 kg/m.................... Tonne	–	–	–	340.14	340.14	476.20	680.28
950703060C	exceeding 150 kg/m............... Tonne	–	–	–	323.13	323.13	452.38	646.26

Rail 2003 Ref		Unit				Specialist Cost	Unit Rate	Unit Rate	Unit Rate
						£	Base Cost £	Green Zone £	Red Zone £
950	**STRUCTURAL METALWORK**								
95090	**OFF-SITE SURFACE TREATMENT**								
9509010	**Surface preparation**								
950901010	**Blast cleaning:**								
950901010A	grit	m^2	–	–	–	4.10	4.10	5.74	8.20
950901010B	shot	m^2	–	–	–	4.10	4.10	5.74	8.20
950901010C	water	m^2	–	–	–	3.79	3.79	5.31	7.58
950901020	**Flame cleaning:**								
950901020A	generally	m^2	–	–	–	5.50	5.50	7.70	11.00
950901030	**Pickling:**								
950901030A	generally	m^2	–	–	–	7.70	7.70	10.78	15.40
950901040	**Mechanical cleaning:**								
950901040A	wire brushing	m^2	–	–	–	6.35	6.35	8.89	12.70
9509020	**Surface pre-treatment**								
950902010	**Wetting oils:**								
950902010A	generally	m^2	–	–	–	3.35	3.35	4.69	6.70
950902020	**Wash primers:**								
950902020A	generally	m^2	–	–	–	3.15	3.15	4.41	6.30
950902030	**Cold phosphates:**								
950902030A	generally	m^2	–	–	–	7.93	7.93	11.10	15.86
950902040	**Hot phosphates:**								
950902040A	generally	m^2	–	–	–	9.35	9.35	13.09	18.70

General Civils

Rail 2003 Ref		Unit	Specialist Cost £	Unit Rate Base Cost £	Unit Rate Green Zone £	Unit Rate Red Zone £
950	**STRUCTURAL METALWORK**					
9509030	**Surface treatment**					
950903010	**Metal spraying:**					
950903010A	generally	m²	– – –	12.91	12.91	18.07	25.82
950903060	**Galvanising:**					
950903060A	generally	m²	– – –	14.75	14.75	20.65	29.50
950903070	**Priming:**					
950903070A	single coat work	m²	– – –	4.85	4.85	6.79	9.70

General Civils

Rail 2003 Ref		Unit	Specialist Cost £	Unit Rate Base Cost £	Unit Rate Green Zone £	Unit Rate Red Zone £
960	**PILES**					
96010	**BORED CAST IN PLACE CONCRETE PILES**					
9601005	Establishment of piling plant					
960100501	Establishment of piling plant for bored cast in place concrete piles; in main piling at general locations; pile diameter:					
960100501A	300 and 350 mm.................	item	–	–	–	6675.00	6675.00	9345.00	13350.00
960100501B	400 and 450 mm.................	item	–	–	–	6720.00	6720.00	9408.00	13440.00
960100501C	500 and 550 mm.................	item	–	–	–	6800.00	6800.00	9520.00	13600.00
960100501D	600 and 750 mm.................	item	–	–	–	6800.00	6800.00	9520.00	13600.00
960100501E	900 and 1050 mm	item	–	–	–	6850.00	6850.00	9590.00	13700.00
960100501F	1200 and 1350 mm	item	–	–	–	6900.00	6900.00	9660.00	13800.00
960100501G	1500 mm......................	item	–	–	–	7025.00	7025.00	9835.00	14050.00
9601010	Bored cast in place concrete piles; reinforced concrete grade C25; 20 mm aggregate; vertical plane					
960101010	**300 mm diameter:**					
960101010A	number of piles; not exceeding 50 nr..	Nr	–	–	–	187.50	187.50	262.50	375.00
960101010B	number of piles; over 50 nr..........	Nr	–	–	–	150.00	150.00	210.00	300.00
960101010C	concreted length...................	m	–	–	–	7.50	7.50	10.50	15.00
960101010D	bored depth 10 m.................	m	–	–	–	18.00	18.00	25.20	36.00
960101010E	bored depth 15 m.................	m	–	–	–	18.00	18.00	25.20	36.00
960101010F	bored depth 20 m.................	m	–	–	–	19.50	19.50	27.30	39.00
960101010G	bored depth 25 m.................	m	–	–	–	20.25	20.25	28.35	40.50
960101020	**450 mm diameter:**					
960101020A	number of piles; not exceeding 50 nr..	Nr	–	–	–	190.00	190.00	266.00	380.00
960101020B	number of piles; over 50 nr..........	Nr	–	–	–	155.00	155.00	217.00	310.00
960101020C	concreted length...................	m	–	–	–	16.75	16.75	23.45	33.50
960101020D	depth bored 10 m.................	m	–	–	–	21.25	21.25	29.75	42.50
960101020E	depth bored 15 m.................	m	–	–	–	21.75	21.75	30.45	43.50
960101020F	bored depth 20 m.................	m	–	–	–	22.60	22.60	31.64	45.20
960101020G	bored depth 25 m.................	m	–	–	–	23.50	23.50	32.90	47.00

General Civils

Rail 2003 Ref	Unit	Specialist Cost £	Unit Rate Base Cost £	Unit Rate Green Zone £	Unit Rate Red Zone £

960 PILES

9601010 Bored cast in place concrete piles; reinforced concrete grade C25; 20 mm aggregate; vertical plane

960101030 600 mm diameter:

Ref	Description	Unit				Base	Green	Red
960101030A	number of piles; not exceeding 50 nr	Nr	–	–	–	195.00	195.00	273.00	390.00
960101030B	number of piles; over 50 nr	Nr	–	–	–	157.50	157.50	220.50	315.00
960101030C	concreted length	m	–	–	–	22.00	22.00	30.80	44.00
960101030D	bored depth 10 m	m	–	–	–	25.00	25.00	35.00	50.00
960101030E	bored depth 15 m	m	–	–	–	25.00	25.00	35.00	50.00
960101030F	bored depth 20 m	m	–	–	–	25.50	25.50	35.70	51.00
960101030G	bored depth 25 m	m	–	–	–	26.00	26.00	36.40	52.00

960101040 900 mm diameter:

960101040A	number of piles; not exceeding 50 nr	Nr	–	–	–	200.00	200.00	280.00	400.00
960101040B	number of piles; over 50 nr	Nr	–	–	–	165.00	165.00	231.00	330.00
960101040C	concreted length	m	–	–	–	23.50	23.50	32.90	47.00
960101040D	bored depth 10 m	m	–	–	–	26.75	26.75	37.45	53.50
960101040E	bored depth 15 m	m	–	–	–	27.25	27.25	38.15	54.50
960101040F	bored depth 20 m	m	–	–	–	28.50	28.50	39.90	57.00
960101040G	bored depth 25 m	m	–	–	–	29.00	29.00	40.60	58.00

960101050 1200 mm diameter:

960101050A	number of piles; not exceeding 50 nr	Nr	–	–	–	210.00	210.00	294.00	420.00
960101050B	number of piles; over 50 nr	Nr	–	–	–	173.50	173.50	242.90	347.00
960101050C	concreted length	m	–	–	–	24.75	24.75	34.65	49.50
960101050D	bored depth 10 m	m	–	–	–	27.50	27.50	38.50	55.00
960101050E	bored depth 15 m	m	–	–	–	29.00	29.00	40.60	58.00
960101050F	bored depth 20 m	m	–	–	–	29.75	29.75	41.65	59.50
960101050G	bored depth 25 m	m	–	–	–	30.50	30.50	42.70	61.00

960101060 1500 mm diameter:

960101060A	number of piles; not exceeding 50 nr	Nr	–	–	–	215.00	215.00	301.00	430.00
960101060B	number of piles; over 50 nr	Nr	–	–	–	180.00	180.00	252.00	360.00
960101060C	concreted length	m	–	–	–	25.00	25.00	35.00	50.00
960101060D	bored depth 10 m	m	–	–	–	28.00	28.00	39.20	56.00
960101060E	bored depth 15 m	m	–	–	–	29.50	29.50	41.30	59.00
960101060F	bored depth 20 m	m	–	–	–	30.00	30.00	42.00	60.00
960101060G	bored depth 25 m	m	–	–	–	31.00	31.00	43.40	62.00

General Civils

Rail 2003 Ref		Unit	Specialist Cost £	Unit Rate Base Cost £	Unit Rate Green Zone £	Unit Rate Red Zone £
960	**PILES**					
96050	**ISOLATED STEEL PILES**					
960500510	Establishment and setting up of piling plant for isolated steel piles; in main piling; at general locations:					
960500510A	provision and establishment.........	Item	– – –	10130.00	10130.00	14182.00	20260.00
960500510B	setting up at each position...........	Nr	– – –	300.00	300.00	420.00	600.00
9605010	**Isolated steel bearing piles; EN 10025 Grade S275; mechanically driven**					
960501010	Mass not exceeding 15 kg/m:					
960501010A	number of piles; not exceeding 5 m long	Nr	– – –	133.50	133.50	186.90	267.00
960501010B	number of piles; 6 – 10 m long.......	Nr	– – –	205.85	205.85	288.19	411.70
960501010C	number of piles; 11 – 15 m long......	Nr	– – –	274.50	274.50	384.30	549.00
960501010D	number of piles; 16 – 20 m long......	Nr	– – –	345.00	345.00	483.00	690.00
960501010E	number of piles; exceeding 20 m long.	Nr	– – –	402.25	402.25	563.15	804.50
960501010F	vertical driven pile	m	– – –	11.75	11.75	16.45	23.50
960501010G	raking driven pile	m	– – –	14.25	14.25	19.95	28.50
960501015	Mass 16 – 30 kg/m:					
960501015A	number of piles; not exceeding 5 m long	Nr	– – –	163.00	163.00	228.20	326.00
960501015B	number of piles; 6 – 10 m long.......	Nr	– – –	252.00	252.00	352.80	504.00
960501015C	number of piles; 11 – 15 m long......	Nr	– – –	318.50	318.50	445.90	637.00
960501015D	number of piles; 16 – 20 m long......	Nr	– – –	427.25	427.25	598.15	854.50
960501015E	number of piles; exceeding 20 m long.	Nr	– – –	501.00	501.00	701.40	1002.00
960501015F	vertical driven pile	m	– – –	11.75	11.75	16.45	23.50
960501015G	raking driven pile	m	– – –	14.25	14.25	19.95	28.50
960501020	Mass 31 – 60 kg/m:					
960501020A	number of piles; not exceeding 5 m long	Nr	– – –	193.25	193.25	270.55	386.50
960501020B	number of piles; 6 – 10 m long.......	Nr	– – –	298.00	298.00	417.20	596.00
960501020C	number of piles; 11 – 15 m long......	Nr	– – –	397.50	397.50	556.50	795.00
960501020D	number of piles; 16 – 20 m long......	Nr	– – –	483.00	483.00	676.20	966.00
960501020E	number of piles; exceeding 20 m long.	Nr	– – –	546.25	546.25	764.75	1092.50
960501020F	vertical driven pile	m	– – –	11.75	11.75	16.45	23.50
960501020G	raking driven pile	m	– – –	14.25	14.25	19.95	28.50

General Civils

Rail 2003 Ref		Unit	Specialist Cost £	Unit Rate Base Cost £	Unit Rate Green Zone £	Unit Rate Red Zone £
960	**PILES**					
9605010	**Isolated steel bearing piles; EN 10025 Grade S275; mechanically driven**					
960501025	Mass 61 – 120 kg/m:					
960501025A	number of piles; not exceeding 5 m long	Nr	–	–	–	212.25	212.25	297.15	424.50
960501025B	number of piles; 6 – 10 m long	Nr	–	–	–	289.00	289.00	404.60	578.00
960501025C	number of piles; 11 – 15 m long	Nr	–	–	–	326.25	326.25	456.75	652.50
960501025D	number of piles; 16 – 20 m long	Nr	–	–	–	451.00	451.00	631.40	902.00
960501025E	number of piles; exceeding 20 m long	Nr	–	–	–	567.75	567.75	794.85	1135.50
960501025F	vertical driven pile	m	–	–	–	12.35	12.35	17.29	24.70
960501025G	raking driven pile	m	–	–	–	14.90	14.90	20.86	29.80
960501030	Mass 121 – 250 kg/m:					
960501030A	number of piles; not exceeding 5 m long	Nr	–	–	–	241.00	241.00	337.40	482.00
960501030B	number of piles; 6 – 10 m long	Nr	–	–	–	316.25	316.25	442.75	632.50
960501030C	number of piles; 11 – 15 m long	Nr	–	–	–	384.00	384.00	537.60	768.00
960501030D	number of piles; 16 – 20 m long	Nr	–	–	–	507.00	507.00	709.80	1014.00
960501030E	number of piles; exceeding 20 m long	Nr	–	–	–	601.00	601.00	841.40	1202.00
960501030F	vertical driven pile	m	–	–	–	12.90	12.90	18.06	25.80
960501030G	raking driven pile	m	–	–	–	15.25	15.25	21.35	30.50
960501035	Mass 251 – 500 kg/m:					
960501035A	number of piles; not exceeding 5 m long	Nr	–	–	–	270.00	270.00	378.00	540.00
960501035B	number of piles; 6 – 10 m long	Nr	–	–	–	342.25	342.25	479.15	684.50
960501035C	number of piles; 11 – 15 m long	Nr	–	–	–	391.70	391.70	548.38	783.40
960501035D	number of piles; 16 – 20 m long	Nr	–	–	–	511.00	511.00	715.40	1022.00
960501035E	number of piles; exceeding 20 m long	Nr	–	–	–	607.50	607.50	850.50	1215.00
960501035F	vertical driven pile	m	–	–	–	13.25	13.25	18.55	26.50
960501035G	raking driven pile	m	–	–	–	15.75	15.75	22.05	31.50
960501040	Mass exceeding 500 kg/m:					
960501040A	number of piles; not exceeding 5 m long	Nr	–	–	–	299.10	299.10	418.74	598.20
960501040B	number of piles; 6 – 10 m long	Nr	–	–	–	371.00	371.00	519.40	742.00
960501040C	number of piles; 11 – 15 m long	Nr	–	–	–	420.75	420.75	589.05	841.50
960501040D	number of piles; 16 – 20 m long	Nr	–	–	–	531.70	531.70	744.38	1063.40
960501040E	number of piles; exceeding 20 m long	Nr	–	–	–	613.43	613.43	858.80	1226.86
960501040F	vertical driven pile	m	–	–	–	13.50	13.50	18.90	27.00
960501040G	raking driven pile	m	–	–	–	16.25	16.25	22.75	32.50

General Civils

Rail 2003 Ref		Unit	Specialist Cost £	Unit Rate Base Cost £	Unit Rate Green Zone £	Unit Rate Red Zone £
965	**PILING ANCILLARIES**					
96510	**CAST IN PLACE CONCRETE PILES**					
9651020	**Backfilling empty bore**					
965102010	Backfilling with material arising from excavations; diameter of bore:					
965102010A	300 mm	m	–	–	–	1.75	1.75	2.45	3.50
965102010B	450 mm	m	–	–	–	1.90	1.90	2.66	3.80
965102010C	600 mm	m	–	–	–	2.00	2.00	2.80	4.00
965102010D	750 mm	m	–	–	–	3.12	3.12	4.37	6.24
965102010E	900 mm	m	–	–	–	4.57	4.57	6.40	9.14
965102010F	1050 mm	m	–	–	–	6.12	6.12	8.57	12.24
965102010G	1200 mm	m	–	–	–	8.00	8.00	11.20	16.00
965102010H	1350 mm	m	–	–	–	10.12	10.12	14.17	20.24
965102010I	1500 mm	m	–	–	–	12.48	12.48	17.47	24.96
965102020	Backfilling with sand; diameter of bore:					
965102020A	300 mm	m	–	–	–	2.60	2.60	3.64	5.20
965102020B	450 mm	m	–	–	–	3.81	3.81	5.33	7.62
965102020C	600 mm	m	–	–	–	5.40	5.40	7.56	10.80
965102020D	750 mm	m	–	–	–	8.43	8.43	11.80	16.86
965102020E	900 mm	m	–	–	–	12.13	12.13	16.98	24.26
965102020F	1050 mm	m	–	–	–	16.52	16.52	23.13	33.04
965102020G	1200 mm	m	–	–	–	21.58	21.58	30.21	43.16
965102020H	1350 mm	m	–	–	–	27.32	27.32	38.25	54.64
965102020I	1500 mm	m	–	–	–	33.72	33.72	47.21	67.44
965102030	Backfilling with concrete; diameter of bore:					
965102030A	300 mm	m	–	–	–	5.15	5.15	7.21	10.30
965102030B	450 mm	m	–	–	–	9.54	9.54	13.36	19.08
965102030C	600 mm	m	–	–	–	15.60	15.60	21.84	31.20

General Civils

Rail 2003 Ref		Unit			Specialist Cost £	Unit Rate Base Cost £	Unit Rate Green Zone £	Unit Rate Red Zone £
965	**PILING ANCILLARIES**							
9651070	**Cutting off surplus lengths**							
965107010	Cut off top of pile including necessary trimming prior to pile cap casting; pile diameter:							
965107010A	300 mm	m	–	–	–	19.00	19.00	26.60	38.00
965107010B	450 mm	m	–	–	–	24.50	24.50	34.30	49.00
965107010C	600 mm	m	–	–	–	32.00	32.00	44.80	64.00
965107010D	750 mm	m	–	–	–	48.00	48.00	67.20	96.00
965107010E	900 mm	m	–	–	–	66.00	66.00	92.40	132.00
965107010F	1050 mm	m	–	–	–	86.00	86.00	120.40	172.00
965107010G	1200 mm	m	–	–	–	96.00	96.00	134.40	192.00
965107010H	1350 mm	m	–	–	–	110.00	110.00	154.00	220.00
965107010I	1500 mm	m	–	–	–	128.00	128.00	179.20	256.00
965107010J	1650 mm	m	–	–	–	146.00	146.00	204.40	292.00

Rail 2003 Ref		Unit	Labour Hours	Labour Cost £	Plant Cost £	Materials Cost £	Unit Rate Base Cost £	Unit Rate Green Zone £	Unit Rate Red Zone £
9651090	**Reinforcement**								
965109010	Straight bars; nominal size:								
965109010A	not exceeding 25 mm	Tonne	26.00	324.48	–	365.37	689.85	965.79	1379.70
965109010B	over 25 mm	Tonne	26.00	324.48	–	352.86	677.34	948.28	1354.68
965109020	Helical bars; nominal size:								
965109020A	not exceeding 25 mm	Tonne	26.00	324.48	–	385.86	710.34	994.47	1420.67
965109020B	over 25 mm	Tonne	26.00	324.48	–	367.32	691.80	968.52	1383.61

Rail 2003 Ref		Unit			Specialist Cost £	Unit Rate Base Cost £	Unit Rate Green Zone £	Unit Rate Red Zone £
96580	**PILE TESTS**							
9658010	**Testing piles**							
965801010	Non working pile:							
965801010A	to one and a half times working load	Nr	–	–	–	1350.00	1350.00	1890.00	2700.00
965801010B	to twice working load	Nr	–	–	–	1800.00	1800.00	2520.00	3600.00
965801020	Working pile							
965801020A	to one and a half times working load	Nr	–	–	–	1687.50	1687.50	2362.50	3375.00
965801020B	to twice working load	Nr	–	–	–	2250.00	2250.00	3150.00	4500.00

General Civils

Rail 2003 Ref		Unit	Labour Hours	Labour Cost £	Plant Cost £	Materials Cost £	Unit Rate Base Cost £	Unit Rate Green Zone £	Unit Rate Red Zone £

970 ROADS & PAVING

97010 SUB-BASES

9701010 DTp Specified Type Sub-bases

970101010 Granular material; Type 1; levelling and compacting with mechanical plant; depth:

970101010A	100 mm	m²	0.07	0.84	0.93	1.41	3.18	4.46	6.36
970101010B	150 mm	m²	0.09	1.08	1.30	2.02	4.40	6.15	8.78
970101010C	200 mm	m²	0.10	1.20	1.66	2.72	5.58	7.80	11.16
970101010D	250 mm	m²	0.15	1.79	1.88	3.43	7.10	9.94	14.20
970101010E	300 mm	m²	0.15	1.79	2.23	4.14	8.16	11.42	16.31
970101010F	400 mm	m²	0.20	2.39	2.79	5.45	10.63	14.88	21.25

970101020 Granular material; Type 2; levelling and compacting with mechanical plant; depth:

970101020A	100 mm	m²	0.07	0.84	0.93	1.37	3.14	4.40	6.28
970101020B	150 mm	m²	0.09	1.08	1.30	1.96	4.34	6.07	8.66
970101020C	200 mm	m²	0.10	1.20	1.66	2.64	5.50	7.69	10.99
970101020D	250 mm	m²	0.15	1.79	1.88	3.32	6.99	9.79	13.99
970101020E	300 mm	m²	0.15	1.79	2.23	4.01	8.03	11.24	16.06
970101020F	400 mm	m²	0.20	2.39	2.79	5.28	10.46	14.65	20.92

970101030 Cement bound granular material; 100 kg dry OPC per m³ granular sub-base; depth:

970101030A	75 mm	m²	0.07	0.84	0.93	7.22	8.99	12.59	17.98
970101030B	100 mm	m²	0.07	0.84	0.93	7.22	8.99	12.59	17.98
970101030C	150 mm	m²	0.09	1.08	1.30	10.32	12.70	17.77	25.38
970101030D	200 mm	m²	0.10	1.20	1.66	13.93	16.79	23.49	33.56

970101040 Lean concrete; Grade 10; 20 mm aggregate; depth:

970101040A	75 mm	m²	0.06	0.72	1.28	3.16	5.16	7.22	10.31
970101040B	100 mm	m²	0.07	0.84	1.46	4.35	6.65	9.30	13.29
970101040C	150 mm	m²	0.09	1.08	1.83	6.33	9.24	12.93	18.46
970101040D	200 mm	m²	0.11	1.32	2.19	8.31	11.82	16.54	23.62

General Civils

Rail 2003 Ref		Unit	Labour Hours	Labour Cost	Plant Cost	Materials Cost	Unit Rate	Unit Rate	Unit Rate
				£	£	£	Base Cost £	Green Zone £	Red Zone £
970	**ROADS & PAVING**								
9701010	**DTp Specified Type Sub-bases**								
970101050	Soil cement; 100 kg dry OPC per m^3; depth:								
970101050A	75 mm	m^2	0.06	0.72	1.28	2.75	4.75	6.64	9.48
970101050B	100 mm	m^2	0.07	0.84	1.46	4.35	6.65	9.30	13.29
970101050C	150 mm	m^2	0.09	1.08	1.83	6.33	9.24	12.93	18.46
970101050D	200 mm	m^2	0.11	1.32	2.19	8.31	11.82	16.54	23.62
9701030	**Hardcore Sub-bases**								
970103010	Clean imported fill; levelling and compacting with mechanical plant; depth:								
970103010A	100 mm	m^2	0.07	0.84	1.11	1.42	3.37	4.71	6.73
970103010B	150 mm	m^2	0.10	1.20	1.48	2.18	4.86	6.80	9.71
970103010C	200 mm	m^2	0.11	1.32	1.84	2.84	6.00	8.39	11.98
970103010D	250 mm	m^2	0.14	1.67	2.04	3.59	7.30	10.23	14.62
970103010E	300 mm	m^2	0.17	2.03	2.42	4.26	8.71	12.18	17.40
970103010F	400 mm	m^2	0.22	2.63	3.16	5.67	11.46	16.04	22.93
9701050	**Ground Stabilising**								
970105010	Terram 4000 geotextile stabilising matting; 150 mm laps; laid:								
970105010A	horizontally	m^2	0.02	0.24	–	2.23	2.47	3.45	4.94

Rail 2003 Ref		Unit				Specialist Cost	Unit Rate	Unit Rate	Unit Rate
						£	Base Cost £	Green Zone £	Red Zone £
9701060	**Flexible Road Bases**								
970106010	Wet mix macadam; laid in single course; depth:								
970106010A	75 mm	m^2	–	–	–	2.45	2.45	3.43	4.90
970106010B	100 mm	m^2	–	–	–	3.20	3.20	4.48	6.40
970106010C	150 mm	m^2	–	–	–	4.65	4.65	6.51	9.30
970106010D	200 mm	m^2	–	–	–	6.25	6.25	8.75	12.50

General Civils

Rail 2003 Ref		Unit	Specialist Cost £	Unit Rate Base Cost £	Unit Rate Green Zone £	Unit Rate Red Zone £
970	**ROADS & PAVING**					
9701060	**Flexible Road Bases**					
970106020	Dry bound macadam; laid in single course; depth:					
970106020A	75 mm	m^2	– – –	2.35	3.29	4.70
970106020B	100 mm	m^2	– – –	3.15	4.41	6.30
970106020C	150 mm	m^2	– – –	4.60	6.44	9.20
970106020D	200 mm	m^2	– – –	6.00	8.40	12.00
97020	**FLEXIBLE ROAD SURFACINGS**					
9702070	**Road Surfacing**					
970207010	Coated macadam, BS 4987; hand laid surfacing:					
970207010A	60 mm thickness of 20 mm aggregate dense base course, 30 mm thickness of 10 mm aggregate close graded wearing course	m^2	– – –	9.91	13.87	19.82
970207010B	50 mm thickness of 20 mm aggregate open graded base course, 20 mm thickness of 6 mm aggregate medium graded wearing course	m^2	– – –	8.78	12.29	17.56
970207010C	50 mm thickness of 20 mm aggregate dense base course, 20 mm thickness of 6 mm aggregate dense wearing course	m^2	– – –	8.15	11.41	16.30
970207010D	45 mm thickness of 20 mm aggregate open graded base course, 15 mm thickness of fine graded wearing course	m^2	– – –	7.03	9.84	14.06
970207020	Coated macadam, BS 4987; machine laid surfacing:					
970207020A	50 mm thickness of 20 mm aggregate dense base course	m^2	– – –	5.21	7.29	10.42
970207020B	100 mm thickness of 40 mm aggregate dense base course	m^2	– – –	8.57	12.00	17.14
970207020C	80 mm thickness of 28 mm aggregate dense base course, 40 mm thickness of 14 mm aggregate close graded wearing course	m^2	– – –	11.22	15.71	22.44
970207020D	60 mm thickness of 20 mm aggregate dense base course, 30 mm thickness of 10 mm aggregate close graded wearing course	m^2	– – –	9.27	12.98	18.54
970207020E	45 mm thickness of 20 mm aggregate dense base course, 20 mm thickness of 6 mm aggregate dense wearing course	m^2	– – –	8.25	11.55	16.50

General Civils

Rail 2003 Ref		Unit	Specialist Cost £	Unit Rate Base Cost £	Unit Rate Green Zone £	Unit Rate Red Zone £
970	**ROADS & PAVING**					
9702070	**Road Surfacing**					
970207020	Coated macadam, BS 4987; machine laid surfacing:					
970207020F	surface dress with cut-back bitumen K1-70 emulsion and 10 mm aggregate single creasing	m²	–	1.46	2.04	2.92
970207020G	surface dress with cut-back bitumen K1-70 emulsion and 10 mm aggregate double creasing	m²	–	2.13	2.98	4.26
970207020H	surface dress with cut-back bitumen K1-70 emulsion and 6 mm aggregate single creasing	m²	–	1.68	2.35	3.36
970207020I	surface dress with cut-back bitumen K1-70 emulsion and 6 mm aggregate double creasing	m²	–	2.52	3.53	5.04

Rail 2003 Ref		Unit	Labour Hours	Labour Cost £	Plant Cost £	Materials Cost £	Unit Rate Base Cost £	Unit Rate Green Zone £	Unit Rate Red Zone £
97040	**CONCRETE PAVEMENTS**								
9704010	**Plain in situ concrete; mix C25P**								
970401010	Beds; poured on or against earth or unblinded hardcore; depth:								
970401010A	100 mm	m²	0.13	1.55	–	5.70	7.25	10.15	14.50
970401010B	150 mm	m²	0.19	2.27	–	8.29	10.56	14.78	21.11
970401010C	200 mm	m²	0.22	2.63	–	10.87	13.50	18.90	27.01
970401010D	250 mm	m²	0.28	3.35	–	13.98	17.33	24.26	34.65
970401010E	300 mm	m²	0.33	3.95	–	16.57	20.52	28.72	41.03
9704020	**Reinforced in situ concrete; mix C25P**								
970402020	Beds; poured on or against earth or unblinded hardcore; depth:								
970402020A	100 mm	m²	0.19	2.30	0.20	5.70	8.20	11.46	16.37
970402020B	150 mm	m²	0.29	3.51	0.29	8.29	12.09	16.92	24.17
970402020C	200 mm	m²	0.34	4.11	0.39	10.87	15.37	21.53	30.75
970402020D	250 mm	m²	0.43	5.20	0.49	13.98	19.67	27.54	39.34
970402020E	300 mm	m²	0.51	6.17	0.59	16.57	23.33	32.65	46.65

General Civils

Rail 2003 Ref		Unit	Labour Hours	Labour Cost	Plant Cost	Materials Cost	Unit Rate	Unit Rate	Unit Rate
				£	£	£	Base Cost £	Green Zone £	Red Zone £
970	**ROADS & PAVING**								
9704030	Reinforced in situ concrete; mix C35P								
970403010	Beds; poured on or against earth or unblinded hardcore; depth:								
970403010A	100 mm	m²	0.19	2.30	0.20	6.37	8.87	12.40	17.71
970403010B	150 mm	m²	0.29	3.51	0.29	9.26	13.06	18.28	26.12
970403010C	200 mm	m²	0.34	4.11	0.39	12.15	16.65	23.32	33.31
970403010D	250 mm	m²	0.43	5.20	0.49	15.63	21.32	29.85	42.63
970403010E	300 mm	m²	0.51	6.17	0.59	18.52	25.28	35.38	50.55
9704040	**Reinforcement**								
970404005	Steel bar reinforcement; high yield, BS 4449; nominal size:								
970404005A	8 mm	Tonne	72.00	898.56	–	464.35	1362.91	1908.06	2725.81
970404005B	10 mm	Tonne	54.00	673.92	–	452.08	1126.00	1576.40	2252.00
970404005C	12 mm	Tonne	44.00	549.12	–	420.39	969.51	1357.32	1939.03
970404005D	16 mm	Tonne	38.00	474.24	–	402.59	876.83	1227.56	1753.65
970404005E	20 mm	Tonne	30.00	374.40	–	376.83	751.23	1051.72	1502.45
970404005F	25 mm	Tonne	26.00	324.48	–	365.37	689.85	965.79	1379.70
970404005G	32 mm	Tonne	23.00	287.04	–	354.01	641.05	897.47	1282.10
970404005H	40 mm	Tonne	20.00	249.60	–	347.57	597.17	836.03	1194.33
970404010	Steel fabric reinforcement, BS 4483; fixed horizontally; ref:								
970404010A	A142	m²	0.03	0.37	–	1.82	2.19	3.07	4.39
970404010B	A193	m²	0.03	0.37	–	2.42	2.79	3.91	5.60
970404010C	A252	m²	0.03	0.37	–	3.13	3.50	4.90	7.00
970404010D	A393	m²	0.05	0.62	–	4.91	5.53	7.75	11.08
970404010E	B196	m²	0.03	0.37	–	3.13	3.50	4.90	7.00
970404010F	B283	m²	0.04	0.50	–	3.74	4.24	5.94	8.49
970404010G	B385	m²	0.05	0.62	–	4.76	5.38	7.54	10.78
970404010H	C283	m²	0.05	0.62	–	6.10	6.72	9.40	13.44
970404010I	C385	m²	0.03	0.37	–	2.88	3.25	4.55	6.51
970404010J	C503	m²	0.04	0.50	–	3.55	4.05	5.67	8.10

General Civils

Rail 2003 Ref		Unit	Labour Hours	Labour Cost £	Plant Cost £	Materials Cost £	Unit Rate Base Cost £	Unit Rate Green Zone £	Unit Rate Red Zone £
970	**ROADS & PAVING**								
97050	**JOINTS IN CONCRETE PAVEMENTS**								
9705010	**Designed joints in in situ concrete**								
970501010	Horizontal plain joints; depth:								
970501010A	100 mm	m	0.68	9.33	–	1.30	10.63	14.87	21.25
970501010B	150 mm	m	0.72	9.88	–	1.66	11.54	16.16	23.08
970501010C	200 mm	m	0.75	10.29	–	2.52	12.81	17.93	25.62
970501010D	300 mm	m	0.82	11.25	–	3.63	14.88	20.84	29.77
970501020	Horizontal keyed joints; depth:								
970501020A	100 mm	m	0.72	9.88	–	1.43	11.31	15.83	22.62
970501020B	150 mm	m	0.77	10.57	–	1.83	12.40	17.35	24.79
970501020C	200 mm	m	0.80	10.98	–	2.85	13.83	19.36	27.65
970501020D	300 mm	m	0.88	12.08	–	3.88	15.96	22.35	31.92
970501030	Flexcell fibreboard compressible joint filler; 10 mm thick:								
970501030A	100 mm	m	0.08	1.10	–	0.83	1.93	2.70	3.86
970501030B	150 mm	m	0.10	1.37	–	1.21	2.58	3.61	5.16
970501030C	200 mm	m	0.13	1.78	–	1.59	3.37	4.72	6.75
970501030D	300 mm	m	0.18	2.47	–	2.35	4.82	6.74	9.63
970501040	Flexcell fibreboard compressible joint filler; 13 mm thick:								
970501040A	100 mm	m	0.09	1.24	–	1.11	2.35	3.28	4.69
970501040B	150 mm	m	0.11	1.51	–	1.61	3.12	4.37	6.25
970501040C	200 mm	m	0.14	1.92	–	2.12	4.04	5.66	8.08
970501040D	300 mm	m	0.18	2.47	–	3.13	5.60	7.84	11.19
970501050	Flexcell fibreboard compressible joint filler; 19 mm thick:								
970501050A	100 mm	m	0.14	1.92	–	1.39	3.31	4.63	6.61
970501050B	150 mm	m	0.16	2.20	–	2.02	4.22	5.89	8.42
970501050C	200 mm	m	0.17	2.33	–	2.65	4.98	6.98	9.97
970501050D	300 mm	m	0.22	3.02	–	3.91	6.93	9.70	13.86

General Civils

Rail 2003 Ref		Unit	Labour Hours	Labour Cost £	Plant Cost £	Materials Cost £	Unit Rate Base Cost £	Unit Rate Green Zone £	Unit Rate Red Zone £
970	**ROADS & PAVING**								
9705010	Designed joints in in situ concrete								
970501060	Flexcell fibreboard compressible joint filler; 25 mm thick:								
970501060A	100 mm	m	0.17	2.33	–	1.66	3.99	5.60	8.00
970501060B	150 mm	m	0.20	2.75	–	2.42	5.17	7.23	10.33
970501060C	200 mm	m	0.21	2.88	–	3.18	6.06	8.48	12.11
970501060D	300 mm	m	0.24	3.29	–	4.69	7.98	11.18	15.97
970501070	Hot poured bituminous rubber compound joint sealant:								
970501070A	10 × 25 mm	m	0.03	0.41	0.14	1.16	1.71	2.39	3.42
970501070B	13 × 25 mm	m	0.03	0.41	0.14	1.50	2.05	2.87	4.10
970501070C	19 × 25 mm	m	0.04	0.55	0.17	2.19	2.91	4.08	5.84
970501070D	25 × 25 mm	m	0.06	0.82	0.21	2.93	3.96	5.54	7.92
970501080	Cold poured polysulphide rubber compound joint sealant:								
970501080A	10 × 25 mm	m	0.03	0.41	–	0.40	0.81	1.14	1.63
970501080B	13 × 25 mm	m	0.03	0.41	–	0.56	0.97	1.36	1.93
970501080C	19 × 25 mm	m	0.04	0.55	–	0.76	1.31	1.83	2.61
970501080D	25 × 25 mm	m	0.06	0.82	–	1.26	2.08	2.92	4.17
97060	**KERBS, EDGINGS & CHANNELS**								
9706010	Precast Concrete Kerbs, Edgings and Channels								
970601010	Kerbs; 125 × 255 mm; bedding and pointing in cement mortar (1:3); haunching with 20 N/mm^2 concrete:								
970601010A	Fig. 2, 5, 7, 8	m	0.37	4.42	0.58	6.61	11.61	16.26	23.22
970601010B	Fig. 2, 5, 7, 8; curved work	m	0.45	5.38	0.58	6.83	12.79	17.91	25.59
970601010C	Fig. 2a, 7a, 8, 9	m	0.37	4.42	0.58	6.61	11.61	16.26	23.22
970601010D	Fig. 2a, 7a, 8, 9; curved work	m	0.45	5.38	0.58	6.93	12.89	18.05	25.79

General Civils

Rail 2003 Ref		Unit	Labour Hours	Labour Cost £	Plant Cost £	Materials Cost £	Unit Rate Base Cost £	Unit Rate Green Zone £	Unit Rate Red Zone £

970 ROADS & PAVING

9706010 Precast Concrete Kerbs, Edgings and Channels

970601020 Kerbs; 150 × 305 mm; bedding and pointing in cement mortar (1:3); haunching with 20 N/mm² concrete:

970601020A	Fig. 1, 4, 6, 8	m	0.39	4.66	0.58	9.69	14.93	20.91	29.86
970601020B	Fig. 1, 4, 6, 8; curved work	m	0.47	5.62	0.58	10.04	16.24	22.73	32.48

970601030 Extra for:

970601030A	droppers; Fig. 16; 255 to 155 mm high; 125 mm wide	Nr	–	–	–	2.28	2.28	3.19	4.56
970601030B	droppers; Fig. 16; 305 to 205 mm high; 150 mm wide	Nr	–	–	–	2.93	2.93	4.10	5.85
970601030C	quadrants; Fig. 14; 450 × 450 × 250 mm.............................	Nr	–	–	–	7.01	7.01	9.82	14.02

970601050 Edgings; 50 × 150 mm; bedding and pointing in cement mortar (1:3); haunching with 10 N/mm² concrete:

970601050A	Fig. 11, 12, 13	m	0.20	2.39	0.58	2.76	5.73	8.02	11.45
970601050B	Fig. 11, 12, 13; curved work	m	0.25	2.99	0.58	2.87	6.44	9.01	12.88

970601060 Channels; 125 × 255 mm; bedding and pointing in cement mortar (1:3); haunching with 20 N/mm² cement:

970601060A	Fig. 2, 8	m	0.33	3.94	0.58	5.53	10.05	14.07	20.10
970601060B	Fig. 2, 8; curved work...............	m	0.42	5.02	0.58	5.68	11.28	15.79	22.56

970601070 Channels; 150 × 305 mm; bedding and pointing in cement mortar (1:3); haunching with 20 N/mm² cement:

970601070A	Fig. 1, 8	m	0.35	4.18	0.58	5.49	10.25	14.36	20.52
970601070B	Fig. 1, 8; curved work...............	m	0.44	5.26	0.58	5.71	11.55	16.17	23.11

General Civils

Rail 2003 Ref		Unit	Labour Hours	Labour Cost £	Plant Cost £	Materials Cost £	Unit Rate Base Cost £	Unit Rate Green Zone £	Unit Rate Red Zone £

970 ROADS & PAVING

9706010 Precast Concrete Kerbs, Edgings and Channels

970601080 Safeticurb surface water drainage kerbs; 165 × 165 mm Junior Block; bedding and pointing in cement mortar (1:3); haunching with 10 N/mm^2 concrete:

970601080A	76 mm bore....................	m	0.45	5.38	–	16.19	21.57	30.19	43.13
970601080B	76 mm bore; curved work	m	0.60	7.17	–	16.88	24.05	33.67	48.10

970601085 Safeticurb surface water drainage kerbs; 248 × 248 mm DBA/1; bedding and pointing in cement mortar (1:3); haunching with 10 N/mm^2 concrete:

970601085A	127 mm bore....................	m	0.58	6.93	–	20.29	27.22	38.10	54.43
970601085B	127 mm; curved work	m	0.70	8.37	–	21.02	29.39	41.13	58.76
970601085C	silt box........................	Nr	1.80	21.51	–	126.53	148.04	207.25	296.07
970601085D	inspection unit	Nr	0.65	7.77	–	30.43	38.20	53.48	76.40

97070 LIGHT DUTY PAVINGS

9707010 Light Duty Surfacings

970701010 Pea shingle dressing; spread and levelled:

970701010A	25 mm thick	m^2	0.10	1.20	0.15	0.30	1.65	2.29	3.29
970701010B	50 mm thick	m^2	0.12	1.43	0.29	0.71	2.43	3.41	4.86

970701020 Precast concrete flag pavings; bed in cement mortar (1:3), point in lime mortar (1:1:6):

970701020A	600 × 450 × 50 mm thick..........	m^2	0.51	6.10	–	13.54	19.64	27.48	39.27
970701020B	600 × 600 × 50 mm thick..........	m^2	0.43	5.14	–	12.71	17.85	24.98	35.70
970701020C	600 × 750 × 50 mm thick..........	m^2	0.39	4.66	–	11.90	16.56	23.19	33.12
970701020D	600 × 900 × 50 mm thick..........	m^2	0.35	4.18	–	11.26	15.44	21.63	30.89
970701020E	600 × 450 × 63 mm thick..........	m^2	0.52	6.21	–	14.45	20.66	28.93	41.34
970701020F	600 × 600 × 63 mm thick..........	m^2	0.44	5.26	–	13.54	18.80	26.31	37.60
970701020G	600 × 750 × 63 mm thick..........	m^2	0.40	4.78	–	12.65	17.43	24.40	34.85
970701020H	600 × 900 × 63 mm thick..........	m^2	0.36	4.30	–	11.95	16.25	22.75	32.50

General Civils

Rail 2003 Ref		Unit	Labour Hours	Labour Cost £	Plant Cost £	Materials Cost £	Unit Rate Base Cost £	Unit Rate Green Zone £	Unit Rate Red Zone £
970	**ROADS & PAVING**								
9707010	**Light Duty Surfacings**								
970701040	Brick pavings; bed and point in cement mortar (1:3):								
970701040A	brick on flat; half bond or 90 degree herringbone	m²	1.07	14.69	–	12.70	27.39	38.33	54.76
970701040B	brick on flat; parquet bond or 45 degree herringbone	m²	1.12	15.37	0.14	12.70	28.21	39.49	56.41
970701040C	brick on edge; half bond or 90 degree herringbone	m²	1.39	19.08	–	17.37	36.45	51.02	72.88
970701040D	brick on edge; parquet bond or 45 degree herringbone	m²	1.44	19.76	0.14	19.37	39.27	54.98	78.54
970701060	Concrete block paving; 200 × 100 mm chamfered blocks; laid and vibrator compacted on 50 mm sand bed:								
970701060A	65 mm thick; half bond or 90 degree herringbone	m²	1.11	15.23	0.07	9.14	24.44	34.22	48.89
970701060B	65 mm thick; parquet bond or 45 degree herringbone	m²	1.21	16.61	0.21	9.48	26.30	36.81	52.57
970701060C	80 mm thick; half bond or 90 degree herringbone	m²	1.11	15.23	0.21	10.70	26.14	36.59	52.27
970701060D	80 mm thick; parquet bond or 45 degree herringbone	m²	1.21	16.61	0.21	10.70	27.52	38.51	55.01
970701080	Granite sett paving; bedding in cement mortar (1:3) on concrete base (measured separately); laid:								
970701080A	100 mm thick; level and to falls	m²	1.75	34.46	–	43.56	78.02	109.22	156.03

General Civils

Rail 2003 Ref	Unit	Specialist Cost £	Unit Rate Base Cost £	Unit Rate Green Zone £	Unit Rate Red Zone £
970	**ROADS & PAVING**				
97080	**ANCILLARIES**				
9708010	**Road Signage**				
970801010	Non-illuminated traffic signs; post mounted; including excavation, disposal and backfilling; general signs:				
970801010A	small.................... Nr	– – –	495.00	495.00	693.00	990.00
970801010B	medium.................. Nr	– – –	570.00	570.00	798.00	1140.00
970801010C	large.................... Nr	– – –	650.00	650.00	910.00	1300.00
9708050	**Surface Markings**				
970805010	Letters and shapes; to tarmadam and asphalt:				
970805010A	direction arrow; 6 m high........... Nr	– – –	20.00	20.00	28.00	40.00
970805010B	'Give Way' triangle................. Nr	– – –	15.00	15.00	21.00	30.00
970805020	Lines; continuous; to tarmac and asphalt; width:				
970805020A	100 mm......................... m	– – –	0.75	0.75	1.05	1.50
970805020B	200 mm......................... m	– – –	1.00	1.00	1.40	2.00
970805030	Lines; intermittent; to tarmac and asphalt; width:				
970805030A	100 mm......................... m	– – –	0.70	0.70	0.98	1.40
970805030B	200 mm......................... m	– – –	0.93	0.93	1.30	1.86

General Civils

Rail 2003 Ref		Unit	Labour Hours	Labour Cost £	Plant Cost £	Materials Cost £	Unit Rate Base Cost £	Unit Rate Green Zone £	Unit Rate Red Zone £

980 MASONRY

98031 Brickwork and Blockwork

9803101 Walls

980310101 Precast concrete blocks, BS 6073, strength 7 N/mm^2, in cement mortar (1:3):

980310101A	100 mm solid block skins	m^2	0.57	11.22	–	7.27	18.49	25.88	36.98
980310101B	215 mm hollow block skins	m^2	0.63	12.40	–	10.56	22.96	32.16	45.93
980310101C	200 mm solid block skins	m^2	0.71	13.98	–	14.41	28.39	39.74	56.77
980310101D	215 mm solid block skins	m^2	0.72	14.18	–	15.85	30.03	42.05	60.06

980310121 Architectural blocks, in gauged mortar (1:1:6); flush pointing all around:

980310121A	100 mm solid; PC £30:00 per m^2	m^2	0.89	17.52	–	32.14	49.66	69.53	99.34
980310121B	220 mm solid; PC £65:00 per m^2	m^2	1.12	22.05	–	68.89	90.94	127.32	181.89

980310141 Facing bricks, PC £250:00 per 1000, in gauged mortar (1:1:6); flush pointing all around:

980310141A	105 mm	m^2	0.79	15.55	–	16.29	31.84	44.58	63.69
980310141B	215 mm	m^2	1.58	31.11	–	33.86	64.97	90.96	129.95

980310161 Class A engineering bricks, BS 3921, in cement mortar (1:3):

980310161A	half brick thick	m^2	0.73	14.37	–	31.00	45.37	63.52	90.75
980310161B	one brick thick	m^2	1.47	28.94	–	62.67	91.61	128.26	183.23
980310161C	one and a half brick thick	m^2	2.19	43.12	–	98.61	141.73	198.43	283.46
980310161D	two brick thick	m^2	2.93	57.69	–	129.61	187.30	262.22	374.60

980310166 Class B engineering bricks, BS 3921, in cement mortar (1:3):

980310166A	half brick thick	m^2	0.71	13.98	–	21.11	35.09	49.13	70.19
980310166B	one brick thick	m^2	1.41	27.76	–	42.90	70.66	98.93	141.31
980310166C	one and a half brick thick	m^2	2.12	41.74	–	67.31	109.05	152.67	218.09
980310166D	two brick thick	m^2	2.82	55.52	–	88.42	143.94	201.52	287.89

General Civils

Rail 2003 Ref		Unit	Labour Hours	Labour Cost	Plant Cost	Materials Cost	Unit Rate	Unit Rate	Unit Rate
				£	£	£	Base Cost £	Green Zone £	Red Zone £

980 MASONRY

9803101 Walls

980310171 Common bricks, BS 3921, in cement mortar (1:3):

980310171A	half brick thick.....................	m^2	0.68	13.39	–	11.62	25.01	35.01	50.03
980310171B	one brick thick.....................	m^2	1.35	26.58	–	24.59	51.17	71.63	102.34
980310171C	one and a half brick thick............	m^2	2.03	39.97	–	37.92	77.89	109.05	155.79
980310171D	two brick thick.....................	m^2	2.71	53.36	–	50.89	104.25	145.94	208.49

9803111 Sloping walls

980311101 Precast concrete blocks, BS 6073, strength 7 N/mm^2, in cement mortar (1:3):

980311101A	100 mm solid block skins	m^2	0.63	12.40	–	7.27	19.67	27.54	39.34
980311101B	150 mm solid block skins	m^2	0.69	13.59	–	10.56	24.15	33.81	48.29
980311101C	200 mm solid block skins	m^2	0.78	15.36	–	14.41	29.77	41.67	59.52
980311101D	100 mm hollow block skins..........	m^2	0.79	15.55	–	15.85	31.40	43.98	62.82

980311121 Architectural blocks, in gauged mortar (1:1:6); flush pointing all round:

980311121A	100 mm solid; PC £30:00 per m^2	m^2	0.98	19.30	–	32.14	51.44	72.01	102.88
980311121B	220 mm solid; PC £65:00 per m^2	m^2	1.23	24.22	–	68.89	93.11	130.35	186.22

980311141 Facing bricks, PC £250:00 per 1000, in gauged mortar (1:1:6); flush pointing all round:

980311141A	half brick thick.....................	m^2	0.87	17.13	–	16.29	33.42	46.78	66.84
980311141B	one brick thick.....................	m^2	1.74	34.26	–	33.86	68.12	95.37	136.25

980311161 Class A engineering bricks, BS 3921, in cement mortar (1:3):

980311161A	half brick thick.....................	m^2	0.81	15.95	–	31.00	46.95	65.73	93.90
980311161B	one brick thick.....................	m^2	1.61	31.70	–	62.67	94.37	132.12	188.74
980311161C	one and a half brick thick............	m^2	2.42	47.65	–	98.61	146.26	204.77	292.51
980311161D	two brick thick.....................	m^2	3.22	63.40	–	129.61	193.01	270.22	386.02

General Civils

Rail 2003 Ref		Unit	Labour Hours	Labour Cost £	Plant Cost £	Materials Cost £	Unit Rate Base Cost £	Unit Rate Green Zone £	Unit Rate Red Zone £
980	**MASONRY**								
9803111	**Sloping walls**								
980311166	Class B engineering bricks, BS 3921, in cement mortar (1:3):								
980311166A	half brick thick.....................	m²	0.78	15.36	–	21.11	36.47	51.06	72.94
980311166B	one brick thick.....................	m²	1.55	30.52	–	42.90	73.42	102.79	146.83
980311166C	one and a half brick thick............	m²	2.33	45.88	–	67.31	113.19	158.46	226.36
980311166D	two brick thick.....................	m²	3.10	61.04	–	88.42	149.46	209.24	298.91
980311171	Common bricks, BS 3921, in cement mortar (1:3):								
980311171A	half brick thick.....................	m²	0.75	14.77	–	11.62	26.39	36.94	52.78
980311171B	one brick thick.....................	m²	1.49	29.34	–	24.59	53.93	75.49	107.85
980311171C	one and a half brick thick............	m²	2.24	44.10	–	37.92	82.02	114.83	164.06
980311171D	two brick thick.....................	m²	2.98	58.67	–	50.89	109.56	153.38	219.13
9803121	**Battering walls**								
980312101	Precast concrete blocks, BS 6073, strength 7 N/mm², in cement mortar (1:3):								
980312101A	100 mm solid block skins	m²	0.69	13.59	–	7.27	20.86	29.19	41.70
980312101B	150 mm solid block skins	m²	0.76	14.96	–	10.56	25.52	35.74	51.05
980312101C	200 mm solid block skins	m²	0.86	16.93	–	14.41	31.34	43.88	62.68
980312101D	100 mm hollow block skins..........	m²	0.87	17.13	–	15.85	32.98	46.18	65.97
980312121	Architectural blocks, in gauged mortar (1:1:6); flush pointing all round:								
980312121A	100 mm solid; PC £30:00 per m²	m²	1.07	21.07	–	32.14	53.21	74.49	106.42
980312121B	220 mm solid; PC £65:00 per m²	m²	1.35	26.58	–	68.89	95.47	133.66	190.95
980312141	Facing bricks, PC £250:00 per 1000, in gauged mortar (1:1:6); flush pointing all round:								
980312141A	105 mm	m²	0.96	18.90	–	16.29	35.19	49.26	70.38
980312141B	215 mm	m²	1.92	37.80	–	33.86	71.66	100.33	143.34

General Civils

Rail 2003 Ref		Unit	Labour Hours	Labour Cost £	Plant Cost £	Materials Cost £	Unit Rate Base Cost £	Unit Rate Green Zone £	Unit Rate Red Zone £

980 MASONRY

9803121 Battering walls

980312161 Class A engineering bricks, BS 3921, in cement mortar (1:3):

980312161A	half brick thick	m²	0.89	17.52	–	31.00	48.52	67.93	97.05
980312161B	one brick thick	m²	1.78	35.05	–	62.67	97.72	136.80	195.43
980312161C	one and a half brick thick	m²	2.66	52.37	–	98.61	150.98	211.38	301.97
980312161D	two brick thick	m²	3.54	69.70	–	129.61	199.31	279.04	398.62

980312166 Class B engineering bricks, BS 3921, in cement mortar (1:3):

980312166A	half brick thick	m²	0.85	16.74	–	21.11	37.85	52.99	75.70
980312166B	one brick thick	m²	1.71	33.67	–	42.90	76.57	107.20	153.13
980312166C	one and a half brick thick	m²	2.56	50.40	–	67.31	117.71	164.80	235.42
980312166D	two brick thick	m²	3.41	67.14	–	88.42	155.56	217.78	311.12

980312171 Common bricks, BS 3821, in cement mortar (1:3):

980312171A	half brick thick	m²	0.82	16.15	–	11.62	27.77	38.87	55.54
980312171B	one brick thick	m²	1.64	32.29	–	24.59	56.88	79.63	113.76
980312171C	one a and half brick thick	m²	2.46	48.44	–	37.92	86.36	120.90	172.72
980312171D	two brick thick	m²	3.28	64.58	–	50.89	115.47	161.65	230.94

9803131 Curved walls

980313101 Precast concrete blocks, BS 6073, strength 7 N/mm², in cement mortar (1:3):

980313101A	100 mm solid block skins	m²	0.86	16.93	–	7.27	24.20	33.88	48.40
980313101B	150 mm solid block skins	m²	0.95	18.71	–	10.56	29.27	40.98	58.53
980313101C	200 mm solid block skins	m²	1.07	21.07	–	14.41	35.48	49.66	70.94
980313101D	100 mm hollow block skins	m²	1.09	21.46	–	15.85	37.31	52.25	74.63

980313121 Architectural blocks, in gauged mortar (1:1:6); flush pointing all round:

980313121A	100 mm solid; PC £30:00 per m²	m²	1.34	26.38	–	32.14	58.52	81.94	117.06
980313121B	220 mm solid; PC £65:00 per m²	m²	1.69	33.27	–	68.89	102.16	143.03	204.34

General Civils

Rail 2003 Ref		Unit	Labour Hours	Labour Cost £	Plant Cost £	Materials Cost £	Unit Rate Base Cost £	Unit Rate Green Zone £	Unit Rate Red Zone £

980 MASONRY

9803131 Curved walls

980313141 Facing bricks, PC £250:00 per 1000, in gauged mortar (1:1:6); flush pointing all round:

980313141A	half brick thick	m²	1.20	23.63	–	16.29	39.92	55.88	79.83
980313141B	215 mm	m²	2.40	47.25	–	33.86	81.11	113.57	162.24

980313161 Class A engineering bricks, BS 3921, in cement mortar (1:3):

980313161A	half brick thick	m²	1.11	21.86	–	31.00	52.86	74.00	105.71
980313161B	one brick thick	m²	1.95	38.39	–	62.67	101.06	141.49	202.13
980313161C	one and a half brick thick	m²	3.32	65.37	–	98.61	163.98	229.57	327.95
980313161D	two brick thick	m²	4.42	87.03	–	129.61	216.64	303.30	433.27

980313166 Class B engineering bricks, BS 3921, in cement mortar (1:3):

980313166A	half brick thick	m²	1.07	21.07	–	21.11	42.18	59.05	84.36
980313166B	one brick thick	m²	2.13	41.94	–	42.90	84.84	118.77	169.67
980313166C	one and a half brick thick	m²	3.20	63.01	–	67.31	130.32	182.44	260.62
980313166D	two brick thick	m²	4.27	84.07	–	88.42	172.49	241.49	344.98

980313171 Common bricks, BS 3921, in cement mortar (1:3):

980313171A	half brick thick	m²	1.03	20.28	–	11.62	31.90	44.66	63.81
980313171B	one brick thick	m²	2.05	40.36	–	24.59	64.95	90.93	129.90
980313171C	one and a half brick thick	m²	3.08	60.64	–	37.92	98.56	137.99	197.13
980313171D	two brick thick	m²	4.10	80.73	–	50.89	131.62	184.26	263.23

9803141 Walls built against other constructions

980314101 Precast concrete blocks, BS 6073, strength 7 N/mm², in cement mortar (1:3):

980314101A	100 mm solid block skins	m²	0.63	12.40	–	7.27	19.67	27.54	39.34
980314101B	150 mm solid block skins	m²	0.69	13.59	–	10.56	24.15	33.81	48.29
980314101C	200 mm solid block skins	m²	0.78	15.36	–	14.41	29.77	41.67	59.52
980314101D	100 mm hollow block skins	m²	0.79	15.55	–	15.85	31.40	43.98	62.82

General Civils

Rail 2003 Ref		Unit	Labour Hours	Labour Cost	Plant Cost	Materials Cost	Unit Rate	Unit Rate	Unit Rate
				£	£	£	Base Cost £	Green Zone £	Red Zone £
980	**MASONRY**								
9803141	Walls built against other constructions								
980314121	Architectural blocks, in gauged mortar (1:1:6); flush pointing all round:								
980314121A	100 mm solid; PC £30:00 per m^2	m^2	0.98	19.30	–	32.14	51.44	72.01	102.88
980314121B	220 mm solid; PC £65:00 per m^2	m^2	1.23	24.22	–	68.89	93.11	130.35	186.22
980314141	Facing bricks, PC £250:00 per 1000, in gauged mortar (1:1:6): flush pointing all round:								
980314141A	105 mm	m^2	0.87	17.13	–	16.29	33.42	46.78	66.84
980314141B	215 mm	m^2	1.74	34.26	–	33.86	68.12	95.37	136.25
980314161	Class A engineering bricks, BS 3921, in cement mortar (1:3):								
980314161A	half brick thick	m^2	0.81	15.95	–	31.00	46.95	65.73	93.90
980314161B	one brick thick	m^2	1.61	31.70	–	62.67	94.37	132.12	188.74
980314161C	one and a half brick thick	m^2	2.42	47.65	–	98.61	146.26	204.77	292.51
980314161D	two brick thick	m^2	3.22	63.40	–	129.61	193.01	270.22	386.02
980314166	Class B engineering bricks, BS 3921, in cement mortar (1:3):								
980314166A	half brick thick	m^2	0.78	15.36	–	21.11	36.47	51.06	72.94
980314166B	one brick thick	m^2	1.55	30.52	–	42.90	73.42	102.79	146.83
980314166C	one and a half brick thick	m^2	2.33	45.88	–	67.31	113.19	158.46	226.36
980314166D	two brick thick	m^2	3.10	61.04	–	88.42	149.46	209.24	298.91
980314171	Common brick, BS 3921, in cement mortar (1:3):								
980314171A	half brick thick	m^2	0.75	14.77	–	11.62	26.39	36.94	52.78
980314171B	one brick thick	m^2	1.49	29.34	–	24.59	53.93	75.49	107.85
980314171C	one and a half brick thick	m^2	2.24	44.10	–	37.92	82.02	114.83	164.06
980314171D	two brick thick	m^2	2.98	58.67	–	50.89	109.56	153.38	219.13

General Civils

Rail 2003 Ref		Unit	Labour Hours	Labour Cost £	Plant Cost £	Materials Cost £	Unit Rate Base Cost £	Unit Rate Green Zone £	Unit Rate Red Zone £
980	**MASONRY**								
9803151	**Cavity walls**								
980315101	Precast concrete blocks, BS 6073, strength 7 N/mm^2, in cement mortar (1:3):								
980315101A	100 mm solid block skins	m^2	0.57	11.22	–	7.27	18.49	25.88	36.98
980315101B	150 mm solid block skins	m^2	0.63	12.40	–	10.56	22.96	32.16	45.93
980315101C	200 mm solid block skins	m^2	0.71	13.98	–	14.41	28.39	39.74	56.77
980315121	Architectural blocks, in gauged mortar (1:1:6); flush pointing all round:								
980315121A	100 mm solid; PC £30:00 per m^2	m^2	0.89	17.52	–	32.14	49.66	69.53	99.34
980315121B	220 mm solid; PC £65:00 per m^2	m^2	1.12	22.05	–	68.89	90.94	127.32	181.89
980315141	Facing bricks, PC £250:00 per 1000, in gauged mortar (1:1:6); flush pointing all round:								
980315141A	105 mm	m^2	0.79	15.55	–	16.29	31.84	44.58	63.69
980315141B	215 mm	m^2	1.58	31.11	–	33.86	64.97	90.96	129.95
980315161	Class A engineering bricks, BS 3921, in cement mortar (1:3):								
980315161A	half brick thick	m^2	0.73	14.37	–	31.00	45.37	63.52	90.75
980315161B	one brick thick	m^2	1.47	28.94	–	62.67	91.61	128.26	183.23
980315161C	one and a half brick thick	m^2	2.19	43.12	–	98.61	141.73	198.43	283.46
980315161D	two brick thick	m^2	2.93	57.69	–	129.61	187.30	262.22	374.60
980315166	Class B engineering bricks, BS 3921, in cement mortar (1:3):								
980315166A	half brick thick	m^2	0.71	13.98	–	21.11	35.09	49.13	70.19
980315166B	one brick thick	m^2	1.41	27.76	–	42.90	70.66	98.93	141.31
980315166C	one and a half brick thick	m^2	2.12	41.74	–	67.31	109.05	152.67	218.09
980315166D	two brick thick	m^2	2.82	55.52	–	88.42	143.94	201.52	287.89

General Civils

Rail 2003 Ref		Unit	Labour Hours	Labour Cost £	Plant Cost £	Materials Cost £	Unit Rate Base Cost £	Unit Rate Green Zone £	Unit Rate Red Zone £
980	**MASONRY**								
9803151	**Cavity walls**								
980315171	Common bricks, BS 3921, in cement mortar (1:3):								
980315171A	half brick thick	m²	0.68	13.39	–	11.62	25.01	35.01	50.03
980315171B	one brick thick	m²	1.35	26.58	–	24.59	51.17	71.63	102.34
980315171C	one and a half brick thick	m²	2.03	39.97	–	37.92	77.89	109.05	155.79
980315171D	two brick thick	m²	2.71	53.36	–	50.89	104.25	145.94	208.49
9803161	**Isolated piers**								
980316101	Precast concrete blocks, BS 6073, strength 7 N/mm², in cement mortar (1:3):								
980316101A	100 mm solid block skins	m²	0.69	13.59	–	7.77	21.36	29.89	42.70
980316101B	150 mm solid block skins	m²	0.75	14.77	–	11.32	26.09	36.51	52.16
980316101C	200 mm solid block skins	m²	0.85	16.74	–	15.45	32.19	45.06	64.37
980316101D	100 mm hollow block skins	m²	0.86	16.93	–	17.01	33.94	47.53	67.89
980316121	Architectural blocks, in gauged mortar (1:1:6); flush pointing all around:								
980316121A	100 mm solid; PC £30:00 per m²	m²	1.06	20.87	–	32.14	53.01	74.22	106.03
980316121B	220 mm solid; PC £65:00 per m²	m²	1.34	26.38	–	68.89	95.27	133.39	190.56
980316141	Facing bricks, PC £250:00 per 1000, in gauged mortar (1:1:6); flush pointing all round:								
980316141A	105 mm	m²	0.95	18.71	–	18.79	37.50	52.49	74.99
980316141B	215 mm	m²	1.90	37.41	–	36.36	73.77	103.28	147.55
980316161	Class A engineering bricks, BS 3921, in cement mortar (1:3):								
980316161A	half brick thick	m²	0.88	17.33	–	35.94	53.27	74.58	106.53
980316161B	one brick thick	m²	1.76	34.65	–	68.28	102.93	144.11	205.88
980316161C	one and a half brick thick	m²	2.63	51.78	–	104.23	156.01	218.41	312.01
980316161D	two brick thick	m²	3.51	69.11	–	141.51	210.62	294.86	421.24

General Civils

Rail 2003 Ref		Unit	Labour Hours	Labour Cost £	Plant Cost £	Materials Cost £	Unit Rate Base Cost £	Unit Rate Green Zone £	Unit Rate Red Zone £
980	**MASONRY**								
9803161	**Isolated piers**								
980316166	Class B engineering bricks, BS 3921, in cement mortar (1:3):								
980316166A	half brick thick....................	m²	0.85	16.74	–	24.41	41.15	57.60	82.29
980316166B	one brick thick....................	m²	1.69	33.27	–	46.86	80.13	112.19	160.28
980316166C	one and a half brick thick............	m²	2.54	50.01	–	71.27	121.28	169.79	242.57
980316166D	two brick thick....................	m²	3.39	66.75	–	97.02	163.77	229.27	327.54
980316171	Common bricks, BS 3921, in cement mortar (1:3):								
980316171A	half brick thick....................	m²	0.81	15.95	–	13.34	29.29	41.00	58.57
980316171B	one brick thick....................	m²	1.63	32.09	–	26.30	58.39	81.75	116.79
980316171C	one and a half brick thick...........	m²	2.44	48.04	–	39.64	87.68	122.75	175.36
980316171D	two brick thick....................	m²	3.25	63.99	–	54.32	118.31	165.62	236.61
9803181	**Extra over general brickwork for fair faced work**								
980318101	Fair facing and flush pointing:								
980318101A	stretcher bond....................	m²	0.06	1.18	–	0.26	1.44	2.01	2.88
980318101B	English bond.....................	m²	0.06	1.18	–	0.26	1.44	2.01	2.88
980318102	Fair facing and struck or weather struck pointing:								
980318102A	stretcher bond....................	m²	0.06	1.18	–	0.26	1.44	2.01	2.88
980318102B	English bond.....................	m²	0.06	1.18	–	0.32	1.50	2.10	3.00
980318103	Fair facing and tooled or keyed pointing:								
980318103A	stretcher bond....................	m²	0.06	1.18	–	0.19	1.37	1.92	2.75
980318103B	English bond.....................	m²	0.07	1.38	–	0.26	1.64	2.29	3.28

General Civils

Rail 2003 Ref		Unit	Labour Hours	Labour Cost £	Plant Cost £	Materials Cost £	Unit Rate Base Cost £	Unit Rate Green Zone £	Unit Rate Red Zone £
980	**MASONRY**								
9803183	Extra over general blockwork for fair faced work								
980318301	Fair facing and flush pointing:								
980318301A	precast concrete blocks	m^2	0.02	0.39	–	0.29	0.68	0.96	1.37
980318301B	architectural blocks	m^2	0.03	0.59	–	0.26	0.85	1.19	1.70
9803186	Sills, bands and features								
980318601	Sills and copings; facing bricks, PC £250:00 per 1000, in gauged mortar (1:1:6); flush pointing all exposed edges:								
980318601A	brick on edge sills; flush; flat top; half brick wide	m^2	0.24	4.73	–	2.69	7.42	10.39	14.84
980318601B	brick on edge sills; flush; flat top; one brick wide	m^2	0.37	7.29	–	3.14	10.43	14.60	20.86
980318601C	brick on edge copings; one brick wide	m^2	0.24	4.73	–	2.82	7.55	10.57	15.09
980318601D	brick on edge copings; one brick wide	m^2	0.37	7.29	–	3.14	10.43	14.60	20.86
980318611	Quoins and reveals; facing bricks PC £250:00 per 1000, in gauged mortar (1:1:6); flush pointing all exposed edges:								
980318611A	oversailing reveals; 25 mm projection toothed in three course bands; 161.25 mm average width	m	0.59	11.62	–	2.76	14.38	20.12	28.75
9803188	Arches								
980318801	Facing bricks, PC £250:00 per 1000, in gauged mortar (1:1:6); flush pointing all round:								
980318801A	brick on edge flat arches; half brick wide (snapped headers)	m	0.32	6.30	–	2.76	9.06	12.68	18.12
980318801B	brick on edge flat arches; one brick wide	m	0.46	9.06	–	3.14	12.20	17.08	24.40
980318801C	brick on edge flat arches; one brick wide	m	0.66	13.00	–	3.14	16.14	22.59	32.28
980318801D	brick on edge segmental arches; half brick wide (snapped headers)	m	0.39	7.68	–	2.82	10.50	14.70	21.00

General Civils

Rail 2003 Ref		Unit	Labour Hours	Labour Cost £	Plant Cost £	Materials Cost £	Unit Rate Base Cost £	Unit Rate Green Zone £	Unit Rate Red Zone £

980   **MASONRY**

9803188   **Arches**

980318801   **Facing bricks, PC £250:00 per 1000, in gauged mortar (1:1:6); flush pointing all round:**

980318801E	brick on edge segmental arches; one brick wide; one course.............	m	0.57	11.22	–	3.14	14.36	20.11	28.74
980318801F	brick on edge segmental arches; one brick wide; two courses.............	m	0.83	16.34	–	8.14	24.48	34.28	48.97
980318801G	brick on end segmental arches; half brick wide	m	0.60	11.81	–	3.14	14.95	20.94	29.92

General Civils

Rail 2003 Ref		Unit	Labour Hours	Labour Cost £	Plant Cost £	Materials Cost £	Unit Rate Base Cost £	Unit Rate Green Zone £	Unit Rate Red Zone £

990 PAINTING

99010 GENERAL PAINTING

9901010 Oil paint; metalwork backgrounds

990101010 Prepare, etching primer, two undercoats and top coat gloss paint; general surfaces:

990101010A	not exceeding 300 mm girth	m	0.15	2.06	–	0.93	2.99	4.18	5.97
990101010B	300 mm – 1 m girth	m²	0.36	4.94	–	2.78	7.72	10.81	15.43
990101010C	over 1 m girth	m²	0.36	4.94	–	2.78	7.72	10.81	15.43

990101020 Prepare, etching primer, two undercoats and top coat gloss paint; railings, fences and gates:

990101020A	not exceeding 300 mm girth	m	0.24	3.29	–	0.93	4.22	5.91	8.44
990101020B	300 mm – 1 m girth	m²	0.60	8.23	–	2.78	11.01	15.42	22.02
990101020C	over 1 m girth	m²	0.60	8.23	–	2.78	11.01	15.42	22.02

990101030 Prepare, etching primer, two undercoats and top coat gloss paint; staircase and balustrades:

990101030A	not exceeding 300 mm girth	m	0.21	2.88	–	0.93	3.81	5.33	7.61
990101030B	300 mm – 1 m girth	m²	0.51	7.00	–	2.78	9.78	13.69	19.55
990101030C	over 1 m girth	m²	0.51	7.00	–	2.78	9.78	13.69	19.55

9901020 Creosote treatment

990102010 Prepare and apply one coat; general surfaces:

990102010A	not exceeding 300 mm girth	m	0.06	0.82	–	0.07	0.89	1.25	1.80
990102010B	300 mm – 1 m girth	m²	0.14	1.92	–	0.22	2.14	3.00	4.28
990102010C	over 1 m girth	m²	0.14	1.92	–	0.22	2.14	3.00	4.28

990102020 Prepare and apply two coats; general surfaces:

990102020A	not exceeding 300 mm girth	m	0.10	1.37	–	0.15	1.52	2.13	3.03
990102020B	300 mm – 1 m girth	m²	0.25	3.43	–	0.44	3.87	5.42	7.74
990102020C	over 1 m girth	m²	0.25	3.43	–	0.44	3.87	5.42	7.74

General Civils

Rail 2003 Ref		Unit	Labour Hours	Labour Cost £	Plant Cost £	Materials Cost £	Unit Rate Base Cost £	Unit Rate Green Zone £	Unit Rate Red Zone £

993 **WATERPROOFING**

99310 **DAMP PROOFING AND TANKING**

9931010 Membranes and compounds

993101010 RIW liquid asphaltic composition; two coats on concrete surfaces; width:

993101010A	horizontal; not exceeding 300 mm	m	0.03	0.36	–	2.08	2.44	3.41	4.87
993101010B	horizontal; over 300 mm	m²	0.08	0.96	–	6.98	7.94	11.11	15.86
993101010C	vertical; not exceeding 300 mm	m	0.03	0.36	–	2.08	2.44	3.41	4.87
993101010D	vertical; over 300 mm	m²	0.09	1.08	–	6.98	8.06	11.28	16.10

993101020 Synthaprufe waterproofing compound; two coats on concrete surfaces; final coat dusted with sand; width:

993101020A	horizontal; not exceeding 300 mm	m	0.03	0.36	–	1.41	1.77	2.48	3.55
993101020B	horizontal; over 300 mm	m²	0.09	1.08	–	4.72	5.80	8.12	11.60
993101020C	vertical; not exceeding 300 mm	m	0.03	0.36	–	1.41	1.77	2.48	3.55
993101020D	vertical; over 300 mm	m²	0.10	1.20	–	4.72	5.92	8.28	11.84

993101030 Synthaprufe waterproofing compound; three coats on concrete surfaces; final coat dusted with sand; width:

993101030A	horizontal; not exceeding 300 mm	m	0.04	0.48	–	2.13	2.61	3.66	5.23
993101030B	horizontal; over 300 mm	m²	0.11	1.32	–	7.04	8.36	11.70	16.72
993101030C	vertical; not exceeding 300 mm	m	0.04	0.48	–	2.14	2.62	3.67	5.24
993101030D	vertical; over 300 mm	m²	0.11	1.32	–	7.04	8.36	11.70	16.72

993101040 1200 gauge polythene sheeting; 150 mm side and end laps; width:

993101040A	horizontal; not exceeding 300 mm	m	0.01	0.12	–	0.69	0.81	1.14	1.62
993101040B	horizontal; over 300 mm	m²	0.04	0.48	–	0.69	1.17	1.64	2.34
993101040C	vertical; not exceeding 300 mm	m	0.01	0.12	–	0.69	0.81	1.14	1.62
993101040D	vertical; over 300 mm	m²	0.05	0.60	–	0.69	1.29	1.81	2.58

993101050 Bituthene bitumen coated 1000 gauged polythene sheeting; sticking to concrete surfaces primed with bituthene primer; width:

993101050A	horizontal; not exceeding 300 mm	m	0.03	0.36	–	2.64	3.00	4.20	6.00
993101050B	horizontal; over 300 mm	m²	0.10	1.20	–	8.77	9.97	13.95	19.93
993101050C	vertical; not exceeding 300 mm	m	0.04	0.48	–	2.78	3.26	4.56	6.52
993101050D	vertical; over 300 mm	m²	0.13	1.55	–	9.18	10.73	15.04	21.48

General Civils

Rail 2003 Ref		Unit	Labour Hours	Labour Cost £	Plant Cost £	Materials Cost £	Unit Rate Base Cost £	Unit Rate Green Zone £	Unit Rate Red Zone £

993   **WATERPROOFING**

9931010   **Membranes and compounds**

993101060   Hyload pitch polymer d.p.c; width:

993101060A	horizontal; not exceeding 300 mm	m	0.15	2.95	–	3.10	6.05	8.47	12.11
993101060B	horizontal; over 300 mm	m²	0.15	2.95	–	10.16	13.11	18.36	26.24
993101060C	vertical; not exceeding 300 mm	m	0.09	1.77	–	3.10	4.87	6.82	9.74
993101060D	vertical; over 300 mm	m²	0.25	4.92	–	10.16	15.08	21.12	30.17

General Civils

Rail 2003 Ref		Unit	Specialist Cost £	Unit Rate Base Cost £	Unit Rate Green Zone £	Unit Rate Red Zone £
995	**MISCELLANEOUS WORK**					
99510	**FENCING**					
9951010	Chain link, BS 1722 Part 1; galvanised mesh; 3 rows barbed wire at top					
995101010	Concrete posts with steel extension arms at 3 m centres; excavation, disposal and filling:					
995101010A	1.80 m high fence	m	–	21.00	29.40	42.00
995101010B	end post	Nr	–	74.00	103.60	148.00
995101010C	corner post	Nr	–	96.00	134.40	192.00
995101010D	gate post	Nr	–	99.00	138.60	198.00
995101010E	2.10 m high fence	m	–	23.00	32.20	46.00
995101010F	end post	Nr	–	78.00	109.20	156.00
995101010G	corner post	Nr	–	100.00	140.00	200.00
995101010H	gate post	Nr	–	103.00	144.20	206.00
995101020	Concrete posts with cranked tops at 3 m centres; excavation, disposal and filling:					
995101020A	1.80 m high fence	m	–	22.00	30.80	44.00
995101020B	end post	Nr	–	74.50	104.30	149.00
995101020C	corner post	Nr	–	96.75	135.45	193.50
995101020D	gate post	Nr	–	98.50	137.90	197.00
995101020E	2.10 m high fence	m	–	24.25	33.95	48.50
995101020F	end post	Nr	–	78.50	109.90	157.00
995101020G	corner post	Nr	–	101.00	141.40	202.00
995101020H	gate post	Nr	–	103.50	144.90	207.00
995101030	Galvanised steel posts with cranked tops at 3 m centres; excavation, disposal and filling:					
995101030A	1.80 m high fence	m	–	18.00	25.20	36.00
995101030B	end post	Nr	–	62.00	86.80	124.00
995101030C	corner post	Nr	–	64.00	89.60	128.00
995101030D	gate post	Nr	–	66.00	92.40	132.00
995101030E	2.10 m high fence	m	–	20.00	28.00	40.00
995101030F	end post	Nr	–	63.00	88.20	126.00
995101030G	corner post	Nr	–	65.00	91.00	130.00
995101030H	gate post	Nr	–	67.00	93.80	134.00

General Civils

Rail 2003 Ref		Unit	Specialist Cost £	Unit Rate Base Cost £	Unit Rate Green Zone £	Unit Rate Red Zone £
995	**MISCELLANEOUS WORK**					
9951020	**Wooden post and rail, BS 1722 Part 7; sawn mortice**					
995102010	Treated softwood posts and rails; posts at 3 m centres; driven:					
995102010A	1.10 m high 3 rail fence	m	– – –	14.00	19.60	28.00
995102010B	end post	Nr	– – –	12.00	16.80	24.00
995102010C	corner post	Nr	– – –	15.00	21.00	30.00
995102010D	intersection post	Nr	– – –	12.00	16.80	24.00
995102010E	gate post	Nr	– – –	15.00	21.00	30.00
995102010F	1.30 m high 4 rail fence	m	– – –	16.00	22.40	32.00
995102010G	end post	Nr	– – –	13.00	18.20	26.00
995102010H	corner post	Nr	– – –	16.00	22.40	32.00
995102010I	intersection post	Nr	– – –	13.00	18.20	26.00
995102010J	gate post	Nr	– – –	16.00	22.40	32.00
9951030	**Wooden palisade, BS 1722 Part 6; nailed**					
995103010	Treated softwood posts and rails; posts at 3 m centres; excavation, disposal and filling:					
995103010A	1.00 m high 3 rail fence	m	– – –	24.00	33.60	48.00
995103010B	end post	Nr	– – –	12.00	16.80	24.00
995103010C	corner post	Nr	– – –	15.00	21.00	30.00
995103010D	gate post	Nr	– – –	15.00	21.00	30.00
995103010E	1.40 m high 4 rail fence	m	– – –	28.00	39.20	56.00
995103010F	end post	Nr	– – –	13.00	18.20	26.00
995103010G	corner post	Nr	– – –	16.00	22.40	32.00
995103010H	gate post	Nr	– – –	16.00	22.40	32.00
995103010I	1.80 m high 6 rail fence	m	– – –	32.00	44.80	64.00
995103010J	end post	Nr	– – –	14.00	19.60	28.00
995103010K	corner post	Nr	– – –	17.00	23.80	34.00
995103010L	gate post	Nr	– – –	17.00	23.80	34.00

Rail 2003 Ref		Unit				Specialist Cost £	Unit Rate Base Cost £	Unit Rate Green Zone £	Unit Rate Red Zone £
995	**MISCELLANEOUS WORK**								
99530	**GATES**								
9953010	**Entrance gates, BS 4092 Part 2**								
995301010	**Treated softwood:**								
995301010A	0.81 × 0.90 m high	Nr	–	–	–	110.00	110.00	154.00	220.00
995301010B	1.02 × 0.90 m high	Nr	–	–	–	126.00	126.00	176.40	252.00
995301010C	2.13 × 0.90 m high	Nr	–	–	–	142.00	142.00	198.80	284.00
9953020	**Metal gates**								
995302010	**Steel framed; 48 mm diameter tubes, welded contruction; 50 mm × 10g plastic coated steel mesh; size:**								
995302010A	0.90 × 1.80 m high	Nr	–	–	–	188.25	188.25	263.55	376.50
995302010B	0.90 × 2.10 m high	Nr	–	–	–	212.00	212.00	296.80	424.00
995302010C	1.20 × 1.80 m high	Nr	–	–	–	235.50	235.50	329.70	471.00
995302010D	1.20 × 2.10 m high	Nr	–	–	–	264.75	264.75	370.65	529.50
995302020	**Steel gate posts; 48 mm diameter tubes; set in concrete; to suit:**								
995302020A	1.80 m high gate	Nr	–	–	–	103.25	103.25	144.55	206.50
995302020B	2.10 m high gate	Nr	–	–	–	111.50	111.50	156.10	223.00

Appendices

Useful Addresses

Associated Society of Locomotive Engineers and Firemen
9 Arkwright Road
Hampstead, London NW3 6AB
United Kingdom
Tel: +44 171 317 8600
Fax: +44 171 794 6406
E-mail: info@aslef.org.uk
URL: http://www.aslef.org.uk/

Association of Consulting Engineers
12 Claxton Street
London SW1H 0QL
United Kingdom
Tel: +44 171 222 6557
Fax: +44 171 222 0750
E-mail: consult@acenet.co.uk
URL: http://www.acenet.co.uk/

Association of Independent Railways
85 Balmoral Road
Gillingham ME7 4QG
United Kingdom
Tel: +44 1634 852672
Fax: +44 1634 852672
URL http://www.hmilburn.easynet.co.uk/main/

Association of Private Railway Wagon Owners
Homelea, Westland Green
Little Hadham, Ware SG11 2AG
United Kingdom
Tel: +44 1279 843487
Fax: +44 1279 842394

British International Freight Association
Redfern House, Browells Lane
Feltham, Middlesex TW13 7EP
United Kingdom
Tel: +44 181 844 2266
Fax: +44 181 890 5546
E-mail: bifasec@msn.com
URL: http://www.bifa.org/

British Safety Council
70 Chancellor's Road
London W6 9RS
United Kingdom
Tel: +44 181 741 1231
Fax: +44 181 741 4555
E-mail: bscl@mail.britishsafetycouncil.co.uk
URL: http://www.britishsafetycouncil.co.uk/

British Safety Industry Federation
St Asaph Business Park, Glascoed Road
St Asaph, Clwyd LL17 7LJ
United Kingdom
Tel: +44 1745 585600
Fax: +44 1745 585800
E-mail: info@bsif.co.uk
URL: http://www.bsif.co.uk/

British Standards Institute (BSI)
389 Chiswick High Road
Chiswick, London W4 4AL
United Kingdom
Tel: +44 181 996 9000
Fax: +44 181 996 7400
E-mail: info@bsi.org.uk
URL: http://www.bsi.org.uk/

British Transport Officers' Guild
Hayes Court, West Common Road
Hayes BR2 7AU
United Kingdom
Tel: +44 181 462 7755
Fax: +44 181 462 4959
E-mail: info@btog.org.uk
URL: http://www.btog.org.uk/

British Valve and Actuator Manufacturers' Association
The McLaren Building, 35 Dale End
Birmingham, B4 7LN
United Kingdom
Tel: +44 121 200 1297
Fax: +44 121 200 1308
E-mail: enquiry@bvama.org.uk
URL: http://www.bvama.org.uk/

Central Rail Users' Consultative Committee
First Floor, Golden Cross House,
Duncannon Street, London WC2N 4JF
United Kingdom
Tel: +44 171 839 7338
Fax: +44 171 925 2228

Chartered Institute of Transport
80 Portland Place
London W1N 4DP
United Kingdom
Tel: +44 171 636 9952
Fax: +44 171 637 0511
E-mail: gen@citrans.org.uk
URL: http://www.citrans.org.uk/

Useful Addresses

College of Railway Technology
London Road
Derby, DE24 8UX
United Kingdom
Tel: +44 1332 262222
Fax: +44 1332 264820
E-mail: hotline.crt@ems.rail.co.uk
URL: http://www.transcend.co.uk/crt/

Community Railways
6 School Lane
Berry Brow HD4 7LT
United Kingdom
Tel: +44 1484 665273
Fax: +44 1484 666974

Confederation of British Industry (CBI)
103 New Oxford Street
London WC1A 1DU
United Kingdom
Tel: +44 171 379 7400
Fax: +44 171 240 1578
URL: http://www.cbi.org.uk/

Confederation of Passenger Transport UK
Sardinia House, 52 Lincoln's Inn Fields
London WC2A 31Z
United Kingdom
Tel: +44 171 831 7546
Fax: +44 171 242 0053

Crown Agents for Overseas Governments and Administrators
St Nicholas House, St Nicholas Road
Sutton SM1 1EL
United Kingdom
Tel: +44 181 643 3311
Fax: +44 181 643 8232
E-mail: enquiries@crownagents.co.uk
URL: http://www.crownagents.com/

Department of the Environment, Transport and the Regions
Eland House, Bressendan Place
London, SW 1E 5DU
United Kingdom
Tel: +44 171 890 3000
URL: http://www.detr.gov.uk/

Engineering Council
10 Maltravers Street
London WC2R 3ER
United Kingdom
Tel: +44 171 240 7891
Fax: +44 171 379 5586
E-mail: info@engc.org.uk
URL: http://www.engc.org.uk/

Engineering Equipment and Material Users Association
14-15 Belgrave Square
London SW1X 8PS
United Kingdom
Tel: +44 171 235 5316
Fax: +44 171 245 6937
URL: http://www.eemua.co.uk

Engineering Industries Association
16 Dartmouth Street, Westminster
London SW1H 9BL
United Kingdom
Tel: +44 171 222 2367
Fax; +44 171 799 2206

European Passenger Services (EPS)
EPS House, Floor 2, Waterloo Station
London SE1 8SE
United Kingdom
Tel: +44 171 928 5151
Fax: +44 171 922 4499

Eurotunnel
1 Canada Square, Canary Wharf
London E14 5DU
United Kingdom
Tel: +44 171 715 6789
Fax: +44 171 715 6666
URL: http://www3.eurotunnel.com/

Federation of Civil Engineering Equipment
Cowdray House, 6 Portugal Street
London WC2A 2HH
United Kingdom
Tel: +44 171 404 4020
Fax: +44 171 242 0256

Health and Safety Executive
Rose Court, 2 Southwark Bridge
London SE1 9HS
Tel: +44 171 717 6000
Fax: +44 171 717 6907
URL: http://www.open.gov.uk/hse/

Useful Addresses

Her Majesty's Railway Inspectorate
Rose Court, 2 Southwark Bridge
London SE1 9HS
United Kingdom
Tel: +44 171 717 6533
Fax: +44 171 717 6547

International Transport Workers' Federation
49-60 Borough Road
London SE1 1DS
United Kingdom
Tel: +44 171 403 2733
Fax: +44 171 357 7871
E-mail: mail@itf.org.uk
URL: http://www.itf.org.uk/

Institution of Civil Engineers
1 Great George Street
London, SW1P 3AA
United Kingdom
Tel: +44 171 222 7722
Fax: +44 171 222 7500
E-mail: webmaster@ice.org.uk
URL: http://www.ice.org.uk/

Institution of Electrical Engineers
Savoy Place
London WC2R 0BL
United Kingdom
Tel: +44 171 240 1871
Fax: +44 171 240 7735
E-mail: postmaster@iee.org.uk
URL: http://www.iee.org.uk/

Institution of Fire Engineers
148 New Walk
Leicester, Leicestershire LE1 7QB
United Kingdom
Tel: +44 116 255 3654
Fax: +44 116 247 1231

Institute of Fire Safety
PO Box 687
Croydon, Surrey CR9 5DD
United Kingdom
Tel: +44 181 655 2582
Fax: +44 181 654 2583

Institution of Diesel and Gas Turbine Engineers
PO Box 43
Bedford MK40 4JB
United Kingdom
Tel: +44 1234 241 340
Fax: +44 1234 355 493
E-mail: secretary@idgte.org/
URL: http://www.idgte.org/

Institution of Electronics and Electrical Incorporation Engineers
Savoy Hill House
Savoy Hill, London WC2R 0BS
United Kingdom
Tel: +44 171 836 3357
Fax: +44 171 497 9006
E-mail: ieeie@dial.pipex.com
URL: http://www.worldserver.pipex.com/ieeie/

Institution of Mechanical Engineers
1 Birdcage Walk
London SW1H 9JJ
United Kingdom
Tel: +44 171 222 7899
Fax: +44 171 222 4557
E-mail: corporate comm@imeche.org.uk
URL: http://www.imeche.org.uk/

Institution of Mechanical Incorporated Engineers
3 Birdcage walk
London SW1H 9JN
United Kingdom
Tel: +44 171 799 1808
Fax: +44 171 799 2243
E-mail: imechie@dial.pipex.com
URL: http://www.engc.org.uk/IMechIE/

Institution of Railway Signal Engineers
3rd Floor, Savoy Hill House
Savoy Hill, London WC2R 0BS
United Kingdom
Tel: +44 171 836 3357
Fax: +44 171 497 9006
E-mail: admin@irse.u-net.com
URL: http://www.irse.org/

Light Rail Transit Association
Albany House, Petty France
London SW1H 9EA
United Kingdom
Tel: +44 171 918 3116
Fax: +44 171 799 1846
URL: http://lrta.org/

Useful Addresses

Locomotive and Carriage Institution
69 Avondale Close
Horley RH6 8BN
United Kingdom
Tel: +44 1293 773239
URL: http://www.lococarriage.org.uk/

Merseyside Passenger Transport Executive
24 Hatton Garden
Liverpool L3 2AB
United Kingdom
Tel: +44 151 227 5181
Fax: +44 151 236 2457
URL: http://merseytravel.gov.uk/

National Council on Inland Transport
5 Pembridge Crescent
London W11 3DT
United Kingdom
Tel: +44 171 727 4689
Fax: +44 171 727 4689

National Union of Rail, Maritime and Transport Workers
Unity House, Euston Road
London NW1 2BL
United Kingdom
Tel: +44 171 387 4771
Fax: +44 171 387 4123
URL: http://www.rmt.org.uk/

Network Rail
PO Box 100, Euston House
24 Eversholt Street
London NW1 1DZ
United Kingdom
Tel: +44 171 928 5151
Fax: +44 171 922 6545

Northern Ireland Department of the
Environment, Transport Division
Clarence Court, 10-18 Adelaide Street
Belfast BT2 8GB
United Kingdom
Tel: +44 1232 540041
Fax: +44 1232 540020

Office of Passenger Rail Franchising
26 Old Queen Street
London SW1H 9HP
United Kingdom
Tel: +44 171 799 8800
Fax: +44 171 799 8100

Office of the Rail Regulator
1 Waterhouse Square, 138 - 142 Holborn
London EC1N 2ST
United Kingdom
Tel: +44 171 282 2000
Fax: +44 171 282 2045
E-mail: orr@dial.pipex.com
URL: http://www.rail-reg.gov.uk/

Permanent Way Institution
4 Reginald Road, Wombwell
Barnsley S73 0HP
United Kingdom
Tel: +44 1226 752605
Fax: +44 1226 754287
URL: http://www.trackbed.com/

Private Wagon Federation
Homelea, Westland Green
Little Hadam, Ware SG11 2AG
United Kingdom
Tel: +44 1279 843487
Fax: +44 1279 842394

Railway Development Society
15 Clapham Road
Lowestoft NR32 1RQ
United Kingdom
Tel: +44 1502 581721

Railway Industry Association
6 Buckingham Gate
London SW1E 6JP
United Kingdom
Tel: +44 171 834 1426
Fax: +44 171 821 1640
E-mail: ria@riagb.org.uk
URL: http://www.riagb.org.uk/

Railway Industry Fire Association
6th Floor, 40 Melton Street,
Euston Square
London NW1 2EE
United Kingdom
Tel: +44 171 557 8509
Fax: +44 171 557 9025

Useful Addresses

Scottish Association for Public Transport
5 St Vincents Place
Glasgow G1 2HT
United Kingdom
Tel: +44 141 639 3697
Fax: +44 141 639 3697
E-mail: sapt_secy@talk21.com
URL: http://www.transformscotland.org.uk/members/text/sapt.html

South Yorkshire Passenger Transport Executive
PO Box 801, Exchange Street
Sheffield S2 5YT
United Kingdom
Tel: +44 114 276 7575
Fax: +44 114 275 9908
E-mail: sypte@sypte.co.uk
URL: http://www.sypte.co.uk/

Strathclyde Passenger Transport Executive
Consort House, 12 West George Street
Glasgow G2 1HN
United Kingdom
Tel: +44 141 332 6811
Fax: +44 141 332 3076
URL: http://www.spt.co.uk

Tramway and Light Railway Society
216 Brentwood Road
Romford RM1 2RP
United Kingdom

Transport Trust
Hobart House, Grosvenor Place
London SW1X 7AE
United Kingdom
Tel: +44 171 201 4233

Transport 2000
Walkden House, 10 Melton Street
London NW1 2EJ
United Kingdom
Tel: +44 171 388 8386
Fax: +44 171 388 2481

Tyne & Wear Passenger Transport Executive
Cuthbert House, All Saints
Newcastle Upon Tyne NE1 2DA
United Kingdom
Tel: +44 191 261 0431
Fax: +44 191 232 1192

University of Leeds, Institute for Transport Studies
Leeds LS2 9JT
United Kingdom
Tel: +44 113 233 5325
Fax: +44 113 233 5334
Url: http://www.its.leeds.ac.uk/

University of Sheffield, Advanced Railway Research Centre
Regent Court, 30 Regent Street
Sheffield S1 4DA
United Kingdom
Tel: +44 114 222 0151
Fax: +44 114 275 5625
URL: http://www.shef.ac.uk/uni/academic/AC/arrc/

University of York, Institute of Railway Studies
Heslington
York, YO1 5DD
United Kingdom
Tel: +44 1904 432990
E-mail: cd11@york.ac.uk
URL: http://www.york.ac.uk/inst/irs/

Wagon Building and Repair Association
48 Clifford Road
Poynton SK12 1HY
United Kingdom
Tel: +44 1625 873012
Fax: +44 1625 859836

West Midlands Passenger Transport Executive
16 Summer lane
Birmingham B19 3SD
United Kingdom
Tel: +44 121 200 2787
Fax: +44 121 200 7010

West Yorkshire Passenger Transport Executive
Wellington House, 40-50 Wellington Street
Leeds LS1 2DE
United Kingdom
Tel: +44 113 244 0988
Fax: +44 113 234 0654

Acronyms and Common Terms

Acronym	Description
AAS	Auxiliary Aspects Sets
ABCL	Automatic Barrier Crossing Locally Monitored
AC	Alternating Current
ACE	Area Civil Engineer
ACI	Automatic Code Insertion
ACM	Assistant Commercial Manager
ADD	Automatic Dropping Device
ADR	International Carriage of Dangerous Goods Regulations
AE	Electronics Appreciation
AFC	Anticipated Final Cost or Approved For Construction
AHB	Automatic Half Barrier
AHBC	Automatic Half-Barrier Crossing
AHBLC	Automatic Half Barrier Level Crossing
AIP	Approval In Principle
ALC	Accommodation Level Crossing
ALC	Automatic Lining Control
AMP	Asset Maintenance Programme
AMS	Austenitic Manganese Steel
AOCL	Automatic Open Crossing Locally Monitored
AOCR	Automatic Open Crossing Remotely Monitored
APC	Automatic Power Control
APM	Assistant Project Manager or Area Production Manager
APM Ops	Assistant Project Manager Operations
APT-P	Advanced Passenger Train - Prototype
APTDM	AP Electronics, Time Division Multiplex
ARI	Alphanumeric Route Indicator
ARS	Automatic Route Setting
ASB	Adjacent SignalBox
ASBPC	Adjacent SignalBox Protocol Converter
ASC	Auxiliary Supply Cabinet
ASL	Adjacent Signalbox Link
ATC	Automatic Train Control
ATO	Automatic Train Operation
ATOC	Association of Train Operating Companies
ATP	Automatic Train Protection
APIS	Automatic Passenger Information System
ATR	Automatic Train Reporting System
ATRE	Automatic Train Reporting by Exception
ATTA	Automatic Tract Top Alignment
AWB	Advanced Warning Board
AWE	Advanced Warning Estimate
AWS	Automatic Warning System
B.S.Spec	British Standards Specification

Acronym	Description
BABT	British Approved Board for Telecommunications
BCWS	Budgeted Cost Work Schedule
BD	Braking Distance
BH	Bull Head Section Rail
bhp	brake horse power
bmep	brake mean effective pressure
BMS	Basic Mechanical Signalling
BP	Break Pipe
BR	British Railways
BRB	British Railways Board
BRIMS	British Rail Incident Monitoring Equipment
BRIS	British Rail Infrastructure Services
BRS	Business Route Sector
BRT	Business Route Telecommunications
BS&TE	Business Signal & Telecommunications Engineer
BST	Basic Signalling Technology
BTOG	British Transport Officers Guild
CA	Commercial Assistant
CAD	Computer Aided Design
CAG	Contract Approval Group
CAMPS	Computer Assisted Maintenance Planning System
CAMS	Cost Allocation Management System
CAPE	Train Cancelled
CAPEX	CAPital EXpenditure
CARINO	TOPS wagon number
CARKND	TOPS wagon type
CB	Circuit Breaker
CB	Central Battery
CCE	Chief Civil Engineer
CCI	Client Change Instruction
CCM	Contractor Control Manual
CCS	Contract Conditions Safety
CCS	Common Control Set
CCTV	Closed Circuit TeleVision
CDI	Closed Door Indicator
CDM	Construction Design - Management
CEC	Contract Execution Change
CER	Community of European Railways
CESM	Civil Engineering Safety Manual
CHIP	Chemicals Hazard Information and Packaging
CIC	Client Instigated Change
CIMAH	Control of Industrial Major Accident Hazard
CIS	Customer Information System
CMD	Central Materials Depot
COCI	Call-Off Contract Instruction
COSHH	Control Of Substances Hazardous to Health

Acronyms and Common Terms

Acronym	Description
CoWD	Cost of Work Done
CP	Contract Package
CPMT	Central Project Management Team
CRA	Call Routing Apparatus
CRG	Commercial Review Group
csc	contractor safety case
CSEG	Control Systems Executive Group
CTI	Computer Telephony Integrated Technology
CTRL	Channel Tunnel Rail Link
CUS	Counter & UR Sets
CWR	Continuous Welded Rail
D&C	Design & Construction
dbhp	drawbar brake horse power
DC	Direct Current
DCA	Data Concentrator Appreciation
DCE	District Civil Engineer
DCF	Discounted Cash Flow
DCFR	Discounted Cash Flow Return
DCM	Duty Construction Manager
DDA	Design and Development Authority
DEMU	Diesel-Electric Multiple-Unit
DEP	Designated Earthing Point
DIADS	Diagram Input And Distribution System
DLR	Docklands Light Railways
DMU	Diesel Multiple Unit
DO	Driver Only
DOG	Out Of Gauge
DOO	Driver Only Operation (Operated)
DPC	Data Protocol Converter
DPI	Dye Penetrant Inspection
DTMF	Dual Tone Multi Frequency
DTS	Dynamic Track Stabiliser
DVT	Driving Van Trailer
DWL	Dynamic Warning Lights
E&G	Edinburgh - Glasgow Line
E&M	Earth & Mark
EAC	Extended Arm Contract
EAS	Entrance Aspect Sets
EAWA	Electricity At Work Act
ECML	East Coast Main Line
ECR	Evaluation of Change Request
ECR	Electric Control Room
EECS	Electrical Engineering & Control Systems
EFL	External Financing Limit
EGP	Ex-Gratia Payment
EI	Electrical Installation
ELR	Engineers Line Reference
ELWI	Electric Line Working Instruction
EMI	Electro Magnetic Interference
EMU	Electrical Multiple Unit
ENS	European Night Services

Acronym	Description
EP	Electronic Principles
EP	Electro-Pneumatic
EPD	Engineering & Production Directorate
EPROM	Erasable Programmable Read Only Memory
EPS	European Passenger Services
EQAP	EQuipment APpreciation
ERDF	European Regional Development Fund
ES	Engineering Supervisor
ESD	ElectroStatic Discharge
ESICOW	Engineering Supervisor In Charge Of Work
ESSD	ElectroStatic Sensitive Device
ETD	Extension Trunk Dialling
eth	electric train heating
ETM	Electric Track Maintenance
ETR	Electronic Train Reporting
EU	European Union
EVI	Earned Value Indicators
EWI	Emergency Warning Indicator
EWS	English Welsh and Scottish Railways Limited
F.Eng	First Engineering
FB	FlatBottom Rail Section or Fringe Box
FDM	Frequency Division Multiplex
FE	Foundation Electronics
FLAWS	Computer Database of Rail FLAWS
FOC	Freight Operating Company
FOC (BT)	Fibre Optic Cable
FORI	Fibre-Optic Route Indicator
FPL	Facing Point Lock
FPL	Facing Point Lock
FRAME	Fault Reporting And Monitoring Equipment
FREDDY	Flange Reading Electronic Detector Designed York
FS	Feeder Station
FT	Fixed Terminations
G&SW	Glasgow - South Wales Line
GEMINI	Twin System of PROMISE and TRUST
GEOGIS	GEOGraphical Information System
GF	Ground Frame
Gij	Glued insulated rail joint
GLW	Gross Laden Weight
GNER	Great North Eastern Railway
GPL	Ground Position Light
GTO	Gate Turn-Off
GWR	Great Western Railway
GWS	GateWay System
H/S	Hand Singalman
HABD	Hot Axle Box Detector

Acronyms and Common Terms

Acronym	Description
HASAW	Health And Safety At Work
HAZCHEM	HAZardous CHEMical
HAZID	HAZard IDentification
HAZOP	HAZard & OPerability Study
HAZPAK	Training Course for Drivers of Vehicles carrying dangerous substances
HMRI	Her Majesty's Railway Inspectorate
HOP	Head Of Projects
HOPO	Head Of Projects Organisation
HR	Human Resources
HRL	High Rail Level
HSE	Health & Safety Executive
HTS	High Tensile Steel
HVI	High Voltage Impulse
IA	Implementation Authority
IB	Intersection Bridge
IBIS	Internal Business Invoicing System
IBJ	Insulated Block Joint
IDC	Insulation Displacement Connector
IDF	Intermediate Distribution Frame
IECC	Integrated Electronic Control Centre
ihp	indicated horse power
ILWS	Inductive Loop Warning System
IMACS	Inventory Management Accounting & Control System
IMC	Infrastructure Maintenance Company
IMDG	International Maritime Dangerous Goods Code
IMU	Infrastructure Maintenance Unit
INTERFRIGO	INTERnational ReFRIGerated Bogied Vehicle
IP	Investment Proposals
IRJ	Insulated Rail Joint
IRSE	Institution of Railway Signal Engineers
IS	Information System
is	intermediate signalling
ISG	Infrastructure Support Group
ISM	IECC System Monitor
ISO	International Standards Organisation
ISRS	International Safety Rating System
ISTP	Intermediate Signalling Technology: Principles
ITS	Infrastructure Testing Services
ITU	International Telecommunications Union
JCN	Job Cost Number
JCT	Pointless Track Circuit or Joint Construction Trades
JS2	Advanced Signalling Cable Jointing & Testing
jsi	Basic Signalling Cable Jointing & Testing
JTC	Jointless Track Circuit
KE	Kinematic Envelope
KPI	Key Performance Indicator
LAN	Local Area Network
LB	Line Blocked
LC	Level Crossing or Locally Controlled Manned Level Crossing or Line
LCAMS	Low Carbon Austenitic Manganese Steel
LCD	Liquid Crystal Display
LCR	London & Continental Railways
LCU	Local Control Unit
LDG	List of Dangerous Goods
LED	Light Emitting Diodes
LEM	Local Estimating Manager
LJU	Line Jack Unit
LLPA	Long Line Public Address
LLR	Level Locking Release
LOC	LOcation Cabinet/Cupboard
LOS	Limit Of Shunt Indicator
LPS	Local Poilicy Statement
LRT	Light Rail Transit
LSL	LineSide Location
LT	London Transport
LUL	London Underground Limited
LVDT	Linear Variable Differential Transformers
M&E	Mechanical & Electrical
M&EE	Mechanical & Electrical Engineer also covers Traction & R. Stock Eng
MAS	Multiple Aspect Signalling
MAS	S&T Management Aspects of Supervision
MAS	Multi Aspect Signalling
MB	Manned Barriers
MCB	Manually Controlled Barriers
MDC	Management Driven Change
MDF	Main Distribution Frame
MER	Main Entrance Registered Relay
MGL	Multi Groove Locking
MIS	Management Information System
ml	mechanical installation practices
MLRI	Multi-Lamp Route Indicator
MODS	Motor Operated Disconnectors
MPD	Major Projects Division
MPI	Magnetic Particle Inspection
MPM	Multi - Processor Module
MRP	Management Report Pack
MS	Motorised Switch
MSP	Measured Shovel Packing
MTBF	Mean Time Between Failures
MTRT	Matisa Track Recording Trolley

Acronyms and Common Terms

Acronym	Description
MWL	Miniature Warning Lights
N&S	Non Stopping & Stopping
NAE	National Accounts Executive
NAPS	National Accounts Payable System
NBA	NRN Base Station Alignment
NCR	Non Conformance Report
NDA	Not Described Alarm
NDT	Non-Destructive Testing
NEQ	Net Explosive Quantity
NG	Normal Grade
NLR	North London Railways
NPS	National Payroll System
NPV	Nett Present Value
NR	Not Registered
NRG	National Records Group
NRM	Nominated Responsible Manager
NRN	National Radio Network
NRNA	National Radio Network Appreciation
NS	Not Supported
NSE	Network South East
NSKT	No Signalman Key Token
NSR	Nor Supported or Registered
NSTR	No Signalman Token Remote
NUT	Number Unobtainable Tone
NWRR	North West Regional Railways Limited
NX	ENtrance-EXit
O&I	Operations & Interface
O&IM	Operations & Interface Manager
O/L	OverLap
OB	OverBridge
oc	open crossing
occ	occupied
ocs	one control switch
ocu	operator's control unit
OE	Outside Edge
OHLE	OverHead Line Electric
OLC	Occupation Level Crossing
OLE	Overhead Line Equipment
OMO	One Man Operation
OPEX	OPerating EXpenditure
OPO	One Person Operation
OPRAF	Office of Passenger RAil Franchising
OPS	Outline Project Specification
OPS	Overlap Point Set
OPSSAP	OPerationS Safety Assessment Panel
ORN	Overlay Radio Network
ORR	Office for the Rail Regulator
OSD	Overhead System Design
OSZ	Operational Safety Zone
OTDR	Optical Time Domain Reflectometer
OTW	One Train Working

Acronym	Description
P.WAY	Permanent WAY
P3	Primavera Project Planning Application
PA	Public Address
PABX	Private Automatic Branch EXchange
PACT	PAved Concrete Track (slab track)
PAF	Project Authority Form
PALADIN	Performance And Loading Analysis Database of INformation
PAX	Private Automatic EXchange
PC	Power Cubicle
PCB	Poly Chlorinated Biphenyis or Printed Circuit Board
PCI	Pre-Connection Inspection
PCM	Project Commercial Manager
PCM	Project Control Manager
PCM	Pulse Code Modulator
PCS	Point Control Set
PCS&TE	Profit Centre Signal & Telecommunications Engineer
PCSE (M)	Profit Centre Signal Engineer (Maintenance)
PCSE (W)	Profit Centre Signal Engineer (Works)
PCTE	Profit Centre Telecommunications Engineer
PD	Project Delivery
PDG	Product Delivery Group
PDG	Procurement Development Group
PDMX	Programmable Digital MultipleXer STC
PE	Point Estimate
PEARLS	PErsonAl Record Local System
PEARS	Paladin Data Extract And Reporting System
PEE-WEE	A Warning Device for On-or About the line
PFTD	Powered Plant Training: Disc Cutters and Class 9 Cutting Off Wheels
PHIS	Performance Historic Information System
PICOP	Person In Charge Of Possession
PICOT	Person In Charge Of Testing
PICOW	Person In Charge Of Work
PIG	Performance Improvement Group
PIMQUIT	Possessions & Isolations Management QUality Improvement Team
PINE	Train Part Cancelled
PIO	Police Incident Officer
PIP	Property Investment Panel
PIS	Point Interlocking Set & Passenger Information System
PLGS	Position Light Ground Signal
PLJI	Position Light Junction Indicator

Acronyms and Common Terms

Acronym	Description
PLOD	Patrolman's LockOut Device
PLRA	Private Locomotive Registration Agreement
PLS	Position Light Signal
PM	Project Manager
PMBX	Private Manual Branch EXchange
PMCS	Project Management Control System
PMI	Project Manager's Instruction
PMO	Project Manager Operations
PMP	Project Management Process
PON	Periodical Operating Notice
POT	Possessions Operation Team
PPE	Personal Protective Equipment
PPM	Panel Processor Module
PPTC	Powered Plant Training: Chain Saw
PRF	Purchase Order Request Form
PRI	Preliminary Routing Indicator
PROM	Programmable Read Ony Memory
PROMISE	Planning and Resources Monitoring System
PSB	Power SignalBox
PSCR	Project Safety Case Requirements
PSO	Public Service Obligation
PSR	Permanent Speed Restriction
PSTN	Public Switched Telephone Network
PTA	Passenger Transport Authority
PTE	Passenger Transport Executive
PTI	Positive Train Identification
PTO	Principal Technical Officer
PTO	Public Telecomms Operator
PTS	Personal Track Safety
PW	Parallel Wing
PW	Permanent Way
PWID	Portable Warning Issuing Device
PWMA	Permanent Way Maintenance Assistant
PWME	Permanent Way Maintenance Engineer
PWRA	Private Wagon Registration Agreement
PWSS	Permanent Way Section Supervisor
QAM	Quality Assurance Manager
QIT	Quality Improvement Team
QRA	Quantative Risk Analysis
QSL	Structured Query Language
R&R	Rationalisation & Resignalling
R/G	Miniature Red/Green Warning Lights
RA	Route Availability
RAI	Right Away Indicator
RAP	Remedial Action Project
RC	Remote Control
RC	Remotely Controlled Manned Level Crossing
RCE	Regional Civil Engineer
RD	GEC Time Division Multiplex, Type RD
REB	Relocatable Equipment Building
REFOS	Running Edge to Face Of Structure
REG	Railtrack Executive Group
REN	Ring Equivalent Number
RER	Regional Express System
RES	Rail Express Systems
RETB	Radio Electronic Token Block
REV	Date of overhaul - International Registered Wagons
RFA	Request For Authority
RfD	Railfreight Distribution
RGS	Railway Group Standard
RHS	Rectangular Hollow Section
RI	Relay Interlocking
RIC	Rail Incident Commander
RID	Reg re International Carriage of Dangerous Goods by rail
RIDDOR	Reportable Injuries Diseases and Dangerous Occurrences Regulations
RIO	Rail Incident Officer
RIP	Railway Investment Panel
RIPR	Railtrack Internal Possessions Request
RIV	Regulations governing the exchange of International Railway Wagons
RLP	Railway Line Procedures
RM	GEC Time Division Multiplex, Type RM
RMT	Rail, Maritime and Transport Workers Union
RoD's	Rectification of Defects
ROLO	Rail Operations Liaison Officer
ROM	Regional Operating Manager
ROR	Rules Of Route
ROSCO	ROlling Stock COmpany(ies)
RRI	Route Relay Interlocking
RRIP	Route Relay Interlocking Processor
RRMS	Railtrack Risk Management System
RRNE	Regional Railways North East Limited
RRR	Remote Relay Room
RSC	Railway Safety Case
RSEG	Railtrack Scotland Executive Group
RSF	Right Side Failure
RSIP	Railtrack Scotland Investment Panel
RSL	Rolling Stock Library
RT	Radiographic Testing
S&C	Switches & Crossings
S&SD	Safety & Standards Directorate

Acronyms and Common Terms

Acronym	Description
S&T	Signalling & Telecommunications
S&TE	Signal & Telecommunications Engineer
safe cess	Footpath Walkway between outer edge of track and boundary fence
SAP	Safety Advisory Panel
SATBUD	S&T Paybill BUDget System
SC	Signalling Centre
SCM	Site Construction Manager
SCUK	Signalling Control UK
SD	Standard Deviation
SDA	SSI Data Appreciation
SDS	Signalling Display System
SG	Special Ground
SICA	Signalling Infrastructure Condition Assessment
SIGTAN	SIGnalling Equipment Technical Advice Notice
SIGWEN	SIGnalling Equipment Workshop Engineering Notice
SIMBIDS	SIMplified BIDirectional Signalling
SIN	Special Inspection Notice
SIN	Substance Identification Number
SINAC	SIN Action Complete
SINMON	SIN MONitor
SIN's	Special Inspection Notices
SIP	Strike-In Point
SIPS	Standard Infrastructure Performance System
SIS	Staff Information System
SIVS	Station Information VDU System
SLR	Selection Level Release
SMA	Statistical Multiplex Appreciation (or Strathclyde Manning Agreement)
SMOS	Structure Mounted Outdoor Switch
SMS	Signalling Maintenance Specification
SMT	Signalling Maintenance Testing
SMTH	Signalling Maintenance Testing Handbook
SPAD	Signal Passed At Danger
SPG	Signals Project Group
SPM	Senior Project Manager
SPT	Signal Post Telephone (or Strathclyde Passenger Transport)
SPWEE	Safety Procedures for Working on Electrical Equipment
SQLA	Select, Quantify, Label & Adjust
SRA	Safety Risk Assessment
SRG	Safety Review Group
SRP	Statutory Report Pack
SRS	Safety Review System
SS	Subsidiary Signal
SSDC	Signalling System Direct Current
SSI	Solid State Interlocking
SSI DWS	Solid State Interlocking Design WorkStations
SSIC	Solid State Interlocking Controller
SSIDES	SSI DESign
SSISIM	SSI SIMulator
SSP	Signalling Supply Point
SSRB	Special Speed Restriction Board
STABS	S&T Area Budget System
STO	Senior Technical Officer
STOFIS	S & T Outturn Processing System
STRC	Scottish Track Renewals Company
SWT	South West Trains
SWTH	Signalling Works Testing Handbook
SWWR	South Wales & West Railways
T&RS	Traction & Rolling Stock
T-COD	Track Circuit Operating Device
T2	Possession - X, D, H, T (Blockage)
T3	Possession
TAPI	Telephone Application Programming Interface
TAC	Telecomms Appreciation Course
TBS	Transmission Based Signalling
TC	Track Circuit
TCA	Track Circuit Actuator
TCAIS	Track Circuit Actuator Interface Detector
TCB	Track Circuit Block
TCI	Track Circuit Interrupter
TD	Train Describer
TD-D	Data Course
TD-T	Transmission Course
TDM	Time Division Multiplex
TE	Tractive Effort
TEN	Trans European Network
TES	Tender Event Schedule
TEU	Twenty-foot or Equivalent Unit
TFC	Telecomms Fault Control
TFM	Trackside Functional Module
TI 2I	Type of Track Circuit
TIB	Time Interval Block
TIC	Technical Investigation Circuit
TIC	Track Inspection Coach
TIC	Tester In Charge
TID	Track IDentity
TIMI	Telecommunication Maintenance Instruction
TIPS	Telecommunications Installation and Procurement Service

Acronyms and Common Terms

Acronym	Description
TIR	Transport International Routier
TIS	Train Information Systems
TLC	Crossing with Telephone Protection Only
TMO	Traincrew Operated Crossing
TO	Team Organiser or Technical Officer
TOC	Train Operating Company
TOFC	Trailer On Flat Car
TOL	Train On Line
TOPS	Train Operations Processing System
TORR	Train Operated Route Release
TOU	Train Operating Unit
TOWS	Train Operated Warning System
TPR	Track Proving Relay
TQ	Technical Query
TR	Track Relay
TRAMM	Track Renewal And Maintenance Machine
TRANSFESA	An Operator of Privately Owned Wagons
TRC	Track Recording Coach
TRE	Track Recording Engineer
TREMCARD	TRansport EMergency CARD
TRG	Technical Review Group
TRTS	Train Ready To Start
TRU	Track Repair Unit
TRU	Track Renewal Unit
TRUST	Train RUnning syzSTem - TOPS
TRV	Track Recording Vehicle
TSC	Track Sectioning Cabin
TSDB	Train Service DataBase
TSF	Task Specification Form
TSL	Track Sectioning Location
TSR	Temporary Speed Restriction
TSSA	Transport Salaried Staff Association
TTP	TimeTable Processor
TTR	Track to Train Radio
TUPE	Transfer of Undertakings (Protection of Employment) Regulations 1981
TWE	Track Welding Engineer
UB	UnderBridge
UBA	Use to Best Advantage
UIC	Union International des Chemins de fer (International Union of Railways)
URFDO	Ultrasonic Rail Flaw Detector Operator
URX	Under Road Crossing
USART	Universal Synchronous / Asynchronous Receiver
UT	Ultrasonic Testing
UTU	Ultrasonic Testing Unit
UTX	Under Track Crossing
UWSF	Unprotected Wrong Side Failure
VAB	Vehicle Acceptance Body
VAT	Value Added Tax
VB	Viaduct Bridge
VDP	Vertical Design Package
VDU	Visual Display Unit
VoWD	Value of Work Done
WAGN	West Anglia Great Northern Railway Limited
WAN	Wide Area Network
WBS	Work Breakdown Structure
WCML	West Coast Main Line
WEAVE	To WEAVE between the tracks (Possession type)
WEN	Weekly Engineering Notice
WETOC	Wembley European Train Operating Centre
WI	Work Instruction
WI	Warning Indicator
WON	Weekly Operating Notice
WR SSTD	WR Solid State Train Describer
WSF	Wrong Side Failure
XS	EXit Sets
ZIP	Zone Investment Panel

Location Factors

The location map factors relate to the amount that the costs of general construction on railway projects vary throughout the UK.

The cost of construction is affected by the factor value indicated in the required location. The cost data contained in this publication assume a location factor of 1.00.

The chart is intended to illustrate general trends rather than detailed precise data.

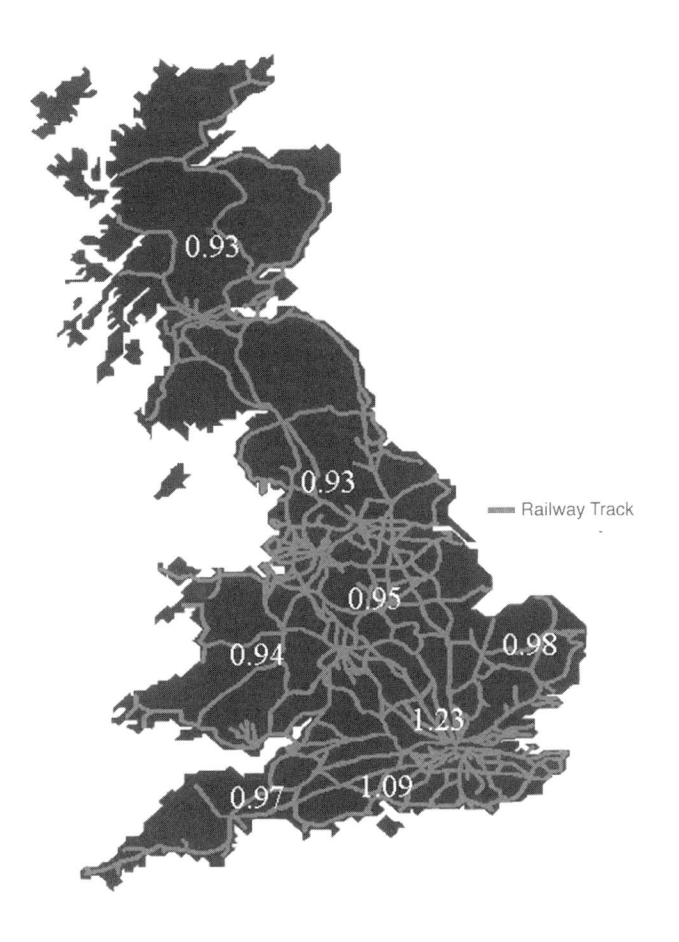

Map of the UK showing location factors:
- 0.93 (Scotland)
- 0.93 (Northern England)
- 0.95 (Midlands)
- 0.94 (Wales)
- 0.98 (East England)
- 1.23 (London)
- 0.97 (South West)
- 1.09 (South East)

Railway Track

Indicative International Location Factors

By the analysis of construction projects throughout various locations around the world the following indicative international location factors have been calculated.

Care should be exercised when applying an international factor to another country's project. Procurement, technology, availability of materials, labour and plant may be different within any individual country, sometimes resulting in a project being factored that cannot actually be built in another part of the world.

These factors are presented as a guide only, a full study should be undertaken to obtain an accurate feasibility report in respect of any proposed project.

(Note: UK = 1.00)

	Mean	Low	High		Mean	Low	High
ARGENTINA	**0.79**	0.62	1.22	Shenzen	0.69	–	–
AUSTRALIA	**1.03**	0.78	1.39	Beijing	0.73	–	–
Adelaide	0.90	–	–	Shanghai	0.75	–	–
Brisbane	0.88	–	–	Macao	0.78	–	–
Darwin	1.17	–	–	**COLOMBIA**	**0.78**	0.47	1.25
Hobart	1.04	–	–	**COTE D'IVOIRE**	**0.88**	0.53	1.40
Perth	0.90	–	–	**CYPRUS**	**0.85**	0.62	1.23
Sydney	1.03	–	–	**CZECH REP**	**0.75**	0.40	1.31
Melbourne	0.97	–	–	Prague	0.75	–	–
Queensland Islands	1.15	–	–	Brno	0.72	–	–
AUSTRIA	**1.19**	0.79	1.52	Liberec	0.70	–	–
BAHRAIN	**0.79**	0.54	1.16	**DENMARK**	**1.24**	0.83	1.67
BANGLADESH	**0.77**	0.29	1.22	Copenhagen	1.24	–	–
BELGIUM	**1.09**	0.84	1.35	Odense	1.33	–	–
BOLIVIA	**0.80**	0.44	1.27	Arhus	1.37	–	–
BRAZIL	**0.76**	0.47	1.19	**EGYPT**	**0.84**	0.50	1.34
Rio de Janeiro	0.76	–	–	Alexandria	0.77	–	–
Sao Paulo	0.77	–	–	Cairo	0.84	–	–
Curitiba	0.77	–	–	Port Said	0.77	–	–
Salvador	0.79	–	–	Suez	0.77	–	–
BULGARIA	**0.79**	0.47	1.26	Other Areas	0.71	–	–
CANADA	**1.02**	0.61	1.63	**EL SALVADOR**	**0.59**	0.35	0.95
Edmonton	0.89	–	–	**FINLAND**	**1.31**	0.79	1.78
Calgary	0.94	–	–	**FRANCE**	**1.16**	0.91	1.43
Halifax	1.07	–	–	Paris	1.16	–	–
Montreal	0.92	–	–	Ardenne	1.12	–	–
Ottowa	0.99	–	–	Picardie	1.11	–	–
Quebec	0.92	–	–	Haute-Normandie	1.08	–	–
Toronto	1.02	–	–	Centre	1.13	–	–
St. Johns	1.14	–	–	Basse Normandie	1.14	–	–
Vancouver	0.97	–	–	Bourgogne	1.17	–	–
Winnipeg	0.89	–	–	Nord-Pas-de-Calais	1.11	–	–
CAPE VERDE	**1.91**	–	–	Lorraine	1.15	–	–
CHILE	**0.77**	0.46	1.22	Alsace	1.16	–	–
CHINA	**0.69**	0.56	1.42	Franche Comte	1.13	–	–

Indicative International Location Factors

	Location Factor				Location Factor		
	Mean	Low	High		Mean	Low	High
GERMANY	**1.22**	0.89	1.56	Hiroshima	1.77	–	–
Baden - Stuttgart	1.29	–	–	Kawasaki	1.82	–	–
Bavaria, Munich	1.28	–	–	Yokohama	1.82	–	–
Berlin	1.46	–	–	Kobe / Kyoto	2.09	–	–
Bremen	1.18	–	–	Nagoya	1.86	–	–
Hamburg	1.18	–	–	Osaka	1.99	–	–
Hannover	1.16	–	–	Sapporo	1.84	–	–
Frankfurt	1.22	–	–	Tokyo	1.90	–	–
Pomerania	1.16	–	–	**KENYA**	**0.68**	0.26	1.10
Cologne	1.24	–	–	**KOREA, SOUTH**	**0.95**	0.57	1.51
Ludwigshaven	1.26	–	–	Incheon	0.93	–	–
Saarbrucken	1.30	–	–	Pusan	0.94	–	–
GHANA	**0.86**	0.51	1.37	Seoul	0.95	–	–
GREECE	**0.91**	0.55	1.46	Taegu	0.92	–	–
HONG KONG	**1.24**	0.65	1.53	**KUWAIT**	**0.75**	0.46	1.23
HUNGARY	**0.88**	0.50	1.29	Kuwait	0.75	–	–
Budapest	0.88	–	–	Al Jahrah	0.81	–	–
INDIA	**0.74**	0.27	1.17	Ad Dawhah	0.85	–	–
INDONESIA	**0.72**	0.36	1.10	Ash Shuaybah	0.94	–	–
Bandung	0.66	–	–	**LIBYA**	**1.04**	0.71	1.17
Jakarta	0.72	–	–	**MALAWI**	**0.70**	0.42	1.13
Medan	0.66	–	–	**MALAYSIA**	**0.79**	0.30	1.13
Surabaya	0.64	–	–	Kuala Lumpur	0.79	–	–
Other Areas	0.57	–	–	Johor Baharu	0.87	–	–
IRELAND	**1.12**	0.89	1.37	**MEXICO**	**0.93**	0.56	1.49
Athlone	1.09	–	–	Ciudad Juarez	1.07	–	–
Cork	1.15	–	–	Guadalajara	0.86	–	–
Dublin	1.12	–	–	Merida	0.82	–	–
Dundalk	1.08	–	–	Mexico City	0.93	–	–
Galway	1.10	–	–	Monterrey	0.89	–	–
Limerick	1.14	–	–	Puebla	0.89	–	–
Waterford	1.13	–	–	**NETHERLANDS**	**1.11**	0.86	1.34
Wexford	1.14	–	–	Arnhem	1.05	–	–
ITALY	**1.00**	0.65	1.43	Amsterdam	1.11	–	–
Bari	0.90	–	–	Apeldorn	1.05	–	–
Bologna	0.98	–	–	Einhoven	1.03	–	–
Cagliari	0.86	–	–	Enschede	1.02	–	–
Florence	1.01	–	–	Groningen	1.01	–	–
Genoa	0.99	–	–	Rotterdam	1.12	–	–
Milan	1.00	–	–	The Hague	1.13	–	–
Naples	0.90	–	–	Utrecht	1.11	–	–
Rome	0.95	–	–	**NEW ZEALAND**	**1.15**	0.87	1.46
Palermo	0.86	–	–	**NIGERIA**	**0.71**	0.49	1.31
Turin	1.01	–	–	Abuja	0.67	–	–
Venice	1.01	–	–	Lagos	0.71	–	–
JAPAN	**1.90**	0.74	2.23	Port Harcourt	0.67	–	–
Fukuoka	1.67	–	–	**NORWAY**	**1.32**	0.86	2.45
Kitakyusha	1.67	–	–	Bergen	1.28	–	–

Indicative International Location Factors

	Location Factor				Location Factor		
	Mean	Low	High		Mean	Low	High
Kristiansand	1.31	–	–	SWEDEN	**1.40**	0.46	1.80
Oslo	1.32	–	–	Gothenburg	1.36	–	–
Stavanger	1.30	–	–	Malmo	1.39	–	–
Trondheim	1.35	–	–	Norrkoping	1.39	–	–
OMAN	**0.76**	0.45	1.21	Orebro	1.37	–	–
PAKISTAN	**0.77**	0.46	1.23	Stockholm	1.40	–	–
PERU	**0.59**	0.35	0.95	Uppsala	1.37	–	–
PHILIPPINES	**1.08**	0.34	1.88	**SWITZERLAND**	**1.40**	1.00	1.83
POLAND	**0.82**	0.48	1.92	**TAIWAN**	**1.18**	0.92	1.95
Lodz	0.78	–	–	**THAILAND**	**0.52**	0.48	1.00
Krakow	0.79	–	–	**TURKEY**	**0.82**	0.49	1.32
Poznan	0.78	–	–	Istanbul	0.82	–	–
Warsaw	0.82	–	–	Ankara	0.78	–	–
Other Areas	0.74	–	–	Other Areas	0.66	–	–
PORTUGAL	**0.83**	0.50	1.32	**UGANDA**	**0.72**	0.43	1.16
Lisbon	0.83	–	–	**UKRAINE**	**0.77**	0.46	1.24
Setubal	0.81	–	–	**UNITED ARAB**			
Coimbra	0.78	–	–	**EMIRATES**	**0.69**	0.42	1.04
Braga	0.74	–	–	**UNITED KINGDOM**	**1.00**	0.60	1.60
Faro	0.95	–	–	USA (New York City)	**1.20**	0.83	1.62
ROMANIA	**0.75**	0.51	1.20	USA (Los Angeles)	**1.08**	0.77	1.10
RUSSIAN FED.	**0.91**	0.61	1.24	USA (Chicago)	**0.96**	0.72	1.00
SAUDI ARABIA	**0.69**	0.57	1.27	USA (New Orleans)	**0.91**	0.71	1.16
Dammam / Dhahran	0.72	–	–	**VENEZUELA**	**0.77**	0.51	1.37
Jeddah	0.65	–	–	**VIETNAM**	**0.96**	0.57	1.53
Mecca	0.65	–	–	**WEST INDIES**	**1.03**	0.62	1.64
Riyadh	0.69	–	–	**ZIMBABWE**	**0.82**	0.49	1.31
Other Areas	0.62	–	–				
SINGAPORE	**1.06**	0.53	1.47				
SLOVENIA	**0.90**	0.54	1.43				
SOUTH AFRICA	**0.72**	0.44	1.31				
Cape Town	0.69	–	–				
Durban	0.70	–	–				
Johannesburg	0.72	–	–				
Pretoria	0.71	–	–				
Port Elizabeth	0.71	–	–				
SPAIN	**0.86**	0.61	1.17				
Barcelona	0.88	–	–				
Bilbao	0.83	–	–				
Las Palmas							
de Gran Canaria	0.82	–	–				
Malaga	0.83	–	–				
Madrid	0.86	–	–				
Palma	0.85	–	–				
Valencia	0.84	–	–				
Seville	0.87	–	–				
SRI LANKA	**0.72**	0.43	1.15				
St. KITTS	**1.10**	0.65	1.23				

Index

AC Electrification
  Cable 54, 62
  Distribution equipment 52, 60
  Distribution equipment – inner city 73
  Distribution equipment – Rural 77
  Distribution equipment – Suburban 75
  Inner city 47-54, 73-74
  Inner city – Repairs and renewal 73
  Inner city – Switchgear 74
  Line installation – Suburban 76
  Overhead line works 47-51, 55-59
  Overhead line works – inner city 72
  Overhead lineworks – Inner city 74
  Repairs and renewals – inner city 72
  Rural 63-72, 77-78
  Rural – Cable 72
  Rural – Distribution equipment 70-71
  Rural – Overhead line works 63-71
  Rural – Testing and commissioning 78
  Rural – Wiring 77
  Suburban 55-62, 75-76
  Suburban – OLW4 75
  Suburban – Switchgear 76
  Switchgear 53, 61
  Testing and commissioning – Rural 78
AC traction 18
Ancillaries – Pipework 127
Ancillaries – Trackwork 125-126
Ancillary items; supply and installation 87
Appendices 393
Asphalt work 202-203, 205-207
Automatic half barrier 235-240
  CCTV 239
  Contractor's testing 239
  PET system 240
  Power supplies 238
  Recoveries 240
  Signals 238
  Treadle equipment 239
Automatic open crossings 241-246
  Barrier equipment 242, 248
  CCTV 245
  Civil works 241
  Contractor's design 247
  Contractor's preliminaries/design 241
  Crossing decks 242
  GATSO equipment 243
  PET system 246
  Power supplies 244
  REB transformers 244
  REBs 243, 249
  Recoveries 246, 252
  Signage 242
  Signals 244, 250
  Telecoms 245, 251
  Treadle equipment 245
Automatic open crossings (AOCL) 247-252
  Civil works 247
  Contractor's preliminaries 247
  Crossing decks 248
  GATSO equipment 249
  PET system 252
  Power supplies 250
  Treadle equipment 251
Automatic warning systems 32

Banner signals 9-10
Barrier equipment 236, 248
  Automatic open crossings 242
  Automatic open crossings (AOCL) 248
Base Construction 97
Bases 8, 38
Beams and columns 277-278
  Precast concrete 277-278
Boreholes 301-302
  General civils 301-302

Brick and block walling 196
Brickwork – Tunnels 291
Brickwork and blockwork 194-200
Brickwork in underpinning 190
Brickwork sundries 191
Bridges
  Concrete ancillaries 269-276
  Concrete work 265-268
  Demolition of structures 253
  Demolitions and removals 253-257
  Earthworks 259-264
  Fencing 256
  Miscellaneous metalwork 285-286
  Pipelines and ducts 257
  Precast concrete 277-280
  Removal of old work 255
  Site clearance 258
  Structural metalwork 281-282, 284

Cable 54, 62
Cable – rural 72
Cable routes 16, 40-42
Cabling 12
Cavity wall insulation 199
CCTV 21-22, 251
  Automatic open crossings 245
Ceiling repairss 229
Ceramic tiling repairs 230
Civil works 235
  Automatic open crossings 241
  Automatic open crossings (AOCL) 247
Clay floor tiling 228
Coloured mastic asphalt flooring 206
Concrete accessories 276
  Concrete ancillaries 276
Concrete ancillaries 269-276, 323-327, 330
  Bridges 269-276
  Concrete accessories 276
  Designed joints in in situ concrete 275
  Formwork 269-271
  General civils 323-327, 330
  Joints in concrete pavings 273-274
  Reinforcement 272-273
Concrete pavements 368
  General civils 368
Concrete work 178-179, 265-268
  Bridges 265-268
  In situ concrete 265-268
Conductor rail 83, 92, 94
Contractor's design 235
  Automatic open crossings (AOCL) 247
Contractor's preliminaries, Automatic open crossings (AOCL) 247
Contractor's preliminaries/design, Automatic open crossings 241
Contractor's testing, Automatic half barrier 239
Copings and sills 280
  Precast concrete 280
Cross wall construction 150
Crossing decks 236
  Automatic open crossings 242
  Automatic open crossings (AOCL) 248
Crossing decks (AOCL) 248
Culverts, ducts and covers 279
  Precast concrete 279

Damp proof courses 192-193, 200
Data cable 13
DC Electrification
  Ancillary items; supply and installation 87
  Conductor rail 83, 92, 94
  Contractor's Testing and Commissioning 96
  High voltage cable 89-90
  Inner city 79-80, 83-87
  Inner city – High voltage cable 80
  Inner city – Labour rates 83

Inner city – Sub stations 79
Labour rates 83, 91
Points heating 83-86
Strip out and recoveries 95
Sub stations 79, 88
Substations 93
Suburban 88-95
Supply and installation 95
Testing and commissioning 96
Track feeder cable 80, 93
DC traction 18
Demolition and site clearance 309-314
  General civils 309-314
Demolition of structures 253, 309
  Bridges 253
  General civils 309
Demolitions and alterations 178
Demolitions and removals 159-167, 169, 253-257
  Bridges 253-257
Designed joints in in situ concrete 275
  Concrete ancillaries 275
Disposal and filling 175-177
Distribution equipment 52, 60
Distribution equipment – inner city 73
Distribution equipment – rural 70-71
Distribution equipment – Suburban 75
Door furniture 212-214
Drilling for grout holes 307
  General civils 307

Earthwork, In situ concrete 320
Earthwork support 173-174
Earthworks 259-264, 315-320
  Bridges 259-264
  Excavation ancillaries 261
  Filling to excavations 262-264
  General excavations 259-260
Erection of frames and other members, Structural metalwork 282
Erection of members for bridges 282
  Structural metalwork 282
Excavation 184-186
Excavation ancillaries 261-264
  Earthworks 261
  General civils 317
Excavation by hand 170
Expanded metal reinforcement 200
External cable 12

Fabrication of main members 281
  Structural metalwork 281
Fabrication of secondary members 353
  General civils 353
Facilities 151-157
Fencing 256-257, 312
  Bridges 256
  General civils 312, 390
Fencing – General civils 390
Fibre optic cable 14
Fibre optic route indicators 9-10
Filling 319
Filling to excavations, Earthworks 262-264
Fixing plates and brackets 33
Fixings 217
Flexible road surfacings 367
Formwork 269-271, 323, 325
  Concrete ancillaries 269-271
Formwork to underpinning 189

Gantries and cantilevers 11
Gates – General civils 392
GATSO equipment 237, 249
  Automatic open crossings 243
  Automatic open crossings (AOCL) 249
General civils
  Concrete ancillaries 323-327, 330

Index

Demolition and site clearance 309-314
Demolition of structures 309
Drilling for grout holes 307
Earthworks 315-320
Fencing 312
General site clearance 313
Geotechnical and other specialist processes 307-308
Ground investigation 297-306
Grout holes 308
Grout materials and injection 308
In situ concrete 321-322
Masonry 376-386
Miscellaneous work 390-392
Painting 387
Pile tests 364
Piles 359-362
Piling ancillaries 363-364
Pipework – fittings and valves 340-341
Pipework – manholes and pipework ancillaries 342-344
Pipework – pipes 334-339
Pipework – supports and protection 345-350
Precast concrete 331-333
Removal of old work 311
Roads and paving 365-375
Structural metalwork 351-358
Waterproofing 388-389
General demolitions 159-161, 167, 169
General excavations 259-260
   Earthworks 259-260
General fixtures and equipment 220
General glazing 231
General items, Tunnels 287
General items – Tunnels 287
General removals 194, 202-203
General site clearance 313
   General civils 313
Geotechnical and other specialist processes 307-308
Glazing 231-232
   Plain glazing 231-232
   Special 232
   Works of removal 231
Granolithic screeds 226-227
Ground investigation 297-306
   General civils 297-306
Grout holes 308
Grout materials and injection 308

High voltage cable 80, 89-90

In situ concrete 179, 265-268, 321-322
   Concrete work 265-268
   Placing of concrete 321
In situ concrete in underpinning 187-188
In situ finishings 223-224
Infrastructure systems 141-147
Interlocking systems 38
Internal finishings 221-230
   Ceiling repairs 229
   Ceramic tiling repairs 230
   Clay floor tiling 228
   Granolithic screeds 226-227
   In situ finishings 223-224
   Labours on rendering 225-226
   Plasterwork to walls repairs 230
   Portland cement finishes 225
   Preparation of surfaces 222
   Removal of old work 221
   Repairs and alterations 229
   Vinyl floor finishes 228
Internal fittings 27-29
Ironmongery 211-214
Ironmongery – Door furniture 212-214

Jointing 15

Joints in concrete pavements 370
Joints in concrete pavings 273-274, 327
   Concrete ancillaries 273-274

Kerbs, edgings and channels 371

Labour rates 91
Labours on rendering 225-226
Lay only: junctions and turnouts 120-124
Lay only; crossovers 115-119
Level crossings
   Automatic half barrier 235-240
   Automatic open crossings 241-246
   Automatic open crossings (AOCL) 247-252
   Barrier equipment 236
   CCTV 21-22
   Civil works 235
   Contractor's design 235
   Contractor's preliminaries 235
   Crossing decks 236
   GATSO equipment 237
   Lifting barriers 19-20
   Lights 20
   REBs 237
   Signage 236
Light duty pavings 373
   General civils 373
Lights 20-21
Line installation – Suburban 76

Manholes, Catchpits etc. 134
Masonry 376-386
Masonry paint 233-234
Mastic asphalt flooring 205
Material supply 105-115
Metalwork 219-220
   General fixtures and equipment 220
   Signage 219
Miscellaneous activities 17
Miscellaneous additional work 32
Miscellaneous equipment 45
Miscellaneous metalwork 285-286
   Bridges 285-286
   Stair units and walkways 285-286
Miscellaneous work 390-392
   General civils 390-392
MK2 rail clamp point locks 30-31

New work 195, 203

Off site surface treatment 284
   Structural metalwork 284
OLW2 – Inner city 73
OLW3 – inner city 74
OLW4 – Suburban 75
Operational systems 140
Overhead line works 47-51, 55-59
Overhead line works – inner city 72
Overhead line works – rural 63-71
Overhead lineworks – Inner city 74

Painting 387
   General civils 387
Painting and decorating 233-234
   Emulsion paint 233
Permanent Way
   Ancillaries 125-127
   Base construction 97
   Lay only: junctions and turnouts 120-124
   Lay only; crossovers 115-119
   Manholes, Catchpits etc. 134
   Material supply 105-115
   Pipe bedding 136-137
   Pipe surrounds 134-135
   Pipework 127-133
   Plain line 99-103

   Removals 97-98
   Soakaways 135
   Switches and crossings 104-124
   Track crossings 135
   Track drainage 128-137
   Track Foundations 97-98
   Trackwork 99, 104, 125-126
PET system
   Automatic open crossings 246
   Automatic open crossings (AOCL) 252
Pile tests 364
Piles 359-362
Piling ancillaries 363-364
Pipe bedding 136-137
Pipe surrounds 134
Pipelines and ducts, Bridges 257
Pipework 127
Pipework – fittings and valves 340-341
Pipework – manholes and pipework ancillaries 342-344
Pipework – pipes 334-339
Pipework – supports and protection 345-350
Plain glazing 231-232
Plain line 99-103
Plasterwork to walls repairs 230
Platforms 149-150
Points heating 83-86
Points operating equipment 29
Portland cement finishes 225
Power and general cable 13-14
Power supplies
   Automatic half barrier 238
   Automatic open crossings 244
   Automatic open crossings (AOCL) 250
Precast concrete 277-280, 331-333
   Beams and columns 277-278
   Bridges 277-280
   Copings and sills 280
   Culverts, ducts and covers 279
   Slabs 278
Preparation of surfaces 222
Property – New Build
   Cross wall construction 150
   Facilities 151-157
   Platforms 149-150
   Security centre 157
   Security facilities 156
   Storage and repair facilities 151
Property – Refurbishment
   Asphalt work 202-203, 205-207
   Brick and block walling 196
   Brickwork and blockwork 194-200
   Brickwork in underpinning 190
   Brickwork sundries 191
   Cavity wall insulation 199
   Coloured mastic asphalt flooring 206
   Concrete work 178-179
   Damp proof courses 192-193, 200
   Demolitions and alterations 178
   Demolitions and removals 165-167
   Disposal and filling 175-177
   Earthwork support 173-174
   Excavation 184-186
   Excavation and earthworks 169-177
   Excavation by hand 170
   Expanded metal reinforcement 200
   Formwork to underpinning 189
   General demolitions 159-161, 167, 169
   General removals 194, 202-203
   Glazing 231-232
   In situ concrete 179
   In situ concrete in underpinning 187-188
   Internal finishings 221-230
   Ironmongery 211-214
   Masonry paint 233-234
   Mastic asphalt flooring 205

Index

Metalwork 219-220
New work 195, 203
Painting and decorating 233-234
Remedial works 207
Removal of building fabric 162
Removal of building fabric finishes 166
Removal of electrical installations 165
Removal of plumbing and heating installations 164-165
Removal of sundry joinery items 164
Removal of windows, doors and frames 163
Repairs and remedial works 206
Roofing 208-210
Service trenches 172
Sills and copings 197
Slate roofing 209
Structural steelwork 215-218
Tile roofing 210
Tile sills and creasings 198
Underpinning 184-193
Work to existing 208

REB Transformers 250
   Automatic open crossings 244
REBs 237, 249
   Automatic open crossings 243
   Automatic open crossings (AOCL) 249
Recoveries 42
   Automatic half barrier 240
   Automatic open crossings 246
   Automatic open crossings (AOCL) 252
Reinforcement 272-273, 326
   Concrete ancillaries 272-273
Relocatable equipment 24-27
Remedial works 207, 291-295
   Brickwork 291
   Tunnels 291-295
Removal of building fabric 162
Removal of building fabric finishes 166
Removal of electrical installations 165
Removal of old work 221, 255, 311
   Bridges 255
   General civils 311
Removal of plumbing and heating installations 164-165
Removal of sundry joinery items 164
Removal of windows, doors and frames 163
Repairs and alterations 229
Repairs and alterations – Ironmongery 211
Repairs and remedial works 206
Repairs and renewals – inner city 72
Roads and paving 365-375
Roofing 208-210
Roofing – work to existing 208

Samples 302
Security centre 157
Security facilities 156
Service trenches 172
Signage 219, 236
   Automatic open crossings 242
Signalling
   AC traction 18
   Automatic warning systems 32
   Axle Counters 9-10
   Banner signals 9-10
   Bases 8, 38
   Bases for structures 11
   Cable 21
   Cable routes 16, 40-42
   Cabling 12
   Cameras 22
   Data cable 13
   DC traction 18
   Equipment/Signals 3
   External cable 12

Fibre optic cable 14
Fibre optic route indicators 9-10
Fixing plates and brackets 33
Gantries and cantilevers 11
Heated mirrors 22
Interlocking systems 38
Internal fittings 27-29
Level crossings 19-20
Lifting barriers 19-20
Location cases 23
Miscellaneous additional work 32
Miscellaneous equipment 18, 45
MK2 rail clamp locks 30-31
Platform 3-8
Points operating equipment 29
Power and general cable 13-16
Power supply equipment 36
PVC pipe 17
Recoveries 42
Relocatable equipment 23-27
Resignalling Works 4-36, 38, 40-42, 45-46
Signalling structures 11
Stripping out and removals 38
Tail cable 14
Telephone cable 13
Terminations 15-16
Testing and commissioning 46
Trackside circuits 18
Train operated warning systems 33-34
Train protection warning systems 35
Treadles 21
Trough 17
Signalling power supply equipment 36
Signalling structures 11
Signals 250
   Automatic half barrier 238
   Automatic open crossings 244
   Automatic open crossings (AOCL) 250
Sills and copings 197
Site clearance 258
   Bridges 258
Slabs, Precast concrete 278
Slate roofing 209
Soakaways 135
Special glazing 232
Stair units and walkways 285-286
   Miscellaneous metalwork 285-286
Storage and repair facilities 151
Strip out and recoveries 95
Stripping out and removals 38
Structural metalwork 281-282, 284, 351-358
   Bridges 281-282, 284
   Erection of frames and other members 282
   Erection of members for bridges 282
   Fabrication of main members 281
   Off site surface treatment 284
Structural steel framing 215-216
Structural steelwork 215-218
   Fixings 217
   Wedging and grouting bases 218
Structural steelwork – works to existing 215
Sub stations 79, 88
Substations 93
Supply and installation 95
Support and stabilisation 290
   Tunnels 290
Switches and crossings 104-124
Switchgear 53, 61
Switchgear – Suburban 76
Systems, cabling and cable routes 139

Tail cable 14
Tee junction 17
Telecommunications
   Cabling 139
   CCTV 143-144

Customer Information Systems 145-146
Equipment 140
Existing line 141
Infrastructure systems 141-147
Operational systems 140
Public Address System 147
Transmission 142
Telecoms
   Automatic half barrier 239
   Automatic open crossings 245
   Automatic open crossings (AOCL) 251
Telephone cable 13
Terminations 15-16
Testing and commissioning 46, 96
Testing and commissioning – Rural 78
Tile roofing 210
Tile sills and creasings 198
Track crossings 135
Track drainage 128-137
Track drainage – Pipe surrounds 134-135
Track drainage – Pipework 128-133
Track feeder cable 93
Track feeder cable and ete 89-90
Track Foundations 97-98
Trackside circuit 18
Trackwork 99, 104, 125-126
Train operated warning systems 33-34
Train protection warning systems 35
Treadle equipment
   Automatic half barrier 239
   Automatic open crossings 245
   Automatic open crossings (AOCL) 251
Treadles 21
Trial holes 297
Tunnel excavation 288
   Tunnels 288
Tunnel lining 289
   Tunnels 289
Tunnels
   Brickwork 291
   Remedial works 291-295
   Sundries 292-295
   Support and stabilisation 290
   Tunnel excavation 288
   Tunnel lining 289

Underpinning 184-193
Vinyl floor finishes 228

Waterproofing 388-389
Wedging and grouting bases 218
Wiring – Rural 77